VOLUME ONE HUNDRED AND EIGHTY FIVE

ADVANCES IN IMAGING AND ELECTRON PHYSICS

VOLUME ONE HUNDRED AND EIGHTY FIVE

ADVANCES IN IMAGING AND ELECTRON PHYSICS

Edited by

PETER W. HAWKES
CEMES-CNRS, Toulouse, France

ELSEVIER

AMSTERDAM • BOSTON • HEIDELBERG • LONDON
NEW YORK • OXFORD • PARIS • SAN DIEGO
SAN FRANCISCO • SINGAPORE • SYDNEY • TOKYO

Academic Press is an imprint of Elsevier

Cover photo credit:
David Agard et al.
Single-Particle Cryo-Electron Microscopy (Cryo-EM): Progress, Challenges, and Perspectives for Further Improvement
Advances in Imaging and Electron Physics (2014) 185, pp. 113–137.

Academic Press is an imprint of Elsevier
32 Jamestown Road, London NW1 7BY, UK
525 B Street, Suite 1800, San Diego, CA 92101-4495, USA
225 Wyman Street, Waltham, MA 02451, USA
The Boulevard, Langford Lane, Kidlington, Oxford OX5 1GB, UK

First edition 2014

British Library Cataloguing in Publication Data
A catalogue record for this book is available from the British Library

Library of Congress Cataloging-in-Publication Data
A catalog record for this book is available from the Library of Congress

ISBN: 978-0-12-800144-8
ISSN: 1076-5670

For information on all Academic Press publications visit our Web site at http://store.elsevier.com/

Printed in the United States of America

CONTENTS

PREFACE

The three chapters in this latest volume span most of the regular themes of the series: an ingenious extension of geometrical optics, electron microscopy and mathematical morphology. We begin with a very complete account of complex geometrical optics by P. Berczynski and S. Marczynski. This variant of traditional geometrical optics allows diffraction phenomena to be studied. It has two forms, one ray-based, the other eikonal-based, and the authors describe these fully. After presenting the underlying theory, the approach is used to study the propagation of Gaussian beams in inhomogeneous media.

This is followed by a summary of the present state of cryo-electron microscopy for the study of unstained biological macromolecules by D. Agard, Y. Cheng, R.M. Glaeser and S. Subramaniam. The subject is not new but a large step foward has recently been made with the introduction of a new type of electron-detection camera. I leave the authors to set out the advantages of this innovation but we can be sure that many valuable new results can be anticipated.

To conclude, we have a long and authoritative account by M. Welk and M. Breuß of the image-adaptive structuring elements known as amoebas. It has been shown that for iterated median filtering with a fixed structuring-element, the process is closely related to a partial differential equation associated with the image in question. Here, the authors examine the relation between discrete amoeba median filtering and their (continuous) counterparts based on partial differential equations. I have no doubt that this clear and very complete account of the subject will be widely appreciated.

Peter Hawkes

FUTURE CONTRIBUTIONS

H.-W. Ackermann
Electron micrograph quality

J. Andersson and J.-O. Strömberg
Radon transforms and their weighted variants

S. Ando
Gradient operators and edge and corner detection

J. Angulo
Mathematical morphology for complex and quaternion-valued images

D. Batchelor
Soft x-ray microscopy

E. Bayro Corrochano
Quaternion wavelet transforms

C. Beeli
Structure and microscopy of quasicrystals

M. Berz (Ed.)
Femtosecond electron imaging and spectroscopy

C. Bobisch and R. Möller
Ballistic electron microscopy

F. Bociort
Saddle-point methods in lens design

K. Bredies
Diffusion tensor imaging

A. Broers
A retrospective

R.E. Burge
A scientific autobiography

A. Carroll
Refelective electron beam lithography

N. Chandra and R. Ghosh
Quantum entanglement in electron optics

A. Cornejo Rodriguez and F. Granados Agustin
Ronchigram quantification

N. de Jonge and D. Peckys
Scanning transmission electron microscopy of whole eukaryotic cells in liquid and in-situ studies of functional materials

J. Elorza
Fuzzy operators

A.R. Faruqi, G. McMullan and R. Henderson
Direct detectors

M. Ferroni
Transmission microscopy in the scanning electron microscope

R.G. Forbes
Liquid metal ion sources

A. Gölzhäuser
Recent advances in electron holography with point sources

J. Grotemeyer and T. Muskat
Time-of-flight mass spectrometry

M. Haschke
Micro-XRF excitation in the scanning electron microscope

M.I. Herrera
The development of electron microscopy in Spain

R. Herring and B. McMorran
Electron vortex beams

M.S. Isaacson
Early STEM development

K. Ishizuka
Contrast transfer and crystal images

C.T. Koch
In-line electron holography

T. Kohashi
Spin-polarized scanning electron microscopy

O.L. Krivanek
Aberration-corrected STEM

M. Kroupa
The Timepix detector and its applications

B. Lencová
Modern developments in electron optical calculations

H. Lichte
New developments in electron holography

M. Matsuya
Calculation of aberration coefficients using Lie algebra

J.A. Monsoriu
Fractal zone plates

L. Muray
Miniature electron optics and applications

M.A. O'Keefe
Electron image simulation

V. Ortalan
Ultrafast electron microscopy

D. Paganin, T. Gureyev and K. Pavlov
Intensity-linear methods in inverse imaging

M. Pap
Hyperbolic wavelets

N. Papamarkos and A. Kesidis
The inverse Hough transform

S.-C. Pei
Linear canonical transforms

P. Rocca and M. Donelli
Imaging of dielectric objects

J. Rodenburg
Lensless imaging

J. Rouse, H.-n. Liu and E. Munro
The role of differential algebra in electron optics

J. Sánchez
Fisher vector encoding for the classification of natural images

P. Santi
Light sheet fluorescence microscopy

C.J.R. Sheppard
The Rayleigh–Sommerfeld diffraction theory

R. Shimizu, T. Ikuta and Y. Takai
Defocus image modulation processing in real time

T. Soma
Focus-deflection systems and their applications

P. Sussner and M.E. Valle
Fuzzy morphological associative memories

J. Valdés
Recent developments concerning the Système International (SI)

G. Wielgoszewski
Scanning thermal microscopy and related techniques

CONTRIBUTORS

David Agard
HHMI and the Department of Biochemistry and Biophysics, University of California, San Francisco, CA 94158, USA

Pawel Berczynski
Institute of Physics, West Pomeranian University of Technology, Szczecin 70-310, Poland

Michael Breuß
Mathematical Image Analysis Group, Saarland University, Campus E1.1, 66041 Saarbrücken (Germany)

Yifan Cheng
Department of Biochemistry and Biophysics, University of California, San Francisco, CA 94158, USA

Robert M. Glaeser
Lawrence Berkeley National Laboratory, University of California, Berkeley, CA 94720, USA

Slawomir Marczynski
Faculty of Mechanical Engineering and Mechatronics, West Pomeranian University of Technology, Szczecin 70-310, Poland

Sriram Subramaniam
Laboratory for Cell Biology, Center for Cancer Research, National Cancer Institute, National Institutes of Health (NIH), Bethesda, MD 20892, USA

Martin Welk
UMIT, Biomedical Image Analysis Division, Eduard-Wallnoefer-Zentrum 1, 6060 HALL (Tyrol), Austria

CHAPTER ONE

Gaussian Beam Propagation in Inhomogeneous Nonlinear Media. Description in Ordinary Differential Equations by Complex Geometrical Optics

Pawel Berczynski[1], Slawomir Marczynski[2]

[1]Institute of Physics, West Pomeranian University of Technology, Szczecin 70-310, Poland
[2]Faculty of Mechanical Engineering and Mechatronics, West Pomeranian University of Technology, Szczecin 70-310, Poland

Contents

Advances in Imaging and Electron Physics, Volume 185
ISSN 1076-5670
http://dx.doi.org/10.1016/B978-0-12-800144-8.00001-X

1. INTRODUCTION

In the traditional understanding, geometrical optics is a method assigned to describe trajectories of rays, along which the phase and amplitude of a wave field can be calculated via diffractionless approximation (Kravtsov & Orlov 1990; Kravtsov, Kravtsov, & Zhu, 2010). Complex generalization of the classical geometrical optics theory allows one to include diffraction processes into the scope of consideration, which characterize wave rather than geometrical features of wave beams (by *diffraction,* we mean diffraction spreading of the wave beam, which results in GB having inhomogeneous waves). Although the first attempts to introduce complex rays and complex incident angles started before World War II, the real understanding of the potential of complex geometrical optics (CGO) began with the work of Keller (1958), which contains the consistent definition of a complex ray. Actually, the CGO method took two equivalent forms: *the ray-based form,* which deals with complex rays—i.e., trajectories in complex space (Kravtsov et al., 2010; Kravtsov, Forbes, & Asatryan 1999; Chapman *et al.* 1999; Kravtsov 1967)—and the *eikonal-based form,* which uses complex eikonal instead of complex rays (Keller & Streifer 1971; Kravtsov et al., 2010; Kravtsov, Forbes, & Asatryan 1999; Kravtsov 1967). The ability of the CGO method to describe the diffraction of GB on the basis of complex Hamiltonian ray equations was demonstrated many years ago in the framework of the ray-based approach. Development of numerical methods in the framework of the ray-based CGO in the recent years allowed for the description of GB diffraction in inhomogeneous media, including GB focusing by localized inhomogeneities (Deschamps 1971; Egorchenkov & Kravtsov 2000) and reflection from a linear-profile layer (Egorchenkov & Kravtsov 2001). The evolution of paraxial rays through optical structures also was studied by Kogelnik and Li (1966), who introduced the concept of a very convenient ray-transfer matrix (also see Arnaud 1976). This method of transformation is known as the *ABCD matrix method* (Akhmediev 1998; Stegeman & Segev 1999; Chen, Segev, & Christodoulides 2012; Agrawal 1989).

The eikonal-based CGO, which deals with complex eikonal and complex amplitude was essentially influenced by quasi-optics (Fox 1964), which is based on the parabolic wave equation (PWE; Fox 1964; Babič & Buldyrev 1991; Kogelnik 1965; Kogelnik & Li 1966; Arnaud 1976; Akhmanov & Nikitin 1997; Pereverzev 1993). In the case of a spatially narrow wave beam concentrated in the vicinity of the central ray, the parabolic equation reduces to the abridged PWE (Vlasov & Talanov 1995; Permitin & Smirnov 1996), which preserves only quadratic terms in small deviations from the central ray. The abridged PWE allows for describing the electromagnetic GB evolution in inhomogeneous and anisotropic plasmas (Pereverzev 1998) and in optically smoothly inhomogeneous media (Permitin & Smirnov 1996). The description of GB diffraction by the abridged PWE is an essential feature of quasi-optical model. It is a convenient simplification, nevertheless it still requires solving of partial differential equations.

The essential step in the development of quasi-optics was done in various studies that analyzed laser beams by introducing a quasi-optical complex parameter q (Kogelnik 1965; Kogelnik and Li 1966), which allows for solving the parabolic equation in a more compact way, taking into account the wave nature of the beams. The obtained PWE solution enables one to determine such GB parameters as beam width, amplitude, and wave front curvature. The quasi-optical approach is very convenient and commonly used in the framework of beam transmission and transformation through optical systems. However, modeling GB evolution by means of the quasi-optical parameter q using the ABCD matrix is effective for GB propagation in free space or along axial symmetry in graded-index optics (on axis beam propagation) when the A,B,C, and D elements of the transformation matrix are known. Thus, the problem of GB evolution along curvilinear trajectories requires the solution of the parabolic equation, which is complicated even for inhomogeneous media (Vlasov & Talanov 1995). In fact, the description of GB evolution along curvilinear trajectories by means of the parabolic equation is limited only to the consideration of linear inhomogeneous media (Pereverzev 1998; Vlasov & Talanov 1995; Permitin & Smirnov 1996). In our opinion, the eikonal-based form of the paraxial CGO seems to be a more powerful and simpler tool involving wave theory, as opposed to quasi-optics based on the parabolic equation, and even the CGO ray-based version based on Hamiltonian equations.

The problem of Gaussian beam self-focusing in nonlinear media was usually studied by solving the nonlinear parabolic equation (Akhmanov, Sukhorukov, & Khokhlov 1968; Akhmanov, Khokhlov, & Sukhorukov 1972). The abberrationless approximation enables to reduce the nonlinear parabolic equation to solving the second-order ordinary differential equation for Gaussian beam width evolution in a nonlinear medium of the Kerr type, but the procedure is complicated. Because of the general refraction coefficient, the CGO method presented in this paper deals with ordinary differential equations; it does not ask to reduce diffraction and self-focusing descriptions starting every time from partial differential equations. The well-known approaches of nonlinear optics, such as the variational method and method of moments, demand that the nonlinear parabolic equation gets solved by complicated integral procedures of theoretical physics, which can be unfamiliar to engineers of optoelectronics, computer modeling, and electron physics. It is worthwhile to emphasize that the variational method and method of moments have been applied to model Gaussian beam evolution in nonlinear graded-index fibers (Manash, Baldeck, & Alfano 1988; Karlsson, Anderson, & Desaix 1992; Paré & Bélanger 1992; Perez-Garcia *et al.* 2000; Malomed 2002; Longhi & Janner 2004). Moreover, analogous solutions can be obtained by the CGO method in a more convenient and illustrative way. The CGO method deals with Gaussian beams, which are convenient and appropriate wave objects to model famous optical solutions (Anderson 1983; Hasegawa 1990; Akhmediev 1998; Stegeman and Segev 1999; Chen, Segev, & Christodoulides 2012) propagating in nonlinear optical fibers (Agrawal 1989).

The CGO method presented in this paper has been applied in the past to describe GB evolution in inhomogeneous media (Berczynski and Kravtsov 2004; Berczynski *et al.* 2006), nonlinear media of the Kerr type (Berczynski, Kravtsov, & Sukhorukov 2010), nonlinear inhomogeneous fibers (Berczynski 2011) and nonlinear saturable media (Berczynski 2012, 2013a,b). In Berczynski (2013c), the CGO method was generalized to describe spatiotemporal effects for Gaussian wave packets propagating in nonlinear media and nonlinear transversely and longitudinally inhomogeneous fibers. In Berczynski (2014), the CGO method was generalized to describe elliptical Gaussian beam evolution in nonlinear inhomogeneous fibers of the Kerr type. To access the accuracy of the CGO method, Berczynski, Kravtsov, and Zeglinski (2010) showed that it demonstrated a great ability to describe GB evolution in graded-index optical fibers reducing the time of numerical calculations by 100 times with a comparable accuracy with the Crank-Nicolson scheme of the beam propagation method (BPM).

The present paper is organized as follows. Section 2 presents the fundamental equations of the CGO method and its boundary applicability. Section 3 presents the analytical CGO solution for paraxial GB propagating in free space, which demonstrates the advantages of the CGO method over the standard approach of diffraction theory, specifically from the Kirchhoff-Fresnel integral. Section 4 presents the problem of propagation of the Gaussian beam in linear inhomogeneous media by the CGO method, which is a good introduction to the problems of graded-index optics and integrated optics. Section 5 presents the natural generalization of the CGO method for nonlinear inhomogeneous media. Section 6 presents an analytical CGO solution for an axially symmetric Gaussian beam propagating in a nonlinear medium of the Kerr type, and it discusses the accuracy of presented CGO results compared to solutions of the nonlinear parabolic equation within aberration-less approximations. Section 7 discusses the influence of beam ellipticity on Gaussian beam propagation in a nonlinear medium of the Kerr type, in the case when elliptical cross section of the beam conserves its orientation with respect to a natural trihedral. We also discuss the interrelations of the CGO method with the standard approaches of nonlinear wave optics. In section 8, we discuss the sophisticated phenomenon of GB rotation in a nonlinear medium. In section 9, we present an orthogonal ray-centered coordinate system that is indispensable to describing the problem of elliptical rotating Gaussian beams propagating along a curvilinear trajectory in nonlinear inhomogeneous media. We also discuss the influence of nonlinear refraction on the evolution of the central ray of the beam in nonlinear inhomogeneous medium, reported first by Berczynski (2013a).

For clarity, in section 10, we derive from the eikonal equation the complex ordinary differential Riccati equation, which models the problem of the rotation of elliptical GBs propagating along a curvilinear trajectory in a nonlinear inhomogeneous medium. In section 11, we derive from the partial transport equation ordinary differential equations for the evolution of complex amplitude and flux conservation principle for a single elliptical rotating GB in a nonlinear medium. In section 12, we present a generalization of the CGO method for *N*-rotating GBs propagating along a helical ray in nonlinear graded-index fiber. We also demonstrate here the matrix form of CGO equations that are convenient for numerical simulations. In section 13, we present and discuss the evolution of a single rotating Gaussian beam propagating along a helical ray in nonlinear graded-index fiber. We also demonstrate the great ability of the CGO method to model explicitly the rotation of the beam intensity and wave-front cross section. In sections 14 and 15, we discuss the interaction of two, three, and four rotating Gaussian beams in nonlinear graded-index fiber. As mentioned previously, the effect of *N*-interacting Gaussian beams in a nonlinear inhomogeneous medium is a new problem, which demands application of a simple, effective tool and fast and accurate numerical algorithms. From a practical point of view, the rotating Gaussian beams can model the properties of rotating elliptical solitons. Furthermore, the existing state of knowledge of the interaction of optical wave objects in nonlinear media is limited to the description of only two copropagating axially symmetric beams or pulses (Pietrzyk 1999, 2001; Jiang *et al.* 2004; Medhekar, Sarkar, & Paltani 2006; Sarkar & Medhekar 2009). Thus, the CGO method presented in this paper is a convenient tool that easily and effectively generalizes the results of previous research (e.g., Pietrzyk 1999, 2001; Jiang *et al.* 2004; Medhekar, Sarkar, & Paltani 2006; Sarkar & Medhekar 2009) on N-rotating Gaussian beams interacting during propagation along a curvilinear trajectory. In our opinion, CGO can be recognized as the simplest method of nonlinear wave optics, which can make it applicable not only to theoretical physicists but also to engineers of electron physics.

2. CGO: FUNDAMENTAL EQUATIONS, MAIN ASSUMPTIONS, AND BOUNDARY OF APPLICABILITY

As is well known, classical geometrical optics represents diffractionless behavior of the wave field (i.e., behavior that does not take into account wave phenomena). In classical geometrical optics, we deal with a quasi-plane wave $u(\mathbf{r}) = A(\mathbf{r})\exp(i\Psi(\mathbf{r}))$, where the real amplitude $A(\mathbf{r})$ and the real local wave vectors $\mathbf{k}(\mathbf{r}) = \nabla\Psi(\mathbf{r})$ vary insignificantly over the wavelength $\lambda(\mathbf{r})$ in the medium. The wave fronts of the quasi-plane wave experience geometrical transformations because of wave focusing or defocusing in inhomogeneous media. The preservation of the quasi-plane wave form of the wave front is the necessary condition of classical geometrical optics applicability. Thus, classical geometrical optics becomes invalid near focal points, where the wave front loses its quasi-planar form. CGO is the generalization of classical geometrical optics. Unlike classical geometrical optics, which deals with real rays and quasi-plane homogeneous waves, the CGO method is involved with quasi-plane inhomogeneous waves (i.e., evanescent waves) in the form

$$u(\mathbf{r}) = A(\mathbf{r})\exp(i\Psi(\mathbf{r})) = A(\mathbf{r})\exp(ik_0\psi(\mathbf{r})). \tag{1}$$

In contrast to classical geometrical optics, the eikonal (optical path) $\psi(\mathbf{r})$ and amplitude $A(\mathbf{r})$ are complex values in the framework of CGO. The direction of wave propagation is determined by the gradient of the real part of the complex eikonal $\nabla\psi_R(\mathbf{r}) = \nabla\mathrm{Re}\{\psi(\mathbf{r})\}$. The direction of exponential decay of the field's magnitude is given principally by the gradient of the imaginary part $\nabla\psi_I(\boldsymbol{r}) = \nabla \mathrm{Im}\{\psi(\boldsymbol{r})\}$. The gradient of complex eikonal determines the ray momentum $\mathbf{p} = \nabla\psi(\boldsymbol{r})$, which satisfies the ray equations in Hamiltonian form:

$$\frac{d\boldsymbol{r}}{d\sigma} = \frac{\boldsymbol{p}}{n}, \qquad \frac{d\boldsymbol{p}}{d\sigma} = \frac{1}{2}\nabla n(\boldsymbol{r}), \tag{2}$$

where n is the refractive index and $d\sigma$ is the elementary arc length. In Eq. (2), we can use also the parameter of relative permittivity $\boldsymbol{\varepsilon}(\boldsymbol{r})$, which for an isotropic nonmagnetic medium, is related with refractive index by the formula

$$\boldsymbol{\varepsilon}(\boldsymbol{r}) = n^2(\boldsymbol{r}). \tag{3}$$

Based on the assumption of a quasi-plane inhomogeneous wave structure of the CGO wave field, it is natural to require (analogously, as in classical geometrical optics) that the real and imaginary parts of the complex amplitude $A(\boldsymbol{r})$ and local wave vector $\mathbf{k} = k_0\mathbf{p} = k_0\nabla\psi(\boldsymbol{r})$ do not change significantly on the scale of $\lambda'(\boldsymbol{r}) = \lambda(\boldsymbol{r})/2\pi = 1/|\mathbf{k}(\boldsymbol{r})|$. That is, the inequalities

$$\lambda'|\nabla Q_R| << |Q_R|, \quad \lambda'|\nabla Q_I| << |Q_I|, \tag{4}$$

where Q stands for $\mathbf{k}_R = k_0\mathbf{p}_R = k_0\nabla\psi_R(\boldsymbol{r}), \mathbf{k}_I = k_0\mathbf{p}_I = k_0\nabla\psi_I(\boldsymbol{r})$, and $A(\boldsymbol{r})$, together with the expression

$$\lambda'|\nabla\boldsymbol{\varepsilon}| << |\boldsymbol{\varepsilon}|, \tag{5}$$

determine the necessary conditions for the validity of CGO in weakly inhomogeneous media, where the wavelength $\lambda'(\boldsymbol{r})$ is equal to $\lambda'(\boldsymbol{r}) = \lambda'_0/n(\boldsymbol{r})$, where $\lambda'_0 = \lambda_0/2\pi$ and λ_0 is the wavelength in free space. By using inequalities in Eqs. (4) and (5), we can introduce the parameters of characteristic scales: L_i^R $(i = 1, 2, 3)$ for quantity $\boldsymbol{\varepsilon}(\mathbf{r})$, real parts of $\mathbf{k} = k_0\mathbf{p} = k_0\nabla\psi(\boldsymbol{r})$, $A(\boldsymbol{r})$, and L_i^I $(i = 1, 2, 3)$ for the imaginary parts, which are equal to

$$\frac{\lambda'_0}{n_R(\boldsymbol{r})} << \frac{|Q_R|}{|\nabla Q_R|} \equiv L_i^R, \quad \frac{\lambda'_0}{n_R(\boldsymbol{r})} << \frac{|Q_I|}{|\nabla Q_I|} \equiv L_i^I. \tag{6}$$

Alternatively, these conditions, stating that $\boldsymbol{\varepsilon}(\mathbf{r})$, $\mathbf{k}(\boldsymbol{r})$, and $A(\boldsymbol{r})$ vary insignificantly within the region of the order of $\lambda'(\boldsymbol{r})$ may be united in a single inequality:

$$\mu_{CGO} = \frac{1}{kL} = \frac{1}{k_0\sqrt{\boldsymbol{\varepsilon}(\mathbf{r})}L} = \frac{\lambda'(\boldsymbol{r})}{L} << 1, \tag{7}$$

where μ_{CGO} is the parameter of smallness in the method of CGO, and L is the smallest of the characteristic lengths of $\boldsymbol{\varepsilon}(\mathbf{r}), \mathbf{k}(\boldsymbol{r})$, and $A(\boldsymbol{r})$; i.e., $L = \min(L_i^R, L_i^I)$. To derive the basic equations of the CGO method, let us take advantage of the Rytov expansion (Kravtsov et al., 2010) of the field in a small dimensionless parameter $\mu = 1/k_0L$, where we assume that $\mu_{CGO} \cong \mu$. Within dimensionless variables $x_I = k_0x$, $y_I = k_0y$, and $z_I = k_0z$, the Helmholtz equation for an inhomogeneous medium takes the form

$$\Delta_I u(\boldsymbol{r}_I) + \boldsymbol{\varepsilon}(\boldsymbol{r}_I)u(\boldsymbol{r}_I) = 0, \tag{8}$$

where $\Delta_I = \frac{\partial^2}{\partial x_I^2} + \frac{\partial^2}{\partial y_I^2} + \frac{\partial^2}{\partial z_I^2}$, and characteristic length L for the change of parameters $\boldsymbol{\varepsilon}, \mathbf{k}$, and A converts into the dimensionless parameter $kL = 1/\mu_{CGO}$. We can introduce this parameter into Eq. (8) by transforming $\boldsymbol{r}_{II} = \mu_{CGO}\boldsymbol{r}_I = \boldsymbol{r}/L$, $n(\boldsymbol{r}_{II}) = n(\mu_{CGO}\boldsymbol{r}_I) = n(\mathbf{r}/L)$ into the following:

$$\Delta_{II} u(\boldsymbol{r}_{II}) + \frac{\boldsymbol{\varepsilon}(\boldsymbol{r}_{II})}{\mu_{CGO}^2} u(\boldsymbol{r}_{II}) = 0, \tag{9}$$

where $\Delta_{II} = \frac{\partial^2}{\partial x_{II}^2} + \frac{\partial^2}{\partial y_{II}^2} + \frac{\partial^2}{\partial z_{II}^2}$. The amplitude is assumed to vary slowly, so $A = A(\mu_{CGO}\boldsymbol{r}_I) = A(\boldsymbol{r}_{II})$. It is also convenient to write the phase in the form $\Psi(\boldsymbol{r}) = \Psi_1(\mu_{CGO}\boldsymbol{r}_I)/\mu_{CGO} = \Psi_1(\boldsymbol{r}_{II})/\mu_{CGO}$, so that the local wave vector as a gradient of phase $\mathbf{k} = \nabla\Psi = k_0\nabla_I\Psi_1(\mu_{CGO}\boldsymbol{r}_I)/\mu_{CGO} = k_0\nabla_2\Psi_1(\boldsymbol{r}_{II})$ would also be assumed to be a function that changes slowly with the coordinates, where $\nabla_I = \partial/\partial\mathbf{r}_I$ and $\nabla_{II} = \partial/\partial\mathbf{r}_{II}$. As a result, the wave field in coordinates $\boldsymbol{r}_I$ and $\boldsymbol{r}_{II}$ has the form

$$u = A(\mu_{CGO}\boldsymbol{r}_I)\exp(i\Psi_1(\mu_{CGO}\boldsymbol{r}_I)/\mu_{CGO}) = A(\boldsymbol{r}_{II})\exp(i\Psi_1(\boldsymbol{r}_{II})/\mu_{CGO}). \tag{10}$$

Substituting Eq. (10) into Eq. (9), we obtain

$$\Delta_{II} u(\boldsymbol{r}_{II}) + \frac{\boldsymbol{\varepsilon}(\boldsymbol{r}_{II})}{\mu_{CGO}^2} u(\boldsymbol{r}_{II}) \equiv \left\{ \frac{1}{\mu_{CGO}^2} (\boldsymbol{\varepsilon} - (\nabla_{II}\Psi_I)^2)A \right.$$
$$\left. + \frac{i}{\mu_{CGO}} (2\nabla_{II}A \cdot \nabla_{II}\Psi_1 + A\Delta_{II}\Psi_I) + \Delta_{II}A \right\} \exp(i\Psi_I/\mu_{CGO}) = 0. \tag{11}$$

Thus, comparison of the components with $1/\mu_{CGO}^2$ in Eq. (11) leads to the eikonal equation

$$(\nabla_{II}\Psi_I)^2 = \boldsymbol{\varepsilon}. \tag{12}$$

Comparing also the components of i/μ_{CGO}, we obtain the transport equation in the form

$$2(\nabla_{II}A\cdot\nabla\Psi_I) + A\Delta_{II}\Psi_I = 0. \tag{13}$$

Strictly speaking, Eq. (13) is the transport equation derived in a zeroth-order geometrical optics approximation (Kravtsov et al., 2010), which enables satisfactory accuracy of wave analysis on the example of Gaussian beam evolution in inhomogeneous media (Berczynski & Kravtsov 2004; Berczynski *et al.* 2006) and nonlinear media (Berczynski, Kravtsov, & Sukhorukov 2010), including optical fibers (Berczynski 2011). In more sophisticated problems, such as wave field reflection from a weak interface (Kravtsov et al., 2010), which requires the description of the entire wave phenomena in the framework of CGO, one can use the expansion of the wave amplitude A in the CGO method in parameter μ_{CGO} (Rytov expansion):

$$u(\mathbf{r}_{II}) = \sum_{m=0}^{\infty} A_m(\mathbf{r}_{II})\left(\frac{\mu_{CGO}}{i}\right)^m \exp(i\Psi_I(\mathbf{r}_{II})/\mu_{CGO}), \tag{14}$$

or a Debye expansion of the field in inverse powers of wave number $1/k_0$ in the form

$$u(\mathbf{r}) = \sum_{m=0}^{\infty}\frac{A_m(\mathbf{r})}{(ik_0)^m}\exp(ik_0\psi(\mathbf{r})). \tag{15}$$

Although these expansions give equivalent results, in our opinion the Rytov expansion in the dimensionless small parameter $\mu_{CGO}\cong 1/k_0L$ has some methodological advantages, which shows that the smallest value of CGO parameter μ_{CGO} is achieved not only when $\lambda'_0\to 0$, but also when the parameter L is much greater than wavelength λ'_0. Limiting ourselves to amplitudes of the zeroth order in the Rytov expansion in Eq. (14) and rewriting Eqs. (12) and (13) in dimensional variables x, y, and z and introducing eikonal $k_0\psi = \Psi_I/\mu_{CGO}$, we obtain CGO equations in the standard form derived for weakly absorptive media in the form of

$$(\nabla\psi)^2 = \varepsilon, \tag{16}$$

$$2(\nabla A\cdot\nabla\psi) + A\Delta\psi = 0. \tag{17}$$

The CGO method deals with small-angle (paraxial) beams, which are localized in the vicinity of the central ray (beam trajectory) satisfying the ray equations in Eq. (2). To satisfy the condition of the CGO method applicability, we introduce the small paraxial parameter, which takes the form

$$\mu_{Parax} = 1/k_0 w_0 = \lambda'_0/w_0 << 1. \tag{18}$$

The CGO paraxial parameter defined in Eq. (18) appears usually in explicit form in problems of nonlinear optics, where the optical description by means of the nonlinear Helmholtz equation reduces to the nonlinear parabolic equation. The propagation of linearly polarized, continuous wave beams in an isotropic Kerr is governed by the scalar nonlinear Helmholtz equation, where variable $\boldsymbol{r}$ has the form

$$\Delta u(\boldsymbol{r}) + k_0^2\left(1 + \varepsilon_{NL}|u(\boldsymbol{r})|^2\right)u(\boldsymbol{r}) = 0. \tag{19}$$

If we model the light propagation along the z-axis in an axially symmetric medium of the Kerr type [introducing cylindrical coordinates (z, r), where $r = \sqrt{x^2 + y^2}$], the wave field $u = u(z, r)$ can be presented in the form

$$u(z, r) = \frac{1}{k_0^2 w_0^2 \varepsilon_{NL}} \varphi(z, r)\exp(ik_0 z), \tag{20}$$

where w_0 is the initial beam width and k_0 is the vacuum wave number. Introducing the next dimensionless variables $r = r/w_0$ and $z' = z/2L_D$, where $L_D = k_0 w_0^2$ is diffraction length, and substituting Eq. (20) into Eq. (19), we obtain the equation for complex envelope evolution in the form

$$\frac{\mu_{Parax}^2}{4}\frac{\partial^2 \varphi}{\partial z'^2} + i\frac{\partial \varphi}{\partial z'} + \Delta_\perp \varphi + \varphi|\varphi|^2 = 0, \tag{21}$$

where $\Delta_\perp = \frac{1}{r'}\frac{\partial}{\partial r'} + \frac{\partial^2}{\partial r'^2}$ is the transverse gradient in cylindrical coordinates. Assuming that $\lambda_0 << w_0$, we obtain that $\mu_{Parax} << 1$ and as a result, $\frac{\mu_{Parax}^2}{4}\frac{\partial^2 \varphi}{\partial z^2} << \frac{\partial \varphi}{\partial z}$. Therefore, the paraxial approximation, which in this case ignores $\frac{\mu_{Parax}^2}{4}\frac{\partial^2 \varphi}{\partial z^2}$, allows one to examine the beam propagation in nonlinear media using the nonlinear parabolic equation. It is shown (Berczynski 2011) that the CGO method supplies solutions for nonlinear inhomogeneous fibers of the Kerr type, which are identical to results obtained by the nonlinear parabolic equation. Eq. (21) shows interrelations between the CGO method and the nonlinear parabolic equation paraxial description in a nonlinear medium of the Kerr type, which is essential for the development of the use of the CGO method for nonlinear saturable media. Berczynski (2011) demonstrated that the CGO method supplies solutions for GB evolution in inhomogeneous and nonlinear Kerr-type fibers in a much simpler way than standard methods of nonlinear optics such as the variational method and the method of moments. The CGO method also reduces essentially the time spent on numerical calculations compared to the beam propagation method (BPM), which was shown in the example of GB propagation in optical graded-index fibers in Berczynski, Kravtsov, and Zeglinski (2010). The Gaussian beam is a self-sustained solution in the framework of the CGO method, and we noticed from our numerical calculations that the necessary condition for GB to preserve its "Gaussian" profile in a nonlinear medium of the Kerr type is that GB width be small enough with respect to the characteristic nonlinear scale. To satisfy this condition, let us introduce the next small, nonlinear parameter of the following form:

$$\mu_{NL} = w_0/L_{NL} << 1, \tag{22}$$

where $L_{NL} = w_0/\sqrt{\varepsilon_{NL}|A_0|^2}$ (with w_0 being the initial beam width and A_0 being the initial complex amplitude), limiting ourselves to low intensities of the beams. One can deduce from numerical calculations that the small parameter in Eq. (22) determines the condition of the applicability of the CGO method in nonlinear saturable media. To generalize the description by the CGO method for nonlinear inhomogeneous media, we should take into account the effect of linear refraction. Thus, introducing the refraction parameter μ_{REF}, we notice that the beam preserves its "Gaussian" form during propagation in an inhomogeneous medium when this refraction parameter is small enough. Thus, we obtain that

$$\mu_{REF} = w_0/L << 1, \tag{23}$$

where $L = |\varepsilon|/|\nabla\varepsilon|$ is the inhomogeneity scale of smoothly inhomogeneous media. Small parameters in Eqs. (22) and (23) determine also the boundary applicability of the abridged PWE (Vlasov & Talanov 1995; Permitin & Smirnov 1996).

3. GAUSSIAN BEAM DIFFRACTION IN FREE SPACE. CGO METHOD AND CLASSICAL DIFFRACTION THEORY

For an axially symmetric wave beam propagating along the z-axis in free space, the CGO method suggests a solution of the form

$$u(r,z) = A\exp(ik_0\psi) = A(z)\exp\left(ik_0\left(B(z)r^2/2 - z\right)\right), \tag{24}$$

where ψ is a complex-valued eikonal, which takes the form

$$\psi = B(z)r^2/2 - z, \tag{25}$$

where $r = \sqrt{x^2 + y^2}$ is the distance from the z-axis (with a radius in cylindrical symmetry) and parameter *B(z)* is the complex curvature of the beam wave front. The eikonal equation in Eq. (16) in coordinates (r, z) takes the form

$$\left(\frac{\partial \psi}{\partial r}\right)^2 + \left(\frac{\partial \psi}{\partial z}\right)^2 = 1, \tag{26}$$

where relative permittivity is equal to unity: $\varepsilon = 1$ in free space. Taking the complex eikonal from Eq. (25) and putting it into the eikonal equation in Eq. (26), we obtain the Riccati equation in the form

$$\frac{dB}{dz} = B^2, \tag{27}$$

which has the following solution:

$$B(z) = \frac{B(0)}{1 - B(0)z}. \tag{28}$$

For the GB, which has the initial width w(0) and with initial wave curvature equal to zero $\kappa(0) = 0$, the initial value of parameter B is equal to $B(0) = i/k_0 w^2(0)$. As a result, we obtain

$$B(z) = \frac{i/k_0(0)}{1 - iz/k_0 w^2(0)} = \frac{i/k_0 w^2(0)}{1 + z/ik_0 w^2(0)}. \tag{29}$$

In the framework of paraxial approximation, where r is a small parameter, amplitude A = A(z) satisfies the transport equation in Eq. (17), which for the axially symmetric beam in cylindrical coordinates, (r, z) takes the following form:

$$\frac{dA^2}{dz}\frac{\partial \psi}{\partial z} + \left(\frac{1}{r}\frac{\partial}{\partial r}\left(r\frac{\partial \psi}{\partial r}\right) + \frac{\partial^2 \psi}{\partial z^2}\right)A^2 = 0. \tag{30}$$

In accordance with Eq. (25), we obtain

$$\frac{\partial \psi}{\partial z} = 1, \quad \frac{1}{r}\frac{\partial}{\partial r}\left(r\frac{\partial \psi}{\partial r}\right) = 2B, \tag{31}$$

and as a result, Eq. (30) reduces to an ordinary differential equation in the form:

$$\frac{dA}{dz} + B(z)A = 0. \tag{32}$$

As a result, the complex amplitude of the axially symmetric GB takes the form

$$A(z) = A(0)\exp\left(-\int B(z)dz\right), \tag{33}$$

where $A(0)$ is the initial amplitude. Putting Eq. (28) into Eq. (33), we obtain the connection between the amplitude of GB and the complex wave front curvature in the form

$$A(z) = \frac{A(0)B(z)}{B(0)}. \tag{34}$$

As a result, the complex amplitude equals

$$A(z) = \frac{A(0)}{1 + z/ik_0 w^2(0)}. \tag{35}$$

Let us compare the obtained CGO results presented in Eq. (28) and Eq. (35) for GB diffraction in free space with solutions of the diffraction theory within Fresnel approximation, for which the diffraction integral has the form

$$E(x, y, z) = \frac{i}{\lambda z} e^{-ik_0 z} \int_{-\infty}^{+\infty}\int E_0(x', y')\exp\left\{\frac{ik_0}{2z}\left((x - x')^2 + (y - y')^2\right)\right\} dx'dy', \tag{36}$$

where (x', y') are the coordinates in the plane of the screen and (x, y) are the coordinates in the observation plane. $E(x, y, z)$ is the envelope of the field in the observation plane, $E_0(x', y')$ denotes the field envelope in the plane of the screen with the aperture, and z is the distance between the screen plane and the observation plane. The integral in Eq. (36) is the approximation of the diffraction integral written in standard form:

$$E(x, y, z) = \frac{i}{\lambda} \int_{-\infty}^{+\infty}\int E_0(x', y')\frac{e^{-ik_0\rho}}{\rho} dx'dy', \tag{37}$$

where $\rho = \sqrt{z^2 + (x - x')^2 + (y - y')^2}$ is the distance between two points on the aperture and observation plane. Remember that Fresnel's approximation describes the diffraction of the paraxial (weakly diverging) optical beams for the inequalities, where $z >> x, y, x', y'$ are well satisfied. The inequalities shown here allow one to write the following approximate expression for parameter ρ: $\rho = z + ((x - x')^2 + (y - y')^2)/2z$. In this way, we can disregard the difference between parameters ρ and z in the denominator of the integrand in Eq. (37). From a physical point of view, the formula $\rho = z + ((x - x')^2 + (y - y')^2)/2z$ obtained within the Fresnel approximation implies substitution of the parabolic surfaces for the spherical wave fronts of the Huygens secondary wavelets. By virtue of the axial symmetry of the GB with initially circular cross sections, we perform further calculations of diffraction by a round aperture, making the transition to the polar coordinates via the following formula:

$$x = r\cos\varphi, \quad y = r\sin\varphi \quad x' = r'\cos\varphi', \quad y' = r'\sin\varphi'. \tag{38}$$

Writing the surface area in the form $d\sigma' = r'dr'd\varphi'$, the integral in Eq. (36) takes the form

$$E(x, y, z) = \frac{i}{\lambda z} e^{-ik_0 z} \int_0^\infty r'dr' \int_0^{2\pi} d\varphi' E_0(r', \varphi') \times \exp\left\{\frac{ik_0}{2z}\left(r'^2 + r^2 - 2rr'\cos(\varphi - \varphi')\right)\right\}. \tag{39}$$

By virtue of the axial symmetry of the field distribution, where $E_0(r', \varphi') = E_0(r')$, and expressing the integral

$$\Phi(r, r') = \int_0^{2\pi} \exp\left\{\frac{ik_0 rr'}{2z}\cos(\varphi - \varphi')\right\} d\varphi' \tag{40}$$

via the Bessel functions of the zeroth order, $J(\alpha)$, Eq. (39) takes the form

$$E(r, z) = \frac{2\pi i}{\lambda z}\exp\left\{ik_0\left(z + \frac{r^2}{2z}\right)\right\} \int_0^\infty E_0(r') J_0\left(\frac{k_0 rr'}{z}\right)' e^{\frac{-k_0 r'^2}{2z}} r'dr'. \tag{41}$$

The field envelope $E_0(r')$ in the plane of the screen with the aperture takes the following form:

$$E_0(r') = E_0\exp\left(-r'^2/2w_0^2\right). \tag{42}$$

Putting Eq. (42) into Eq. (41), we obtain the diffraction field, which on the observation plane has the form

$$E(r,z)=\frac{E_0\exp(-ik_0z)}{1+z/ik_0w_0^2}\exp\left(-\frac{r^2/2w_0^2}{1+z/ik_0w_0^2}\right). \tag{43}$$

One can notice that the solution in Eq. (43) is in total agreement with the CGO wave field in the form $u(r,z)=A\exp(ik_0\psi)$, where the complex amplitude A for GB propagating in free space is presented in Eq. (35); and the complex eikonal ψ is shown in Eq. (29). Thus, using the CGO method as applied to Gaussian beam propagating and diffracting in free space, we obtain in a simple and illustrative way the same result as can be obtained in the standard way within a Fresnel approximation to the Kirchoff integral, taking into account the fact that the wave function $u(r,z)$ used in the CGO method plays the same role as the field envelope $E(r,z)$ used within classical diffraction theory, and the CGO quantity $A(0)$ is equivalent to parameter E_0 used in Eq. (43).

4. ON-AXIS PROPAGATION OF AN AXIALLY SYMMETRIC GAUSSIAN BEAM IN SMOOTHLY INHOMOGENEOUS MEDIA

4.1 First-Order Ordinary Differential Equation for Complex Parameter B

For an axially symmetric wave beam in an axially symmetric inhomogeneous linear medium, we use a Gaussian ansatz of the form

$$u(r,z)=A\exp(ik_0\psi)=A(z)\exp\left(ik_0\left(z+B(z)r^2/2\right)\right). \tag{44}$$

Analogously, as before, ψ denotes a complex eikonal, which has the form

$$\psi=z+B(z)r^2/2. \tag{45}$$

In Eq. (45), complex parameter $B=B(z)$ can be associated with the complex wave front curvature used in quasi-optics and resonator optics. The general eikonal equation in Eq. (16) for an axially symmetric inhomogeneous medium with coordinates (*r*, z) takes the form

$$\left(\frac{\partial\psi}{\partial r}\right)^2+\left(\frac{\partial\psi}{\partial z}\right)^2=\varepsilon(z,r). \tag{47}$$

In accordance with the paraxial approximation, where we assume that GB is localized in the vicinity of axis z, radius *r* should be small enough. Therefore, relative permittivity in Eq. (47) can be expanded in the Taylor series in r in the vicinity of symmetry axis z:

$$\varepsilon(z,r) = \varepsilon(r=0) + \left(\frac{\partial\varepsilon}{\partial r}\Big|_{r=0}\right) r + \left(\frac{\partial^2\varepsilon}{\partial r^2}\Big|_{r=0}\right)\frac{r^2}{2}. \tag{48}$$

Putting Eqs. (48) and (45) into the eikonal equation in Eq. (48) and comparing the next coefficients of r^0, r and r^2, we obtain the following relations:

$$\varepsilon(r=0) = 1, \quad \frac{\partial\varepsilon}{\partial r}\Big|_{r=0} = 0 \tag{49}$$

and the Riccati equation for complex curvature B, which for a smoothly inhomogeneous medium has the form

$$\frac{dB}{dz} + B^2 = \beta. \tag{50}$$

In Eq. (50), parameter $\beta = \frac{\partial^2\varepsilon}{\partial r^2}\Big|_{r=0}$ describes the linear refraction. When we describe GB diffraction in a homogeneous medium where $\varepsilon = const$, this parameter equals zero. At this point, let us determine the physical meaning of complex parameter B. The real and imaginary parts of parameter $B = \mathrm{Re}B + i\mathrm{Im}B$ determine the real curvature κ of the wave front and the beam width w correspondingly:

$$\mathrm{Re}B(z) = \kappa(z), \; \mathrm{Im}B(z) = \frac{1}{k_0 w^2(z)}. \tag{51}$$

Putting Eq. (51) into Eq. (44), we obtain the Gaussian beam of the form

$$u(r,z) = A(z)\exp\left(-\frac{r^2}{2w^2(z)}\right)\exp\left(ik_0\left(z + \kappa(z)\frac{r^2}{2}\right)\right). \tag{52}$$

The expression in Eq. (52) reflects the general feature of the CGO method, which in fact deals with the Gaussian beams.

4.2 The Second-Order Ordinary Differential Equation for GB Width Evolution in an Inhomogeneous Medium

The Riccati equation in Eq. (50) is equivalent to the set of two equations for the real and imaginary parts of the complex parameter B:

$$\begin{cases} \dfrac{d\mathrm{Re}B(z)}{dz} + (\mathrm{Re}B(z))^2 - (\mathrm{Im}B(z))^2 = \beta(z) \\[2ex] \dfrac{d\mathrm{Im}B(z)}{dz} + 2(\mathrm{Re}B(z))(\mathrm{Im}B(z)) = 0 \end{cases}. \tag{53}$$

Substituting Eq. (51) into Eq. (53), we obtain the expression

$$\frac{d}{dz}\left(\frac{1}{w^2}\right) = -\frac{2\kappa}{w^2}, \tag{54}$$

which leads to the known relation (Kogelnik 1965) between the beam width w and the wave front curvature κ in the form

$$\kappa = \frac{1}{w}\frac{dw}{dz}. \tag{55}$$

Putting the relation in Eq. (55) into a system of equations in Eq. (53), we obtain an ordinary differential equation of the second order:

$$\frac{d^2w}{dz^2} - \beta w = \frac{1}{k_0^2 w^3}, \tag{56}$$

which describes the influence of linear refraction on GB diffraction in an inhomogeneous linear medium. Remember that the refraction parameter β is the same as in the Riccati equation in Eq. (50). An identical equation was obtained within quasi-optics dealing with an abridged PWE (Vlasov & Talanov 1995; Permitin & Smirnov 1996).

4.3 The First-Order Ordinary Differential Equation for the GB Complex Amplitude

As previously discussed, now we describe within the CGO method paraxial GBs that are now localized in the vicinity of symmetry axis z. Thus, in the framework of paraxial approximation, radius r is a small parameter and amplitude A = A(z) is complex-valued. It satisfies the transport equation in Eq. (17), which for an axially symmetric beam in cylindrical coordinates (r, z) takes the following form:

$$\frac{dA^2}{dz}\frac{\partial\psi}{\partial z} + \left(\frac{1}{r}\frac{\partial}{\partial r}\left(r\frac{\partial\psi}{\partial r}\right) + \frac{\partial^2\psi}{\partial z^2}\right)A^2 = 0. \tag{57}$$

In accordance with Eq. (45), we obtain that

$$\frac{\partial\psi}{\partial z} = 1, \quad \frac{1}{r}\frac{\partial}{\partial r}\left(r\frac{\partial\psi}{\partial r}\right) = 2B. \tag{58}$$

As a result, Eq. (57) reduces to an ordinary differential equation:

$$\frac{dA}{dz} + B(z)A = 0. \tag{59}$$

Eq. (59) for GB complex amplitude and the Riccati equation for complex curvature parameter B are the basic CGO equations, which in fact reduce the problem of GB diffraction to the domain of an ordinary differential equation. Having calculated the complex parameter B from the Riccati equation in Eq. (50), we can readily find complex amplitude A by integrating Eq. (59). As a result, the complex amplitude of axially symmetric GB propagating in inhomogeneous medium takes the form

$$A(z) = A(0)\exp\left(-\int B(z)dz\right), \tag{60}$$

where $A(0)$ is the initial amplitude.

4.4 The Energy Flux Conservation Principle in GB Cross Section

The absolute value of the complex amplitude in Eq. (60) equals

$$|A(z)| = |A(0)|\exp\left(-\int \mathrm{Re}B(z)dz\right). \tag{61}$$

Integrating the second equation of the system in Eq. (53), we get

$$\mathrm{Im}B = \mathrm{Im}B(0)\exp\left(-2\int \mathrm{Re}B(z)dz\right). \tag{62}$$

Comparing Eqs. (53) and (62), we obtain the following relation:

$$w^2|A|^2 = w^2(0)|A(0)|^2, \tag{63}$$

which represents the energy flux conservation principle in a GB cross section.

5. GENERALIZATION OF THE CGO METHOD FOR NONLINEAR INHOMOGENEOUS MEDIA

For clarity, let us start our analysis with the case of on-axis beam propagation in the simplest nonlinear medium—namely, a medium with (cubic) Kerr-type nonlinearity. In such a medium, the relative permittivity depends on the beam intensity $|u|^2$ in the form

$$\varepsilon(\mathbf{r}) = \varepsilon_{LIN}(\mathbf{r}) + \varepsilon_{NL}|u(\mathbf{r})|^2, \tag{64}$$

where coefficient ε_{NL} is assumed to be positive ($\varepsilon_{NL} > 0$) when we consider the self-focusing nonlinear effect. Putting the wave field in Eq. (52) into Eq. (64), we obtain

$$\varepsilon(r,z) = \varepsilon_{LIN}(r,z) + \varepsilon_{NL}|A(z)|^2 \exp\left(-\frac{r^2}{w(z)^2}\right). \tag{65}$$

Thus, when the CGO method deals with Gaussian beams, we can notice that this method treats a formally nonlinear medium of the Kerr type as a smoothly inhomogeneous medium whose profile is modulated by GB parameters w and A. This is the simplest explanation why the CGO method presented earlier in this chapter for the linear inhomogeneous case also can be applicable for nonlinear media of the Kerr type. In accordance with relation in Eq. (63), we can present Eq. (65) as

$$\varepsilon(r,z) = \varepsilon_{LIN}(r,z) + \frac{\varepsilon_{NL}|A(0)|^2 w^2(0)}{w^2(z)} \exp\left(-\frac{r^2}{w^2(z)}\right), \tag{66}$$

Thus, the beta parameter in Eqs. (50) and (56) can be given as

$$\beta = \alpha + \chi, \tag{67}$$

where the first term answers to the linear medium:

$$\alpha = \frac{1}{2}\frac{d^2\varepsilon_{LIN}}{dr^2}\Big|_{r=0}; \tag{68}$$

and the second term accounts for the nonlinear one:

$$\chi = \frac{1}{2}\frac{\varepsilon_{NL}|A(0)|^2 w^2(0)}{w^2(z)}\frac{d^2}{dr^2}\exp\left(-\frac{r^2}{w^2(z)}\right)\Big|_{r=0} = -\frac{\varepsilon_{NL}|A(0)|^2 w^2(0)}{w^4(z)}. \tag{69}$$

Substituting the relations in Eqs. (67) and (68) for the ones in Eq. (50), we obtain the Riccati equation, generalized now for the case of on-axis GB propagation in a nonlinear inhomogeneous medium of the Kerr type:

$$\frac{dB}{dz} + B^2 = \alpha + \chi = \frac{1}{2}\frac{d^2\varepsilon_{LIN}}{dr^2}\Big|_{r=0} - \frac{\varepsilon_{NL}|A(0)|^2 w^2(0)}{w^4(z)}. \tag{70}$$

6. SELF-FOCUSING OF AN AXIALLY SYMMETRIC GAUSSIAN BEAM IN A NONLINEAR MEDIUM OF THE KERR TYPE. THE CGO METHOD AND SOLUTIONS OF THE NONLINEAR PARABOLIC EQUATION

In this section, let us describe the classical example of on-axis GB diffraction and self-focusing in a nonlinear medium of the Kerr type without any influence of linear refraction. In conditions when the contribution of the linear term in Eq. (68) is negligibly small, the Riccati equation in Eq. (70) takes the form

$$\frac{dB}{dz} + B^2 = \chi = -\frac{\varepsilon_{NL}|A(0)|^2 w^2(0)}{w^4(z)}. \tag{71}$$

Following analogously, as in the previous section, Eq. (56) for the beam width evolution in the medium of the Kerr type takes the form

$$\frac{d^2 w}{dz^2} + \frac{\varepsilon_{NL}|A(0)|^2 w^2(0)}{w^3} = \frac{1}{k_0^2 w^3}. \tag{72}$$

Introducing a dimensionless beam width $f = w/w(0)$, Eq. (72) can be rewritten as

$$\frac{d^2 f}{dz^2} = -\frac{1}{f^3}\left(\frac{1}{L_{NL}^2} - \frac{1}{L_D^2}\right), \tag{73}$$

where $L_D = k_0 w^2(0)$ is the diffraction length and $L_{NL} = w(0)/\sqrt{\varepsilon_{NL}|A(0)|^2}$ is the characteristic nonlinear scale. Eq. (73) also can be presented as

$$\frac{d^2 f}{d\zeta^2} = -\frac{1}{f^3}\left(\frac{L_D^2}{L_{NL}^2} - 1\right), \tag{74}$$

where dimensionless distance $\zeta = z/L_D$ is involved in this description. It can be proved that

$$\frac{L_D^2}{L_{NL}^2} = \frac{P}{P_{crit}}, \tag{75}$$

where $P = \frac{1}{8}c\sqrt{\varepsilon_0}w^2(0)|A(0)|^2$ is the total beam power and $P_{crit} = \frac{1}{8}\frac{c\sqrt{\varepsilon_0}}{k_0^2\varepsilon_{NL}}$ is the critical power. As a result, the equation for GB width evolution in a nonlinear medium of the Kerr type takes the form

$$\frac{d^2 f}{d\zeta^2} + \frac{1}{f^3}\left(\frac{P}{P_{crit}} - 1\right) = 0. \tag{76}$$

Integrating once Eq (76) and assuming that $\frac{df}{dz}\Big|_{z=0} = 0$, which corresponds to the GB with an initial wave front [see Eq. (55)], we obtain the following solution:

$$f^2 = 1 - \frac{z^2}{L_D^2}\left(\frac{P}{P_{crit}} - 1\right). \tag{77}$$

This CGO result is in total agreement with the solution of the nonlinear parabolic equation presented in several studies (e.g., Akhmanov, Sukhorukov, & Khokhlov 1968; Akhmanov, Khokhlov, & Sukhorukov 1972). Thus, the CGO method reproduces the classical results of nonlinear optics, but in a more simple and illustrative way (Berczynski, Kravtsov, & Sukhorukov 2010). In addition, let us analyze the CGO solution in Eq. (77) for three cases:

1. Under-critical power: $P < P_{crit}$. In this case, the beam width increases without limits, and in accordance with Eq. (63), the beam amplitude tends to zero at $z \to \infty$.
2. Critical power: $P = P_{crit}$. In this case, we obtain a stationary solution.
3. Over-critical power: $P > P_{crit}$, and the beam width decreases to zero at a finite propagation distance. In accordance with Eq. (63), the wave amplitude increases to infinity over such a distance. In this case, the collapse phenomenon takes place (Akhmanov, Sukhorukov, & Khokhlov 1968; Akhmanov, Khokhlov, & Sukhorukov 1972).

7. SELF-FOCUSING OF ELLIPTICAL GB PROPAGATING IN A NONLINEAR MEDIUM OF THE KERR TYPE

In realistic optical systems of integrated nonlinear optics, we should include beam ellipticity in the description. To analyze the problem of rotating elliptical beam propagating along a curvilinear trajectory in an inhomogeneous nonlinear medium, let us consider first the case where an elliptical beam conserves its orientation in transverse Cartesian coordinates x, y when it propagates along the symmetry axis in a nonlinear medium of the Kerr type. As Berczynski (2014) did, we model by the CGO method the elliptical GB propagating along the z-axis in the form

$$u = A(z)\exp\left(ik_0\left(z + B_1(z)x^2/2 + B_2(z)y^2/2\right)\right). \tag{78}$$

As in Berczynski (2014), we obtain two coupled ordinary differential equations for principal GB widths in the following form:

$$\frac{d^2w_1}{dz^2} + \frac{P/P_{crit}}{k_0^2 w_1^2 w_2} = \frac{1}{k_0^2 w_1^3} \quad \text{and} \tag{79}$$

$$\frac{d^2 w_2}{dz^2} + \frac{P/P_{crit}}{k_0^2 w_1 w_2^2} = \frac{1}{k_0^2 w_2^3}, \tag{80}$$

where

$$P = \frac{1}{8} c\sqrt{\varepsilon_0} w_1(0) w_2(0) |A(0)|^2 \text{ and } P_{crit} = \frac{1}{8} \frac{c\sqrt{\varepsilon_0}}{k_0^2 \varepsilon_{NL}} \tag{81}$$

denote the total beam power of an elliptical beam and the critical power of an axially symmetric GB. We can present Eqs. (79) and (80) in the following forms:

$$\frac{d^2 w_1}{dz^2} = \frac{1}{k_0^2 w_1^3}\left(1 - \frac{P}{P_{crit}} \frac{w_1}{w_2}\right) \quad \text{and} \tag{82}$$

$$\frac{d^2 w_2}{dz^2} = \frac{1}{k_0^2 w_2^3}\left(1 - \frac{P}{P_{crit}} \frac{w_2}{w_1}\right). \tag{83}$$

Using differential identity $(w_i^2)'' = 2((w'_i)^2 + w_i w''_i)$ (i=1,2), we can present Eqs. (82) and (83) in the form

$$\frac{d^2 w_1^2}{dz^2} = 2\left(\left(\frac{dw_1}{dz}\right)^2 + \frac{1}{k_0^2 w_1^2}\left(1 - \frac{P}{P_{crit}} \frac{w_1}{w_2}\right)\right) \quad \text{and} \tag{84}$$

$$\frac{d^2 w_2^2}{dz^2} = 2\left(\left(\frac{dw_2}{dz}\right)^2 + \frac{1}{k_0^2 w_2^2}\left(1 - \frac{P}{P_{crit}} \frac{w_2}{w_1}\right)\right). \tag{85}$$

Summing Equations (84) and (85), we obtain

$$\frac{d^2\left(w_1^2 + w_2^2\right)}{dz^2} = 2\left(\left(\frac{dw_1}{dz}\right)^2 + \left(\frac{dw_2}{dz}\right)^2 + \frac{1}{k_0^2 w_1^2} + \frac{1}{k_0^2 w_2^2} - 2\frac{P}{P_{crit}} \frac{1}{k_0^2 w_1 w_2}\right). \tag{86}$$

One can notice that the component

$$\left(\frac{dw_1}{dz}\right)^2 + \left(\frac{dw_2}{dz}\right)^2 + \frac{1}{k_0^2 w_1^2} + \frac{1}{k_0^2 w_2^2} - 2\frac{P}{P_{crit}} \frac{1}{k_0^2 w_1 w_2} = C = const \tag{87}$$

is the invariant. Thus, Eq. (86) takes the form

$$\frac{d^2\left(w_1^2 + w_2^2\right)}{dz^2} = 2C \tag{88}$$

and has the following solution:

$$w_1^2 + w_2^2 = \left(\frac{1}{k_0^2 w_{01}^2} + \frac{1}{k_0^2 w_{02}^2} - 2 \frac{P}{P_{crit}} \frac{1}{k_0^2 w_{01} w_{02}} \right) z^2 + w_{01}^2 + w_{02}^2, \quad (89)$$

when the initial wave front curvatures are equal to zero $[\kappa_{01}(0) = 0, \kappa_{02}(0) = 0]$. In the solution presented in Eq. (89), we can distinguish three subcases:

1. When $P < \frac{P_{crit}}{2}\left(\theta + \frac{1}{\theta}\right)$, where θ denotes the ratio of the GB width along the large axis to the GB width along the small axis of the elliptical cross section $\theta = w_{02}/w_{01}$, the combination of principal widths $w_1^2 + w_2^2$ increase like parabola.
2. When $P = \frac{P_{crit}}{2}\left(\theta + \frac{1}{\theta}\right)$, the self-trapping effect takes place.
3. When $P > \frac{P_{crit}}{2}\left(\theta + \frac{1}{\theta}\right)$, the combination of the widths $w_1^2 + w_2^2$ decreases to zero at a finite (self-focusing) distance, and the GB collapses.

The accuracy of the CGO method as compared to solutions of the nonlinear parabolic equations for elliptical GB propagating in a nonlinear medium of the Kerr type, as well as the problem of the influence of initial beam divergence and convergence, were discussed by Berczynski (2014).

8. ROTATING ELLIPTICAL GAUSSIAN BEAMS IN NONLINEAR MEDIA

In the previous section, we considered the problem of elliptical GB propagating along the symmetry axis in a nonlinear medium of the Kerr type. The form of a complex eikonal,

$$\psi = z + B_1(z)x^2/2 + B_2(z)y^2/2, \quad (90)$$

means that both the beam intensity and wave-front cross section conserve their orientation with respect to the transverse Cartesian coordinates x, y. As shown by Goncharenko *et al.* (1991), this is not the only possible solution for the nonlinear parabolic equations for elliptical GB propagating in a nonlinear medium. A more general and sophisticated case is when the wave field rotates during propagation. In CGO language, we can express such a situation by a complex eikonal in the following form:

$$\psi = z + B_{11}(z)x^2/2 + B_{12}(z)xy + B_{22}(z)y^2/2, \quad (91)$$

where $B_{ij}(i = 1, 2)$ are complex-valued functions that constitute a symmetric matrix where $B_{12} \equiv B_{21}$. Generalizing the regularities presented in

Eq. (52) for two-dimensional (2-D) GB for the case of a three-dimensional (3-D) beam, one can notice that real parts of the functions $R_{ij} = \mathrm{Re}B_{ij}$ determine the principal wave front curvatures and imaginary parts $I_{ij} = \mathrm{Im}B_{ij}$ determine the widths of elliptical cross sections of the beam. Thus, the complex eikonal in Eq. (91) can be written as

$$\begin{aligned}\psi = z + R_{11}(z)x^2/2 + R_{12}(z)xy + R_{22}(z)y^2/2 + i\big(I_{11}(z)x^2/2 \\ + I_{12}(z)xy + I_{22}(z)y^2\big),\end{aligned} \tag{92}$$

where principal widths and principal wave front curvatures are equal to

$$w_{1,2}^2 = \frac{2}{k_0\left(I_{11} + I_{22} \pm \sqrt{(I_{11} - I_{22})^2 + 4I_{12}^2}\right)},$$

$$\kappa_{1,2} = \frac{R_{11} + R_{22} \pm \sqrt{(R_{11} - R_{22})^2 + 4R_{12}^2}}{2}, \tag{93}$$

As shown by Berczynski (2014), we can transform CGO equations for the parameter B_{ij} into a rotating coordinate system related with a Cartesian one in the following way:

$$\eta_1 = x\cos\varphi + y\sin\varphi,\ \eta_2 = x\cos\varphi - y\sin\varphi, \tag{94}$$

where

$$tg2\varphi = \frac{2I_{12}}{I_{11} - I_{22}}. \tag{95}$$

The parameter I_{ij} of a beam cross section ellipse $I_{11}x^2 + 2I_{12}xy + I_{22}y^2 = 1$ is connected with principal widths w_1 and w_2 of a rotating coordinate system by the following relations:

$$w_1^{-2} = k_0^2\left(I_{11}\cos^2\varphi + I_{12}\sin 2\varphi + I_{22}\sin^2\varphi\right) \quad \text{and} \tag{96}$$

$$w_2^{-2} = k_0^2\left(I_{22}\cos^2\varphi - I_{12}\sin 2\varphi + I_{11}\sin^2\varphi\right). \tag{97}$$

Without the loss of generality, we can assume the following initial conditions for parameters R_{ij} and I_{ij} in the form

$$\begin{aligned}R_{11}(0) = 0,\ R_{12}(0) = R_{12}^0,\ R_{22}(0) = 0,\ I_{11}(0) = I_{11}^0, \\ I_{12}(0) = 0,\ I_{22}(0) = I_{22}^0.\end{aligned} \tag{98}$$

Following analogously, as in Berczynski (2014), we can derive the set of equations for principal widths w_1 and w_2 in the form

$$\frac{d^2 w_1}{dz^2} = \frac{1}{k_0^2 w_1^3} - \frac{P/P_{crit}}{k_0^2 w_1^2 w_2} + \frac{\sigma \cdot w_2}{w_1 \sqrt{2}} \left(\frac{1 + 3w_1/w_2}{\left(1 - (w_2/w_1)^2\right)^3} \right) / k_0^2 w_1^2 w_2 \quad (99)$$

$$\frac{d^2 w_2}{dz^2} = \frac{1}{k_0^2 w_2^3} - \frac{P/P_{crit}}{k_0^2 w_2^2 w_1} + \frac{\sigma \cdot w_1}{w_2 \sqrt{2}} \left(\frac{1 + 3w_1/w_2}{\left(1 - (w_1/w_2)^2\right)^3} \right) / k_0^2 w_2^2 w_1, \quad (100)$$

where

$$\sigma = \frac{R_{12}^0}{I_{11}^0} - \frac{R_{12}^0}{I_{22}^0}. \quad (101)$$

The ratio of P/P_{crit} in Eq. (99) can be calculated using Eq. (81). The set of Eqs. (99) and (100) is identical to that derived by Goncharenko *et al.* (1991) using nonlinear parabolic equations and an aberrationless approximation. Moreover, we can calculate the characteristic power for the self-trapping effect for rotating GB as follows:

$$P = \frac{P_{crit}}{2} \left(\theta + \frac{1}{\theta} + \frac{\left(R_{12}^0\right)^2}{I_{11}^0 I_{22}^0} \left(\theta + \frac{1}{\theta} \right) \right), \quad (102)$$

where $\theta = w_{02}/w_{01}$ as discussed previously. One can see that when $\left(R_{12}^0\right)^2 / I_{11}^0 I_{22}^0 = N$, the power for the self-trapping effect for rotating GB can be presented as

$$P = P_{oe} + \frac{N P_{crit}}{2} \left(\theta + \frac{1}{\theta} \right), \quad (103)$$

where P_{oe} is the self-trapping power for an ordinary elliptical beam (i.e., an elliptical beam that conserves its orientation relative to transverse Cartesian coordinates x, y), as analyzed previously. Strictly speaking, the result presented in Eq. (103) is one of the most important conclusions on the evolution of rotating GB propagating along a rectilinear trajectory in Cartesian coordinates. Namely, as the degree of symmetry of GB decreases (GB starts to rotate with respect to transverse Cartesian coordinates x, y), the self-trapping power for self-focusing increases. A much more complex problem is the elliptical GB rotating in a nonlinear saturable medium propagating along a curvilinear trajectory in an inhomogeneous nonlinear medium. This indicates the need for a simple method effectively describing the wave motion in curvilinear differential geometry, computationally efficient and convenient for the implementation in typical environments for numerical computations (eg Matlab or MathCAD). The CGO method meets these requirements. It need solving only ordinary differential equations instead much more complicated partial differential ones. Therefore, computational algorithms are relatively simple. Suitable implementations are available and can be run in software environments well known to engineers. Therefore CGO method should be preferred way to treat the self-focussing and the diffraction of Gaussian beams, especially in nonlinear systems.

9. ORTHOGONAL RAY-CENTERED COORDINATE SYSTEM FOR ROTATING ELLIPTICAL GAUSSIAN BEAMS PROPAGATING ALONG A CURVILINEAR TRAJECTORY IN A NONLINEAR INHOMOGENEOUS MEDIUM

CGO is a method based on geometrical optics, nevertheless it describes accurately wave phenomena related to Gaussian beam propagation (along the central ray of the beam). Because CGO is a paraxial method, we describe spatially narrow beams, which are localized in the vicinity of central rays. In inhomogeneous media, such central rays are curvilinear due to a nonzero gradient of the relative medium permittivity and can be described using Hamiltonian equations in the following form:

$$\frac{d\mathbf{r}}{d\tau}=\mathbf{p},\ \frac{d\mathbf{p}}{d\tau}=\frac{1}{2}\nabla\varepsilon(\mathbf{r}), \tag{104}$$

where $\mathbf{r}$ is the location vector, $\mathbf{p}$ is the generalized momentum, and τ is a parameter that changes along the curvilinear central ray of the beam and is related to arc-length by the following formula:

$$ds=\sqrt{\varepsilon}d\tau. \tag{105}$$

As mentioned previously, the CGO method deals with paraxial (spatially narrow) beams that are localized along central rays described by the Hamiltonian equations shown in Eq. (104). Thus, the natural choice of the 3-D reference system is when the longitudinal axis is located on the central ray of the beam, which is curvilinear for inhomogeneous media, and two remaining transverse axis are orthogonal to one another. In this way, we constructed a ray-centered coordinate system. From fundamental differential geometry, we know that with each curve, we can associate three characteristic unit vectors, $\mathbf{l}$, $\mathbf{n}$ and $\mathbf{b}$. Vector $\mathbf{l}$ is tangent to the curve and vectors $\mathbf{n}$ and $\mathbf{b}$ are perpendicular to the tangent vector $\mathbf{l}$. The normal vector is $\mathbf{n}$, and the binormal one is **b**. In accordance with Serret-Frenet formulas, we can describe the evolution of such vectors in 3-D geometry in the following form:

$$\frac{d\mathbf{l}}{ds} = \kappa \mathbf{n}, \tag{106}$$

$$\frac{d\mathbf{n}}{ds} = -\kappa \mathbf{l} + \chi \mathbf{b}, \quad \text{and} \tag{107}$$

$$\frac{d\mathbf{b}}{ds} = -\chi \mathbf{n}, \tag{108}$$

where parameter s is the arc-length of the curve, κ is its curvature, and χ denotes the torsion (which makes the wave problem more difficult). Namely, analyzing Eqs. (106)–(108), one can see that the ray-centered coordinate system related to the natural trihedral $(\mathbf{l}, \mathbf{n}, \mathbf{b})$ is not orthogonal because unit vectors **n** and **b** undergo local rotation around the central ray of the beam, with angular velocity proportional to its torsion χ. Thus, it is extremely important to construct an orthogonal ray-centered coordinate system that allows for describing beam wave motion in an unequivocal way. So, the unit vectors

$$\mathbf{e}_1 = \mathbf{n}\cos\varphi + \mathbf{b}\sin\varphi, \ \mathbf{e}_2 = \mathbf{b}\cos\varphi - \mathbf{n}\sin\varphi \tag{109}$$

undergo rotation relative to the base **n** and **b**, with angular velocity equal to

$$d\varphi/ds = -\chi_c. \tag{110}$$

Alternatively, using CGO parameter τ, we can present this in the form

$$d\varphi/d\tau = -\sqrt{\varepsilon_c}\chi_c, \tag{111}$$

where φ is the rotation angle. From the other side, as mentioned previously, the base $(\mathbf{l}, \mathbf{n}, \mathbf{b})$ rotates itself relative to the central ray in accordance with the Serret-Frenet formulas in Eqs. (106)–(108). As a result, vectors $\mathbf{e}_1$ and $\mathbf{e}_2$ are transported in such a way that they do not rotate along **l**. Thus, the unit vectors $\mathbf{e}_1$ and $\mathbf{e}_2$, together with the tangent **l**, form the base $(\mathbf{l}, \mathbf{e}_1, \mathbf{e}_2)$ of an orthogonal ray-centered coordinate system. Such an orthogonal base was introduced into optics by a number of studies (e.g., Popov 1969, 1977, 1982; Popov and Pšenčik 1978a, b) is now widely applied both in geophysics and in optics. In Babič and Buldyrev (1991) and Babič and Kirpichnikova (1980), such an orthogonal base was applied to the problem of paraxial beams described by the abridged PWE. Such a base is considered essential for the description of seismic rays (Červený 2001). One also finds applications in quasi-optics (Permitin & Smirnov 1996), CGO of inhomogeneous media (Berczynski *et al.* 2006), and CGO of nonlinear inhomogeneous media (Berczynski 2013a). If ξ denotes a vector lying in the plane perpendicular to the central ray $\mathbf{r}_c(\tau)$ in the form

$$\xi = \xi_1 \mathbf{e}_1 + \xi_2 \mathbf{e}_2, \tag{112}$$

then the location vector of the orthogonal ray-centered coordinate system (τ, ξ_1, ξ_2) takes the following form:

$$\mathbf{r} = \mathbf{r}_c(\tau) + \xi_1 \mathbf{e}_1 + \xi_2 \mathbf{e}_2. \tag{113}$$

Lamé coefficients of such a ray-centered coordinate system are equal to the findings of previous studies (e.g., Popov 1969, 1977, 1982; Popov and Pšenčik 1978a, b; Babič and Kirpichnikova 1980; Červený 2001):

$$h_\tau = h = \sqrt{\varepsilon(\mathbf{r})}\left(1 - \frac{(\xi \nabla_\perp)\varepsilon(\mathbf{r})}{2\varepsilon(\mathbf{r})}\right)_{\mathbf{r}=\mathbf{r}_c}, \; h_{\xi_1} = h_{\xi_2} = 1, \tag{114}$$

where $\nabla_\perp = \frac{\partial}{\partial \xi}$.

As was shown by Berczynski (2013a), the central ray of a symmetric GB is not subjected to nonlinear refraction caused by a nonlinear part of relative permittivity. Therefore, the trajectory of the central ray in inhomogeneous nonlinear media coincides with the central ray in linear inhomogeneous media. This means that low-powered and high-powered beams propagate along the same trajectory.

10. COMPLEX ORDINARY DIFFERENTIAL RICCATI EQUATIONS FOR ELLIPTICAL ROTATING GB PROPAGATING ALONG A CURVILINEAR TRAJECTORY IN A NONLINEAR INHOMOGENEOUS MEDIUM

Now let us consider the propagation of a monochromatic scalar Gaussian beam in a smoothly inhomogeneous isotropic and nonlinear saturable medium, with a permittivity profile in the form

$$\varepsilon = \varepsilon_{LIN}(\boldsymbol{r}) + \varepsilon_S g(I(\boldsymbol{r})). \tag{115}$$

In above equation, g is an arbitrary function of the beam intensity $I(\boldsymbol{r})$ and ε_S is saturating permittivity, and $\varepsilon(\tau) = \varepsilon(\mathbf{r}_c)$ and $\mathbf{r}_c = (\tau, 0, 0)$ is the radius-vector for the central ray in (τ, ξ_1, ξ_2) coordinates. In the framework of the CGO method, the eikonal consists of two summands: $\psi_c(\tau)$ is the eikonal on the central ray, while $\varphi(\tau, \xi_1, \xi_2)$ is a small deviation from ψ_c in the form

$$\psi(\tau, \xi_1, \xi_2) = \psi_c(\tau) + \varphi(\tau, \xi_1, \xi_2), \tag{116}$$

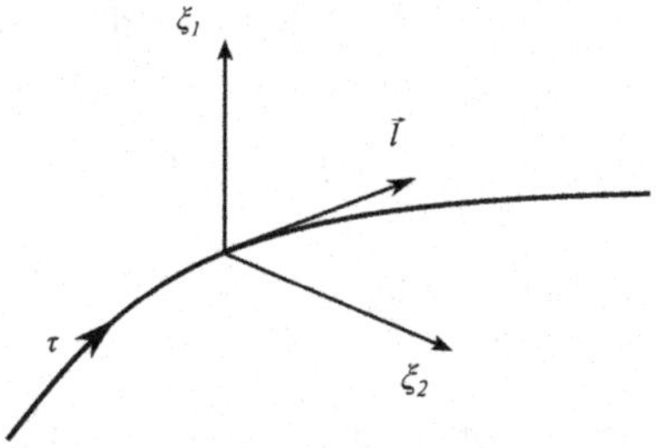

Figure 1 Ray-centered coordinate system.

where τ is the parameter along the central ray and $\xi_{1,2}$ are coordinates orthogonal to the ray in a ray centered-reference system (Figure 1). The complex eikonal φ describes both the curvature of the *phase front* of the beam and its *intensity* profile. Within paraxial approximation, the deviation φ for GB can be presented as a quadratic form:

$$\varphi(\tau,\xi_1,\xi_2)=\frac{1}{2}B_{ij}(\tau)\xi_i\xi_j. \tag{117}$$

In what follows, $i=1,2$ and summation over repeated indices is implied; $B_{ij}=R_{ij}+iI_{ij}$, where $R_{ij}\equiv\mathrm{Re}B_{ij}$ and $I_{ij}\equiv\mathrm{Im}B_{ij}$ are complex-valued functions, which constitute a symmetric tensor with $B_{12}\equiv B_{21}$. The real parts of these functions characterize the curvatures of the GB phase front, whereas the imaginary parts determine the elliptical cross section of the GB. In view of the extreme properties of the central ray, the linear in ξ_i terms does not contribute to Eq. (117) in an isotropic medium. The total complex eikonal ψ satisfies Eq. (16), which, for the case of nonlinear saturable medium, has the form

$$(\nabla\psi)^2=\varepsilon_{LIN}(\boldsymbol{r})+\varepsilon_S g(I(\boldsymbol{r})), \tag{118}$$

where

$$I(\boldsymbol{r})=I(\tau,\xi_i)=\frac{c}{4\pi}|A(\tau)|^2\exp\left(-\frac{k_0}{2}I_{ij}(\tau)\xi_i\xi_j\right). \tag{119}$$

By virtue of Eqs. (116) and (117), the eikonal equation in Eq. (118) in the parallel transport coordinates (τ,ξ_1,ξ_2) takes the following form:

$$\frac{1}{h^2}\left\{\left(\frac{d\psi_c}{d\tau}\right)^2+\frac{d\psi_c}{d\tau}\frac{dB_{ij}}{d\tau}\xi_i\xi_j+\left(\frac{1}{2}\frac{dB_{ij}}{d\tau}\xi_i\xi_j\right)^2\right\}+(B_{1i}\xi_i)^2+(B_{2i}\xi_i)^2$$
$$=\varepsilon(\boldsymbol{r}), \tag{120}$$

where $\varepsilon(\boldsymbol{r})$ is defined in Eq. (115). For the paraxial beam, we can expand the next permittivity ε in Eq. (120) in a Taylor series in small deviation ξ:

$$\varepsilon(\mathbf{r}) = \varepsilon_{LIN}(\mathbf{r}_c) + \varepsilon_{NL}g(I(\mathbf{r}_c)) + ((\xi\nabla_\perp)\varepsilon_{LIN})_{\mathbf{r}=\mathbf{r}_c} + \left(\frac{1}{2}(\xi\nabla_\perp)^2\varepsilon_{LIN} + \frac{\varepsilon_S}{2}(\xi\nabla_\perp)^2 g(I(\mathbf{r}))\right)_{\mathbf{r}=\mathbf{r}_c}. \tag{121}$$

By solving Eq. (120), we obtain the eikonal along the central ray

$$\frac{d\psi_c}{d\tau} = \varepsilon_c(\tau), \tag{122}$$

and the equation

$$\frac{dB_{ij}}{d\tau}\xi_i\xi_j + (B_{1i}\xi_i)^2 + (B_{2i}\xi_i)^2 = \left(\frac{1}{2}\{(\xi\nabla_\perp)^2\varepsilon_{LIN}(\mathbf{r}) + \varepsilon_S(\xi\nabla_\perp)^2 g(I(\boldsymbol{r}))\} - \frac{3(\xi\nabla_\perp\varepsilon_{LIN}(\mathbf{r}))^2}{4\varepsilon_{LIN}(\mathbf{r})}\right)_{\mathbf{r}=\mathbf{r}_c}. \tag{123}$$

Eq. (123) is the quadratic form in ξ_i. In order to satisfy this equation, one should equal all the coefficients in the left and right sides of this equation. As a result, we obtain a tensor Riccati-type equation for $B_{ij}(\tau)$:

$$\frac{dB_{ij}}{d\tau} + B_{ik}B_{kj} = \alpha_{ij} + \chi_{ij}, \tag{124}$$

which is a system of three equations for the complex parameters $B_{ij}(\tau)$:

$$\frac{dB_{11}}{d\tau} + \left(B_{11}^2 + B_{12}^2\right) = \alpha_{11} + \chi_{11}, \tag{125a}$$

$$\frac{dB_{12}}{d\tau} + B_{12}(B_{11} + B_{22}) = \alpha_{12} + \chi_{12}, \quad \text{and} \tag{125b}$$

$$\frac{dB_{22}}{d\tau} + \left(B_{22}^2 + B_{12}^2\right) = \alpha_{22} + \chi_{22}. \tag{125c}$$

The quadratic terms B_{ij} in Eq. (124) are responsible for diffraction in the homogeneous medium. The right-side term α_{ij} in Eq. (124) describes the influence of linear refraction on GB diffraction, whereas quantities χ_{ij} take into account the influence of the self-focusing of GB. These parameters take the following forms:

$$\alpha_{ij}(\tau) = \left(\frac{1}{2}\frac{\partial^2\varepsilon(\mathbf{r})}{\partial\xi_i\partial\xi_j} - \frac{3}{4\varepsilon(\mathbf{r})}\frac{\partial\varepsilon(\mathbf{r})}{\partial\xi_i}\frac{\partial\varepsilon(\mathbf{r})}{\partial\xi_j}\right)_{\mathbf{r}=\mathbf{r}_c} \quad \text{and} \tag{126}$$

$$\chi_{ij}(\tau) = \left(\frac{\varepsilon_{NL}}{2} \frac{\partial g}{\partial I} \frac{\partial^2 I(\mathbf{r})}{\partial \xi_i \partial \xi_j} \right)_{\mathbf{r}=\mathbf{r}_c}. \tag{127}$$

Eq. (125) contains the basic equations for the description of GB diffraction in smoothly inhomogeneous and nonlinear media. It is worth noting that these equations are ordinary differential ones, which are very useful for the analysis and numerical simulations. The complex equations in Eq. (124) can be presented as a system of six real equations for quantities $R_{ij} \equiv \mathrm{Re}B_{ij}$ and $I_{ij} \equiv \mathrm{Im}B_{ij}$:

$$\frac{dR_{ij}}{d\tau} + R_{ik}R_{kj} - I_{ik}I_{kj} = \alpha_{ij} + \chi_{ij}$$

$$\frac{dI_{ij}}{d\tau} + R_{ik}I_{kj} + I_{ik}R_{kj} = 0, \tag{128}$$

or, in the components:

$$\frac{dR_{11}}{d\tau} + R_{11}^2 + R_{12}^2 - I_{11}^2 - I_{12}^2 = \alpha_{11} + \chi_{11} \tag{129a}$$

$$\frac{dR_{12}}{d\tau} + R_{12}(R_{11} + R_{22}) - I_{12}(I_{11} + I_{22}) = \alpha_{12} + \chi_{12} \tag{129b}$$

$$\frac{dR_{22}}{d\tau} + R_{22}^2 + R_{12}^2 - I_{22}^2 - I_{12}^2 = \alpha_{22} + \chi_{22} \tag{129c}$$

$$\frac{dI_{11}}{d\tau} + 2R_{11}I_{11} + 2R_{12}I_{12} = 0 \tag{129d}$$

$$\frac{dI_{12}}{d\tau} + R_{12}(I_{11} + I_{22}) + I_{12}(R_{11} + R_{22}) = 0 \tag{129e}$$

$$\frac{dI_{22}}{d\tau} + 2R_{22}I_{22} + 2R_{12}I_{12} = 0. \tag{129f}$$

If we denote the eigenvalues of tensors $R_{ij}(\tau)$ and $I_{ij}(\tau)$ as $R_i(\tau)$ and $I_i(\tau)$, thus principal curvatures of the wave front, κ_i, and the principal beam widths, w_i, can be presented in the form:

$$R_i = \sqrt{\varepsilon_c} \cdot \kappa_i, \quad I_i = \frac{1}{k_0 w_i^2}. \tag{130}$$

For components R_{ij} and I_{ij} we can derive principal beam widths and principal curvatures of the wave front, which are equal to

$$w_{1,2}^2 = \frac{2}{k_0\left(I_{11} + I_{22} \pm \sqrt{(I_{11} - I_{22})^2 + 4I_{12}^2}\right)},$$

$$\kappa_{1,2} = \frac{R_{11} + R_{22} \pm \sqrt{(R_{11} - R_{22})^2 + 4R_{12}^2}}{2\sqrt{\varepsilon_c}}. \tag{131}$$

11. ORDINARY DIFFERENTIAL EQUATION FOR THE COMPLEX AMPLITUDE AND FLUX CONSERVATION PRINCIPLE FOR A SINGLE ROTATING ELLIPTICAL GB PROPAGATING IN A NONLINEAR MEDIUM

In the ray-centered coordinates (τ, ξ_1, ξ_2), the transport equation in Eq. (17) takes the form

$$\frac{1}{h^2}\frac{dA^2}{d\tau}\frac{\partial\psi}{\partial\tau} + \left(\frac{1}{h}\frac{\partial}{\partial\tau}\left(\frac{1}{h}\frac{\partial\psi}{\partial\tau}\right) + \frac{\partial^2\psi}{\partial\xi_1^2} + \frac{\partial^2\psi}{\partial\xi_2^2}\right)A^2 = 0, \tag{132}$$

where parameter h is defined in Eq. (114). By using paraxial approximation and introducing new amplitude $\widetilde{A} = \varepsilon_c^{1/4} \cdot A$, Eq. (132) can be reduced to the following form:

$$\frac{d\widetilde{A}^2}{d\tau} + TrB_{ij}\widetilde{A}^2 = 0, \tag{133}$$

where $TrB_{ij} \equiv B_{ii} = B_{11} + B_{22}$. It admits an explicit solution:

$$\widetilde{A}^2 = \widetilde{A}_0^2 \exp\left(-\int TrB_{ij}d\tau\right), \tag{134}$$

where $\widetilde{A}_0 = \widetilde{A}(0)$ is the initial amplitude of the beam. Thus, having calculated the tensor B_{ij}, one can readily determine the complex amplitude A as well. The absolute value of $\widetilde{A}$ equals

$$|\widetilde{A}|^2 = |\widetilde{A}_0|^2 \exp\left(-\int TrR_{ij}d\tau\right). \tag{135}$$

It follows from Eqs. (129d)–(129f) that the combination $S \equiv I_{11}I_{22} - I_{12}^2$ obeys the equation

$$\frac{dS}{d\tau} + 2TrR_{ij}S = 0. \tag{136}$$

Therefore,

$$\int \mathrm{Tr}R_{ij}\, d\tau = -\ln\sqrt{\frac{D}{D_0}}, \tag{137}$$

where $D_0 \equiv D(0)$. Then, Eqs. (136) and (137) yield

$$|\widetilde{A}|^2 = |\widetilde{A}_0|^2\sqrt{\frac{S}{S_0}}. \tag{138}$$

Note that $1/\sqrt{D}$ can be interpreted as the area of the GB cross section: $1/\sqrt{S} \propto w_1 w_2$. Therefore, Eq. (138) takes the form

$$w_1 w_2 |\widetilde{A}|^2 = w_1(0) w_2(0) |\widetilde{A}_0|^2. \tag{139}$$

Eq. (139) express the conservation of the energy flux in the GB cross section of a single beam.

12. GENERALIZATION OF THE CGO METHOD FOR *N*-ROTATING GBs PROPAGATING ALONG A HELICAL RAY IN NONLINEAR GRADED-INDEX FIBER

In this section, let us consider an axially symmetrical focusing medium in cylindrical coordinates (r, φ, z) with relative electric permittivity:

$$\varepsilon = \varepsilon_{LIN} + g(X) = \varepsilon_0 - \frac{r^2}{L^2} + \frac{X}{1+X}, \tag{140}$$

where

$$X = \sum_{n=1}^{N} \alpha_n I_n / N \tag{141}$$

and $r = \sqrt{x^2 + y^2}$ is the distance from the z-axis (radius in cylindrical symmetry); $I_n = u_n u_n^*$ beam intensity and α_n is the nonlinear coefficient of the nth beam; $L \sim \varepsilon_{LIN}/|\nabla \varepsilon_{LIN}|$ is the characteristic inhomogeneity scale of the fiber, which is related to fiber core radius r_c by the relation $L = r_c/\delta$, where δ is the difference of the constant refractive indexes between core and cladding; and ε_0 is the permittivity along the symmetry axis. The GB, which incidences on the fiber's core is defined by a unit tangent vector $\mathbf{l}$ with respect to the central ray, which poses a component in the azimuthal direction $\mathbf{l}_\varphi$ and along the fiber symmetry axis $\mathbf{l}_z$. As shown by Berczynski (2013a), where

$$\frac{\varepsilon_{NL}}{2} \nabla g(I(\mathbf{r})) = 0, \tag{142}$$

the nonlinearity does not influence the central ray evolution; rather, it influences only the amplitude, the wave front curvature, and the beam width. As a result, GB propagates along a helical ray like in a linear inhomogeneous medium with a constant radius r_c, which equals

$$r_c = -l_\varphi^2 \frac{2\varepsilon}{\varepsilon'}\bigg|_{r=r_c}, \tag{143}$$

where l_φ is the axial component of unit tangent vector **l** and the prime symbol stands for the derivatives with respect to r. Substituting Eq. (140) into Eq. (143), we obtain that the radius of the central ray equals

$$r_c = \frac{l_\varphi\sqrt{\varepsilon_0}}{\sqrt{1+l_\varphi^2}}L, \tag{144}$$

and the torsion of this helical ray takes the form

$$\chi = \frac{l_z l_\varphi}{r_c} = \frac{\sqrt{1-l_\varphi^4}}{\sqrt{\varepsilon_0}L} = \text{const}, \tag{145}$$

where $l_z = \sqrt{1-l_\varphi^2}$, since $l_r = 0$. The normal and binormal to the ray are connected to the unit vectors of the cylindrical coordinates as

$$\mathbf{n} \equiv -\mathbf{e}_r, \quad \mathbf{b} \equiv -\left(l_z\mathbf{e}_\varphi - l_\varphi\mathbf{e}_z\right). \tag{146}$$

By using Eq. (109), we introduce a coordinate system corresponding to the parallel transport:

$$\begin{aligned} \mathbf{e}_1 &= -\mathbf{e}_r \cos\varphi - \left(l_z\mathbf{e}_\varphi - l_\varphi\mathbf{e}_z\right)\sin\varphi, \\ \mathbf{e}_2 &= -\left(l_z\mathbf{e}_\varphi - l_\varphi\mathbf{e}_z\right)\cos\varphi + \mathbf{e}_r\sin\varphi, \end{aligned} \tag{147}$$

where

$$\varphi = -\frac{\sqrt{1-l_\varphi^2}}{L}\tau, \tag{148}$$

is the rotation angle of the normal **n** around the ray. To solve Eq. (125), it is necessary to determine functions $\alpha_{ij}(\tau)$ and $\chi_{ij}(\tau)$ in a coordinate system (τ, ξ_1, ξ_2) with unit vectors in Eq. (90). Taking into account that $\partial/\partial\xi_i = (\mathbf{e}_i\nabla)$, and that for an axially inhomogeneous medium, $\nabla = \mathbf{e}_r d/dr$, one can rewrite Eq. (126) as

$$\alpha_{ij} = (\mathbf{e}_i\mathbf{e}_r)(\mathbf{e}_j\mathbf{e}_r)\left(\frac{\varepsilon''_{LIN}}{2} - \frac{3\varepsilon'^2_{LIN}}{4\varepsilon_{LIN}}\right), \tag{149}$$

where the prime symbol stands for the derivative with respect to r. In view of Eqs. (146) and (147), the α parameters take the form

$$\alpha_{11} = -\frac{1+3l_\varphi^2}{L^2}\cos^2\varphi, \quad \alpha_{12} = \alpha_{21} = \frac{1+3l_\varphi^2}{L^2}\sin\varphi\cos\varphi,$$
$$\alpha_{22} = -\frac{1+3l_\varphi^2}{L^2}\sin^2\varphi. \tag{150}$$

From Eq. (140), we can also calculate coefficients $\chi_{ij}(\tau)$, which for the case of N-rotating GBs have the form

$$\chi_{ij} = \sum_{n=1}^{N}\left(\frac{\varepsilon_S}{2}\frac{\partial^2 I_n(\mathbf{r})}{\partial\xi_i\partial\xi_j}\frac{(1-X)}{(1+X)^2}\right)_{\mathbf{r}=\mathbf{r}_c}. \tag{151}$$

As a result, Eq. (124) can be presented for the case of N-rotating GBs in the matrix form

$$^nB'_{ij} + {}^nB_{ik}{}^nB_{kj} = \alpha_{ij} + \chi_{ij}. \tag{152}$$

Thus, Eq. (152) describes N-coupled matrixes, where prime denotes derivative $d/d\tau$. The transport equation for N GBs reduces to first-order differential-matrix equations in the form

$$({}^nA^2)' + Tr{}^nB_{ij}{}^nA^2 = 0. \tag{153}$$

Thus, having calculated parameters $^nB_{ij}$ from Eq. (152), we can determine amplidtude of the nth GB in the form

$$^nA^2 = {}^nA_0^2\exp\left(-\sum_{n=1}^{N}\int Tr{}^nB_{ij}d\tau\right), \tag{154}$$

where $^n\widetilde{A}_0 = {}^n\widetilde{A}(0)$ is the initial amplitude of the beam. The absolute value of $^n\widetilde{A}$ equals

$$|{}^n\widetilde{A}|^2 = |{}^n\widetilde{A}_0|^2\exp\left(-\int Tr{}^nR_{ij}d\tau\right). \tag{155}$$

The combination $^nS \equiv {}^nI_{11}{}^nI_{22} - {}^nI_{12}^2$ obeys the equation

$$\frac{d({}^nS)}{d\tau} + 2Tr{}^nR_{ij}({}^nS) = 0. \tag{156}$$

Therefore,

$$|^n\widetilde{A}|^2 = |^n\widetilde{A}_0|^2\sqrt{\frac{^nS}{^nS_0}}. \tag{157}$$

Note that $1/\sqrt{^nS} \propto {}^nw_1{}^nw_2$ can be interpreted as the area of the cross section of the nth GB, so that Eq. (157) leads to conservation principles in the matrix form

$$^nw_1{}^nw_2|^nA|^2 = {}^nw_1(0){}^nw_2(0)|^nA_0|^2, \tag{158}$$

where

$$^nw_{1,2}^2 = \frac{4}{k_0\left({}^nI_{11} + {}^nI_{22} \pm \sqrt{({}^nI_{11} - {}^nI_{22})^2 + 4{}^nI_{12}^2}\right)}. \tag{159}$$

13. SINGLE-ROTATING GB. EVOLUTION OF BEAM CROSS SECTION AND WAVE-FRONT CROSS SECTION

For the case of a single-rotating GB, we present numerical simulations based on CGO equations in sections 10 and 11 for the evolution of the GB cross section, evolution of GB widths, evolution of GB wave front curvatures, and evolution of the wave front cross section, as shown in Figs. 2–29.

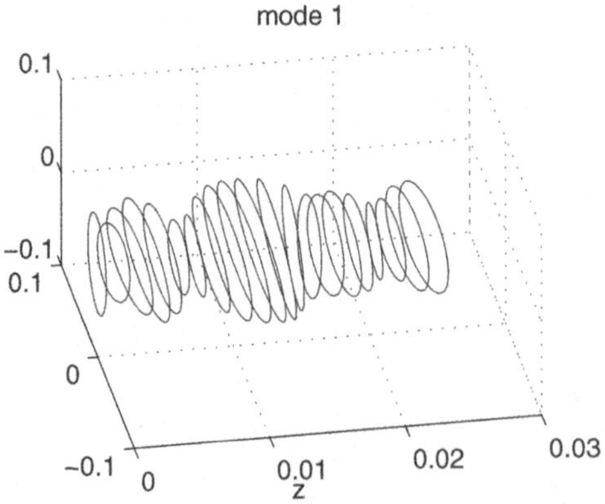

Figure 2 Evolution of GB cross section. Parameters: $\alpha l_0 = 10^{-3}$, $w1_{\min}(0) = 10^{-5}$, $w1_{\max}(0) = 2 \times 10^{-5}$, $R_{12}(0) = -397$, $I_{11}(0) = 1592$, $I_{22}(0) = 397$.

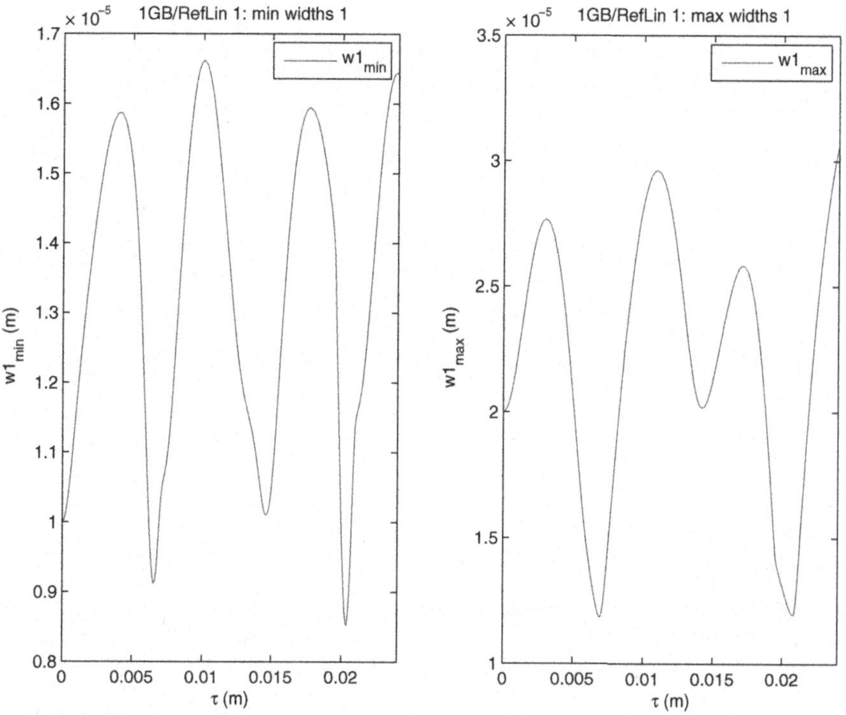

Figure 3 Evolution of GB widths. Parameters: $\alpha l_0 = 10^{-3}$, $w1_{\min}(0) = 10^{-5}$, $w1_{\max}(0) = 2 \times 10^{-5}$, $R_{12}(0) = -397$, $I_{11}(0) = 1592$, $I_{22}(0) = 397$.

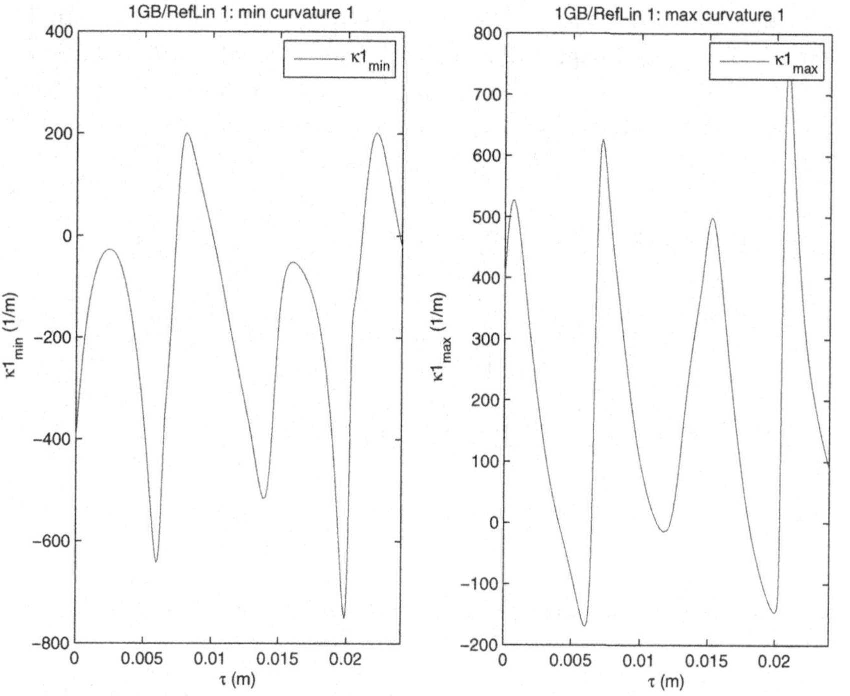

Figure 4 Evolution of GB wave front curvatures. Parameters: $\alpha l_0 = 10^{-3}$, $w1_{\min}(0) = 10^{-5}$, $w1_{\max}(0) = 2 \times 10^{-5}$, $R_{12}(0) = -397$, $I_{11}(0) = 1592$, $I_{22}(0) = 397$.

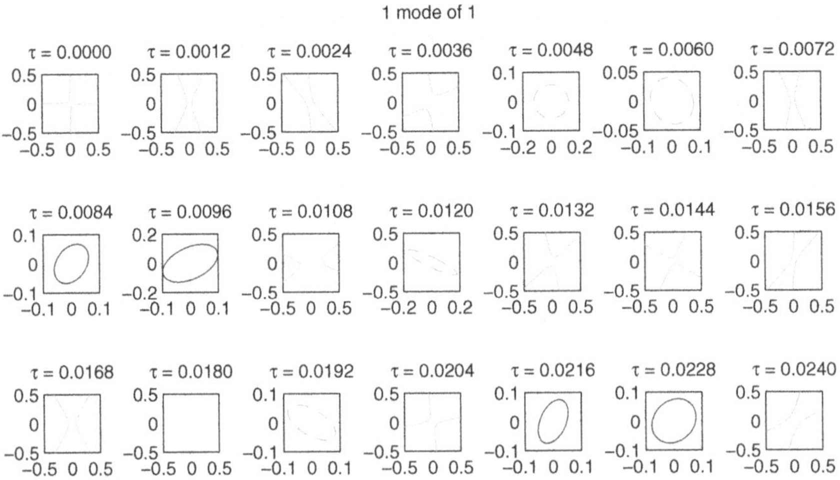

Figure 5 Evolution of wave front cross section. Parameters: $\alpha l_0 = 10^{-3}$, $w1_{\min}(0) = 10^{-5}$, $w1_{\max}(0) = 2 \times 10^{-5}$, $R_{12}(0) = -397$, $l_{11}(0) = 1592$, $l_{22}(0) = 397$.

Figure 2 shows that the GB cross section rotates clockwise, and after the distance of few diffraction lengths, it turns counterclockwise. In Figure 5, it can be seen that the wave front cross section also rotates clockwise and after the distance of few diffraction lengths, it turns counterclockwise. Figures 3 and 4 show the evolution of GB width and GB curvature calculated with the same parameters as in Figures 2 and 5. In Figures 6 and 9, it can be observed that the GB cross section and the wavefront cross section rotate counterclockwise only. Figures 7 and 8 present the evolution of GB widths and GB

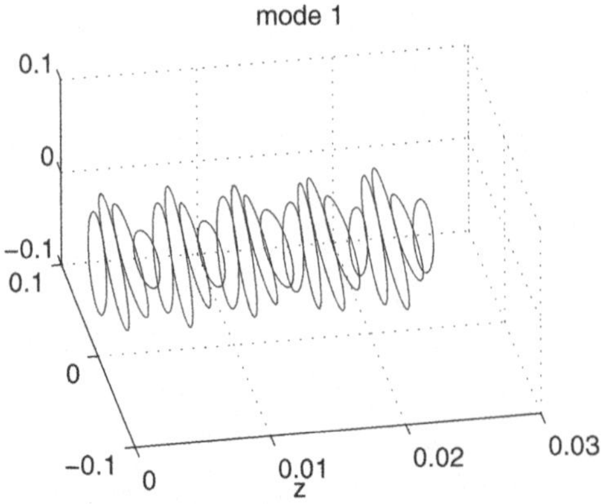

Figure 6 Evolution of GB cross section. Parameters: $\alpha l_0 = 10^{-2}$, $w1_{\min}(0) = 10^{-5}$, $w1_{\max}(0) = 2 \times 10^{-5}$, $R_{12}(0) = 397$, $l_{11}(0) = 1592$, $l_{22}(0) = 397$.

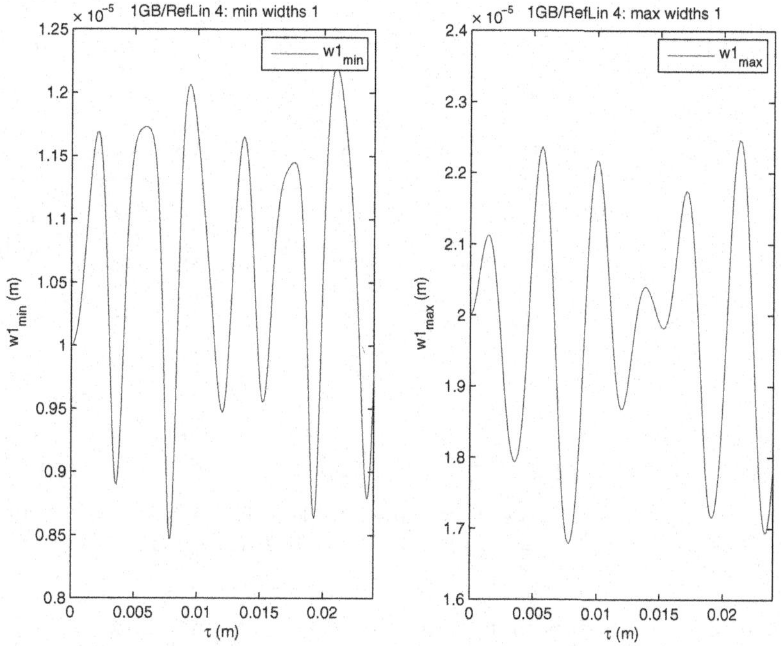

Figure 7 Evolution of GB widths. Parameters: $\alpha l_0 = 10^{-2}$, $w1_{\min}(0) = 10^{-5}$, $w1_{\max}(0) = 2 \times 10^{-5}$, $R_{12}(0) = 397$, $l_{11}(0) = 1592$, $l_{22}(0) = 397$.

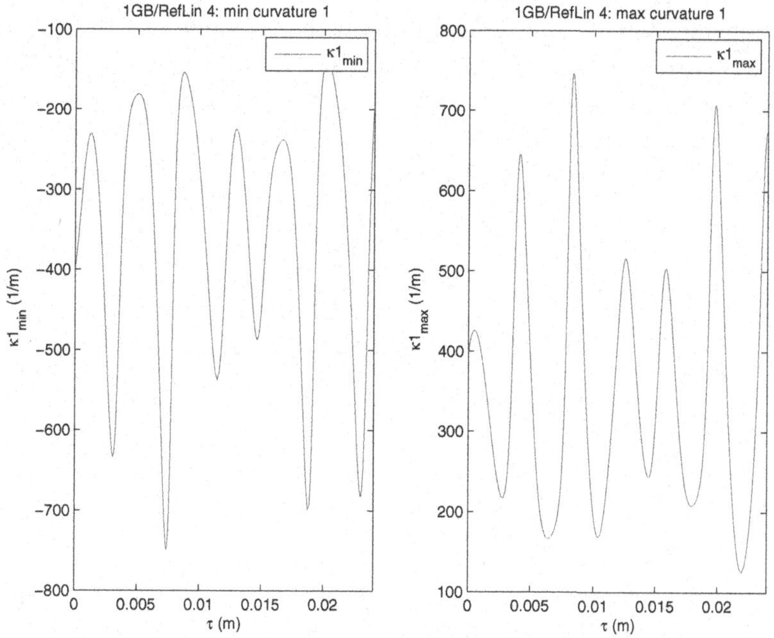

Figure 8 Evolution of GB wave front curvatures. Parameters: $\alpha l_0 = 10^{-2}$, $w1_{\min}(0) = 10^{-5}$, $w1_{\max}(0) = 2 \times 10^{-5}$, $R_{12}(0) = 397$, $l_{11}(0) = 1592$, $l_{22}(0) = 397$.

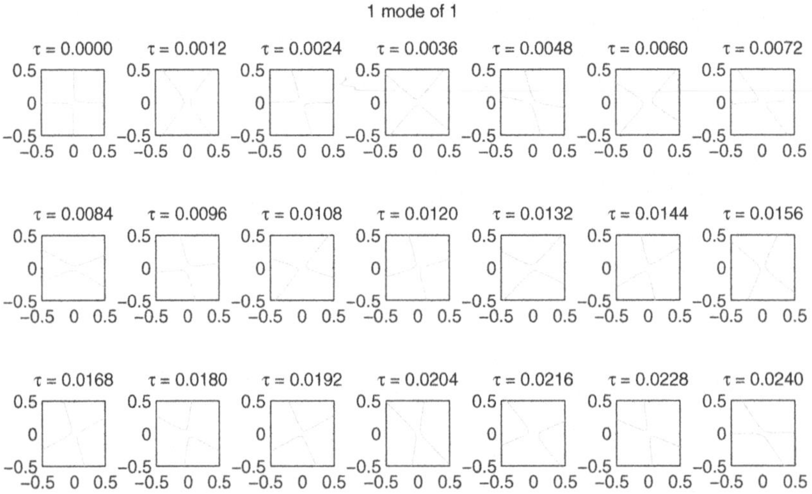

Figure 9 Evolution of wave front cross section. Parameters: $\alpha l_0 = 10^{-2}$, $w1_{\min}(0) = 10^{-5}$, $w1_{\max}(0) = 2 \times 10^{-5}$, $R_{12}(0) = 397$, $I_{11}(0) = 1592$, $I_{22}(0) = 397$.

curvatures calculated with the same parameters as in Figures 6 and 9. Figure 10 illustrates that the GB cross section rotates counterclockwise only. In Figure 13, it can be observed that the wave front cross section rotates clockwise, and after a distance of 10 diffraction lengths, it turns in the opposite direction. Figures 11 and 12 show the evolution of GB width and GB curvature calculated with the same parameters as in Figures 10 and 13. In Figure 14, it can be seen that the GB cross section rotates clockwise only. In

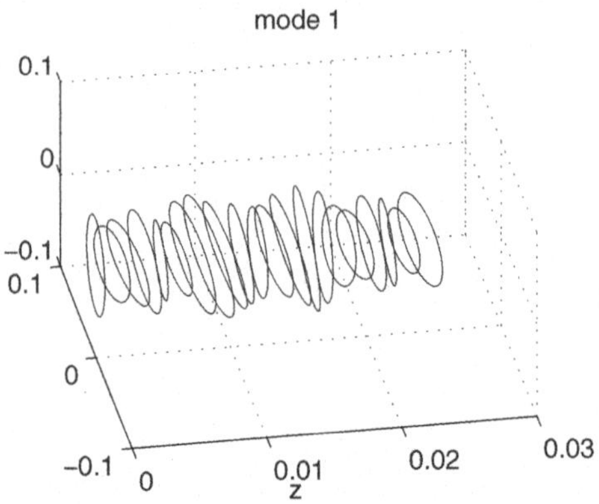

Figure 10 Evolution of GB cross section. Parameters: $\alpha l_0 = 10^{-2}$, $w1_{\min}(0) = 10^{-5}$, $w1_{\max}(0) = 2 \times 10^{-5}$, $R_{11}(0) = 707$, $R_{12}(0) = -397$, $I_{11}(0) = 1592$, $I_{22}(0) = 397$.

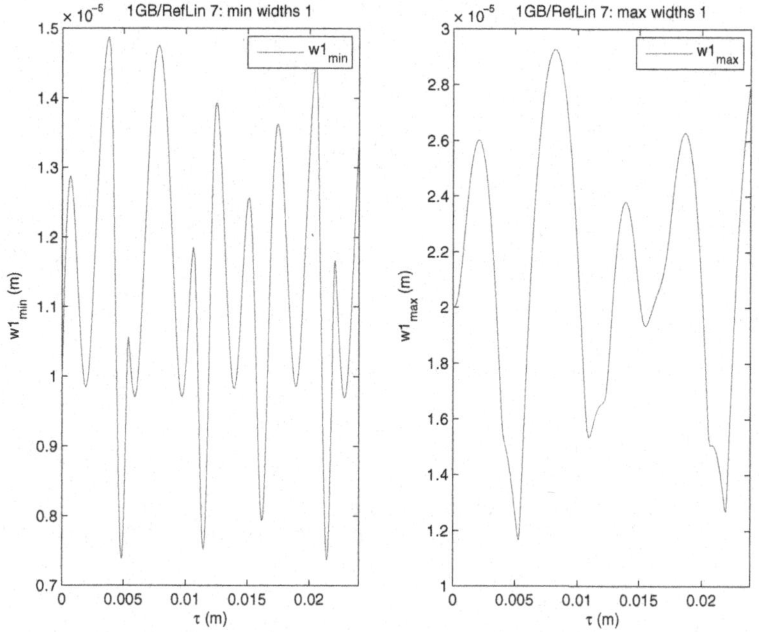

Figure 11 Evolution of GB widths. Parameters: $\alpha l_0 = 10^{-2}$, $w1_{\min}(0) = 10^{-5}$, $w1_{\max}(0) = 2 \times 10^{-5}$, $R_{11}(0) = 707$, $R_{12}(0) = -397$, $I_{11}(0) = 1592$, $I_{22}(0) = 397$.

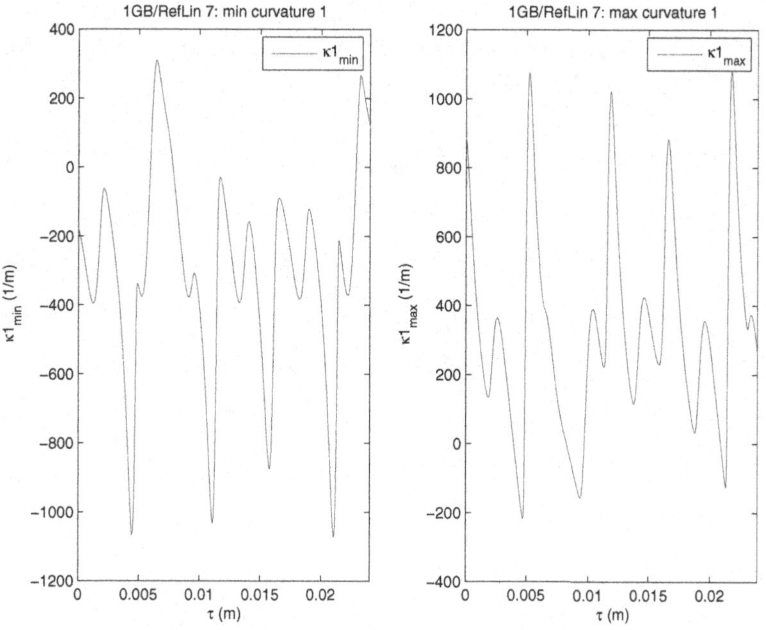

Figure 12 Evolution of GB wave front curvatures. Parameters: $\alpha l_0 = 10^{-2}$, $w1_{\min}(0) = 10^{-5}$, $w1_{\max}(0) = 2 \times 10^{-5}$, $R_{11}(0) = 707$, $R_{12}(0) = -397$, $I_{11}(0) = 1592$, $I_{22}(0) = 397$.

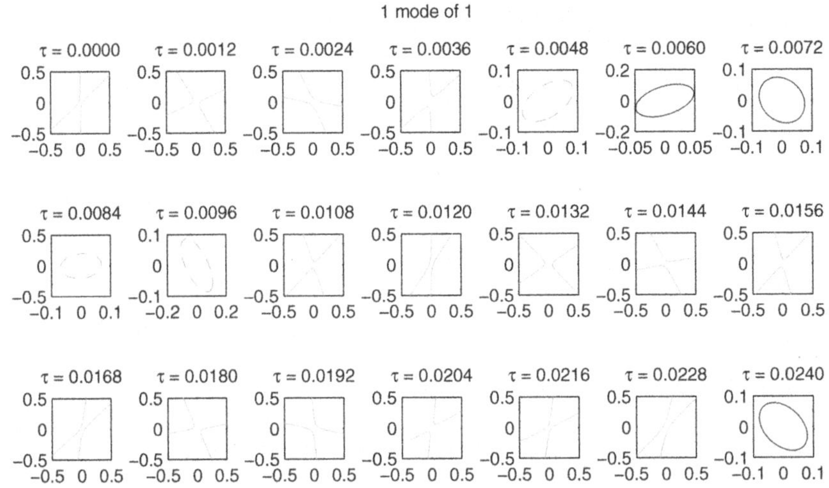

Figure 13 Evolution of wave front cross section. Parameters: $\alpha l_0 = 10^{-2}$, $w1_{\min}(0) = 10^{-5}$, $w1_{\max}(0) = 2 \times 10^{-5}$, $R_{11}(0) = 707$, $R_{12}(0) = -397$, $l_{11}(0) = 1592$, $l_{22}(0) = 397$.

Figure 17, it can be observed that the wave front cross section rotates clockwise and after the distance of a few diffraction lengths, it starts to rotate in the opposite direction. Figures 15 and 16 show the evolution of GB width and GB curvature calculated with the same parameters as in Figures 14 and 17. In Figure 18, it can be seen that the GB cross section rotates counter-clockwise only. In Figure 21, it can be observed that the wave front cross

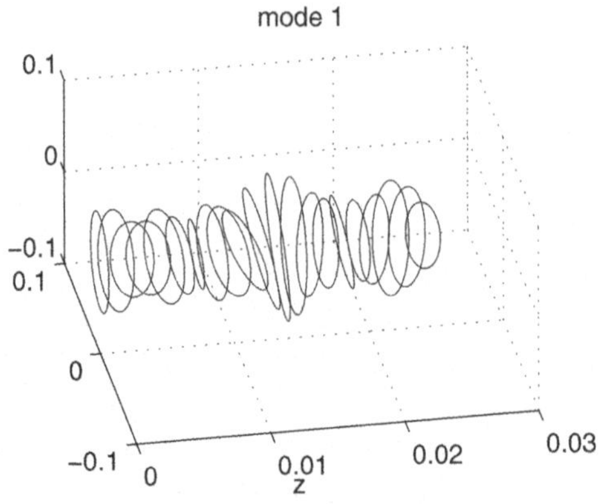

Figure 14 Evolution of GB cross section. Parameters: $\alpha l_0 = 10^{-2}$, $w1_{\min}(0) = 10^{-5}$, $w1_{\max}(0) = 2 \times 10^{-5}$, $R_{11}(0) = 707$, $R_{12}(0) = -397$, $R_{22}(0) = 400$, $l_{11}(0) = 1592$, $l_{22}(0) = 397$.

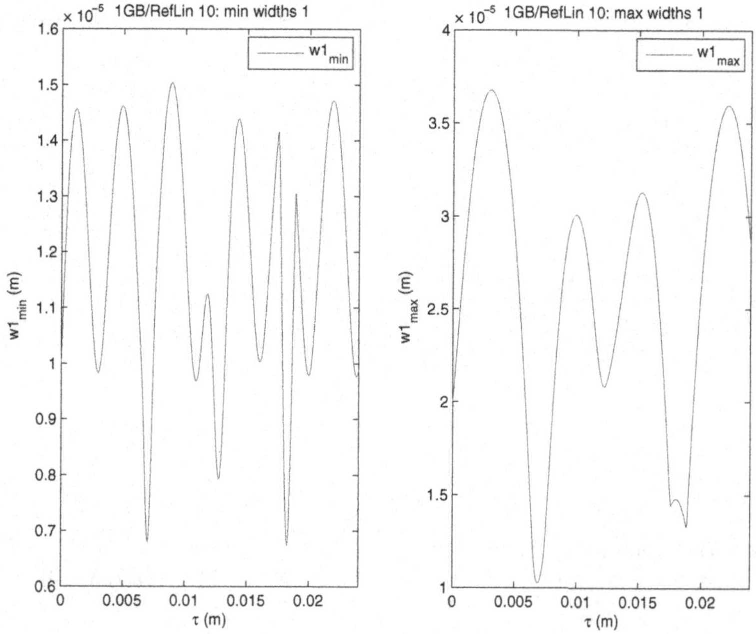

Figure 15 Evolution of GB widths. Parameters: $\alpha l_0 = 10^{-2}$, $w1_{\min}(0) = 10^{-5}$, $w1_{\max}(0) = 2 \times 10^{-5}$, $R_{11}(0) = 707$, $R_{12}(0) = -397$, $R_{22}(0) = 400$, $I_{11}(0) = 1592$, $I_{22}(0) = 397$.

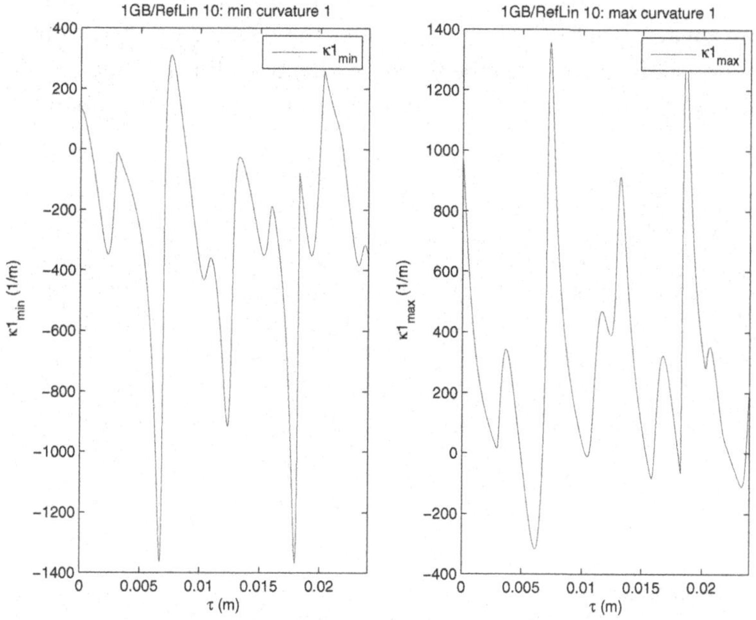

Figure 16 Evolution of GB wave front curvatures. Parameters: $\alpha l_0 = 10^{-2}$, $w1_{\min}(0) = 10^{-5}$, $w1_{\max}(0) = 2 \times 10^{-5}$, $R_{11}(0) = 707$, $R_{12}(0) = -397$, $R_{22}(0) = 400$, $I_{11}(0) = 1592$, $I_{22}(0) = 397$.

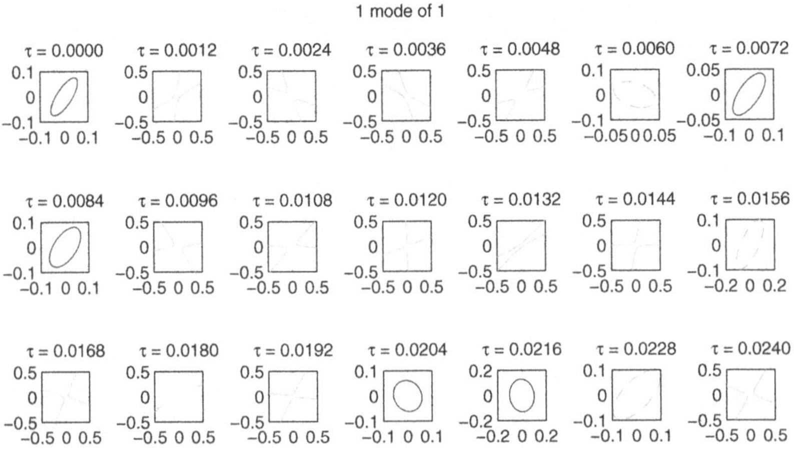

Figure 17 Evolution of wave front cross section. Parameters: $\alpha l_0 = 10^{-2}$, $w1_{\min}(0) = 10^{-5}$, $w1_{\max}(0) = 2 \times 10^{-5}$, $R_{11}(0) = 707$, $R_{12}(0) = -397$, $R_{22}(0) = 400$, $I_{11}(0) = 1592$, $I_{22}(0) = 397$.

section rotates clockwise only. Figures 19 and 20 present the evolution of GB width and GB curvature calculated with the same parameters as in Figures 18 and 21. Figure 22 shows that the GB cross section rotates clockwise only. In Figure 25, it can be observed that the wave front cross section rotates counterclockwise, and after the distance of a few diffraction lengths, it starts to rotate in the opposite direction. It can be observed that after around 10 diffraction lengths, it rotates clockwise, and after 12

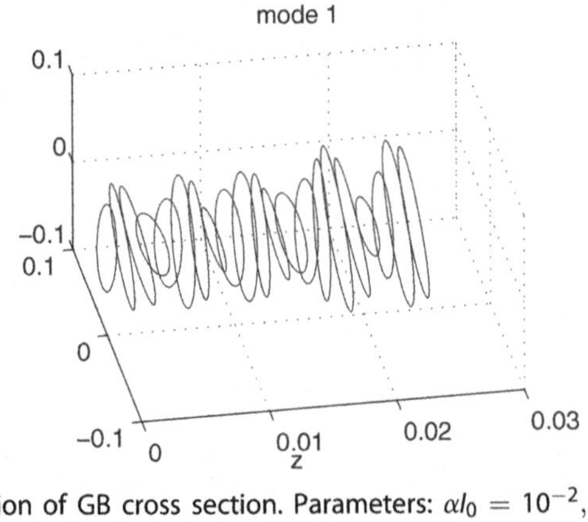

Figure 18 Evolution of GB cross section. Parameters: $\alpha l_0 = 10^{-2}$, $w1_{\min}(0) = 10^{-5}$, $w1_{\max}(0) = 2.4 \times 10^{-5}$, $R_{11}(0) = 0$, $R_{12}(0) = 397$, $R_{22}(0) = 0$, $I_{11}(0) = 1592$, $I_{12}(0) = 397$, $I_{22}(0) = 397$.

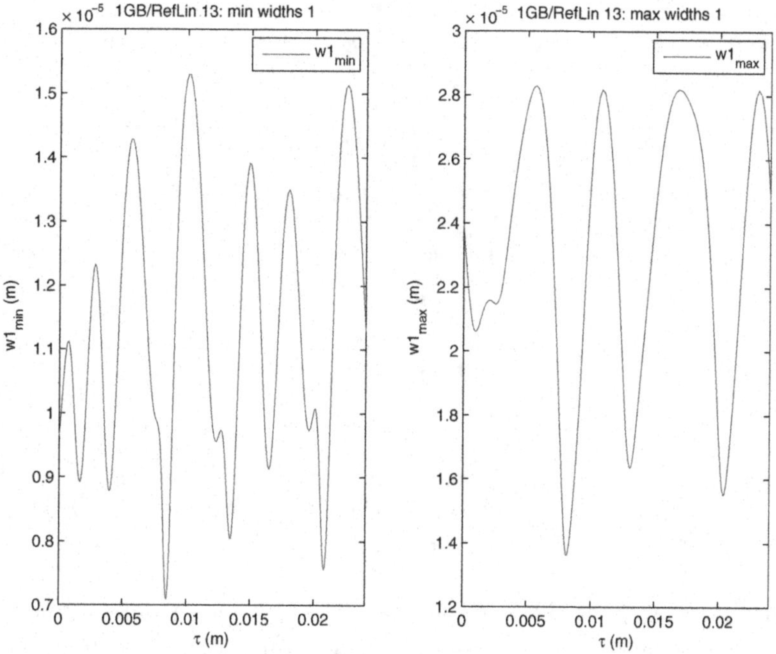

Figure 19 Evolution of GB widths. Parameters: $\alpha I_0 = 10^{-2}$, $w1_{min}(0) = 10^{-5}$, $w1_{max}(0) = 2.4 \times 10^{-5}$, $R_{11}(0) = 0$, $R_{12}(0) = 397$, $R_{22}(0) = 0$, $I_{11}(0) = 1592$, $I_{12}(0) = 397$, $I_{22}(0) = 397$.

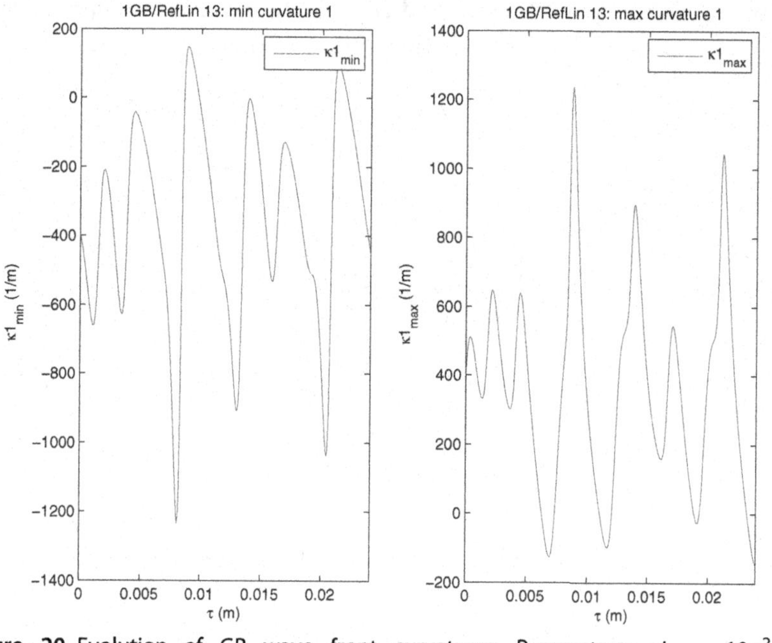

Figure 20 Evolution of GB wave front curvatures. Parameters: $\alpha I_0 = 10^{-2}$, $w1_{min}(0) = 10^{-5}$, $w1_{max}(0) = 2.4 \times 10^{-5}$, $R_{11}(0) = 0$, $R_{12}(0) = 397$, $R_{22}(0) = 0$, $I_{11}(0) = 1592$, $I_{12}(0) = 397$, $I_{22}(0) = 397$.

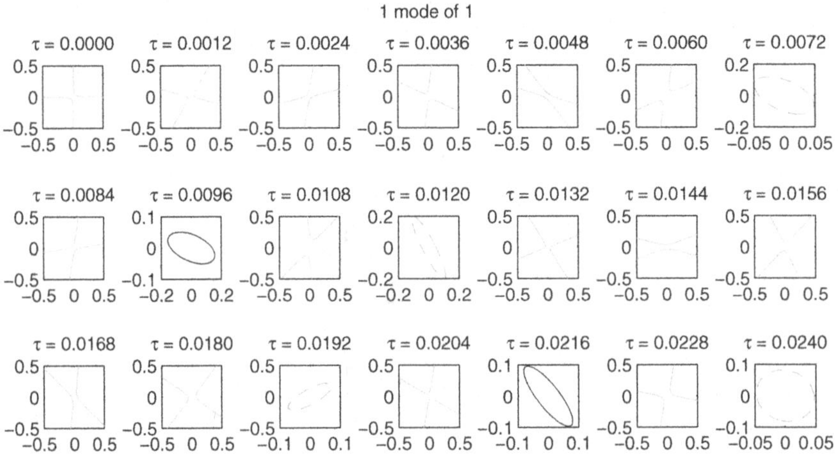

Figure 21 Evolution of wave front cross section. Parameters: $\alpha l_0 = 10^{-2}$, $w1_{\min}(0) = 10^{-5}$, $w1_{\max}(0) = 2.4 \times 10^{-5}$, $R_{11}(0) = 0$, $R_{12}(0) = 397$, $R_{22}(0) = 0$, $I_{11}(0) = 1592$, $I_{12}(0) = 397$, $I_{22}(0) = 397$.

diffraction lengths, it starts to rotate counterclockwise. Figures 23 and 24 show the evolution of GB width and GB curvature calculated with the same parameters as in Figures 22 and 25. In Figure 26, it can be seen that the GB cross section rotates counterclockwise only. In Figure 29, it can be observed that the wave front cross section rotates clockwise only. Figures 27 and 28 show the evolution of GB width and GB curvature calculated with the same parameters as on the two previously mentioned figures.

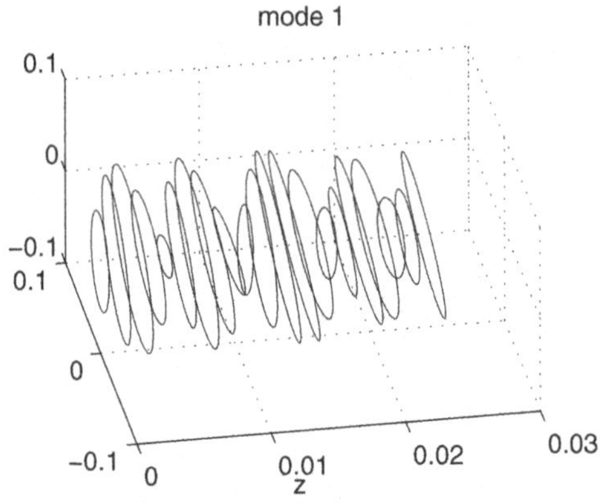

Figure 22 Evolution of GB cross section. Parameters: $\alpha l_0 = 10^{-2}$, $w1_{\min}(0) = 10^{-5}$, $w1_{\max}(0) = 2 \times 10^{-5}$, $R_{11}(0) = -707$, $R_{12}(0) = 397$, $R_{22}(0) = 400$, $I_{11}(0) = 1592$, $I_{12}(0) = 0$, $I_{22}(0) = 397$.

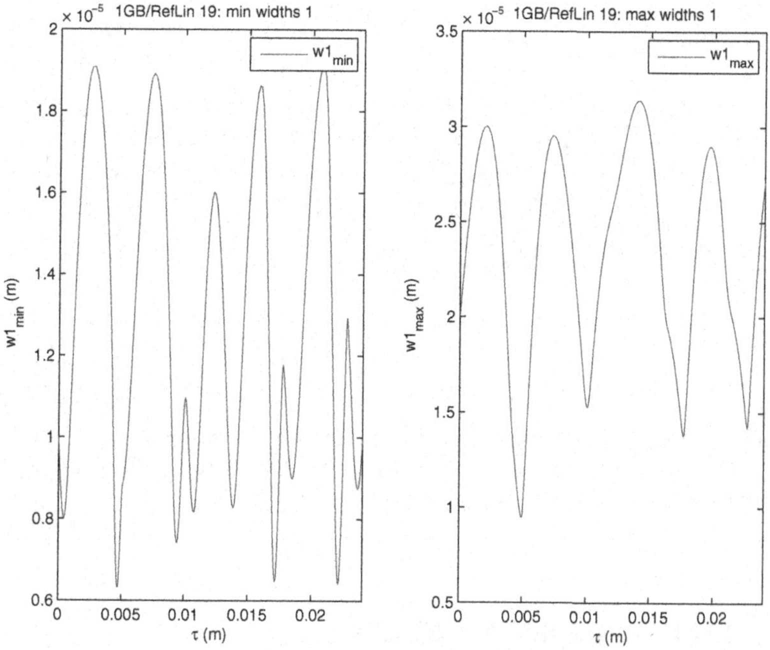

Figure 23 Evolution of GB widths. Parameters: $\alpha l_0 = 10^{-2}$, $w1_{\min}(0) = 10^{-5}$, $w1_{\max}(0) = 2 \times 10^{-5}$, $R_{11}(0) = -707$, $R_{12}(0) = 397$, $R_{22}(0) = 400$, $l_{11}(0) = 1592$, $l_{12}(0) = 0$, $l_{22}(0) = 397$.

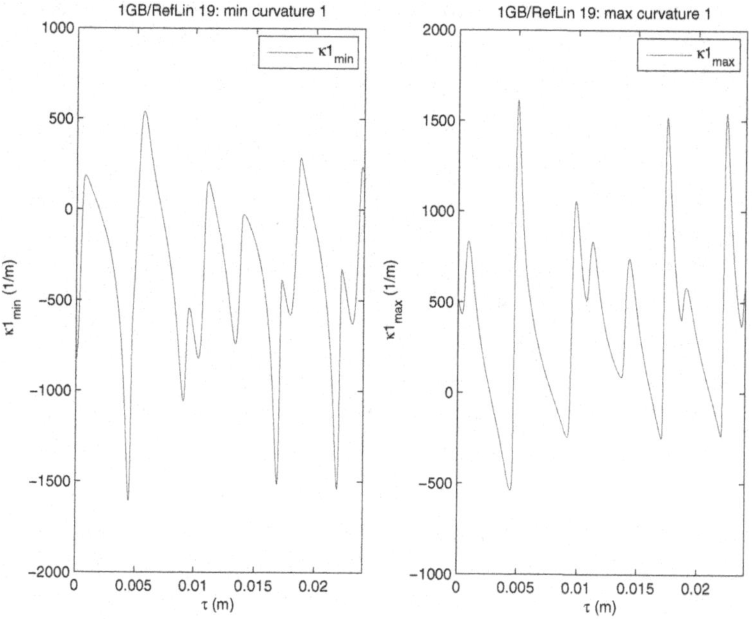

Figure 24 Evolution of GB wave front curvatures. Parameters: $\alpha l_0 = 10^{-2}$, $w1_{\min}(0) = 10^{-5}$, $w1_{\max}(0) = 2 \times 10^{-5}$, $R_{11}(0) = -707$, $R_{12}(0) = 397$, $R_{22}(0) = 400$, $l_{11}(0) = 1592$, $l_{12}(0) = 0$, $l_{22}(0) = 397$.

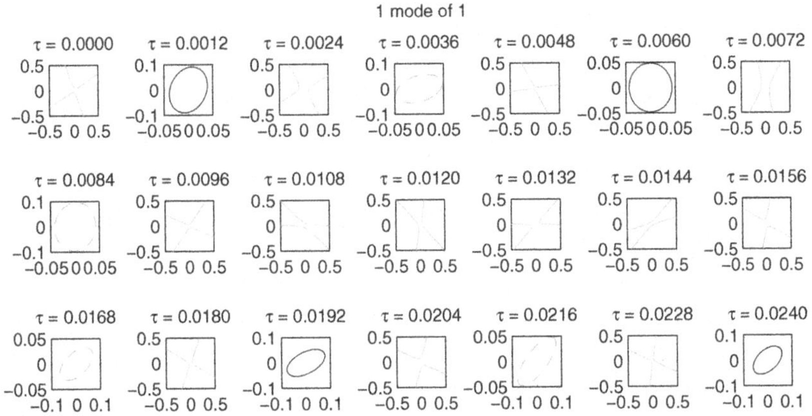

Figure 25 Evolution of wave front cross section. Parameters: $\alpha l_0 = 10^{-2}$, $w1_{\min}(0) = 10^{-5}$, $w1_{\max}(0) = 2 \times 10^{-5}$, $R_{11}(0) = -707$, $R_{12}(0) = 397$, $R_{22}(0) = 400$, $I_{11}(0) = 1592$, $I_{12}(0) = 0$, $I_{22}(0) = 397$.

14. PAIR OF ROTATING GBs

For the case of a pair of rotating GBs, we present numerical simulations based on CGO equations in section 12 for the evolution of GB cross section, evolution of GB widths, evolution of GB spots, evolution of GB wave front curvatures, and evolution of wave front cross section in Figures 30–69.

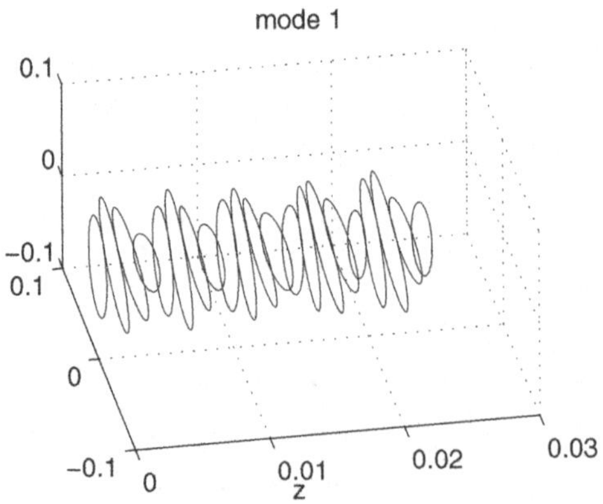

Figure 26 Evolution of GB cross section. Parameters: $\alpha l_0 = 10^{-2}$, $w1_{\min}(0) = 10^{-5}$, $w1_{\max}(0) = 2 \times 10^{-5}$, $R_{11}(0) = 0$, $R_{12}(0) = 397$, $R_{22}(0) = 0$, $I_{11}(0) = 1592$, $I_{12}(0) = 0$, $I_{22}(0) = 397$.

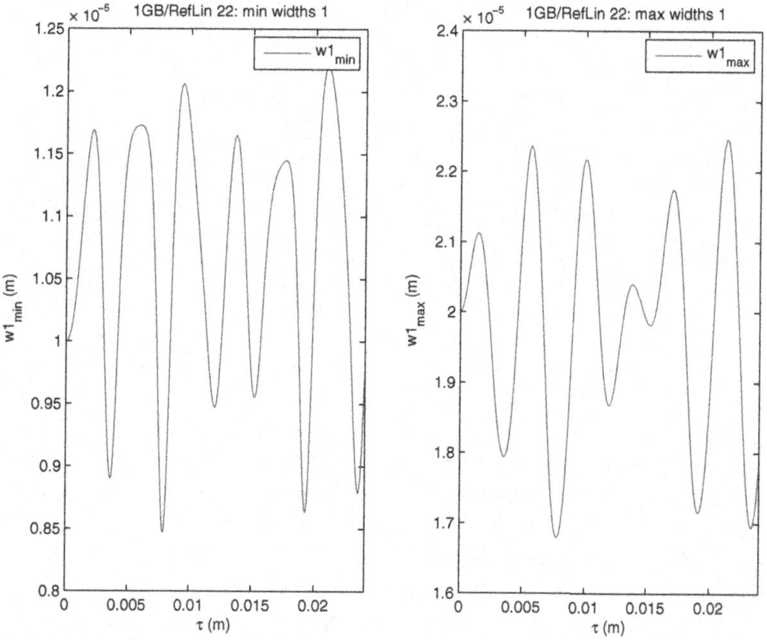

Figure 27 Evolution of GB widths. Parameters: $\alpha l_0 = 10^{-2}$, $w1_{\min}(0) = 10^{-5}$, $w1_{\max}(0) = 2 \times 10^{-5}$, $R_{11}(0) = 0$, $R_{12}(0) = 397$, $R_{22}(0) = 0$, $I_{11}(0) = 1592$, $I_{12}(0) = 0$, $I_{22}(0) = 397$.

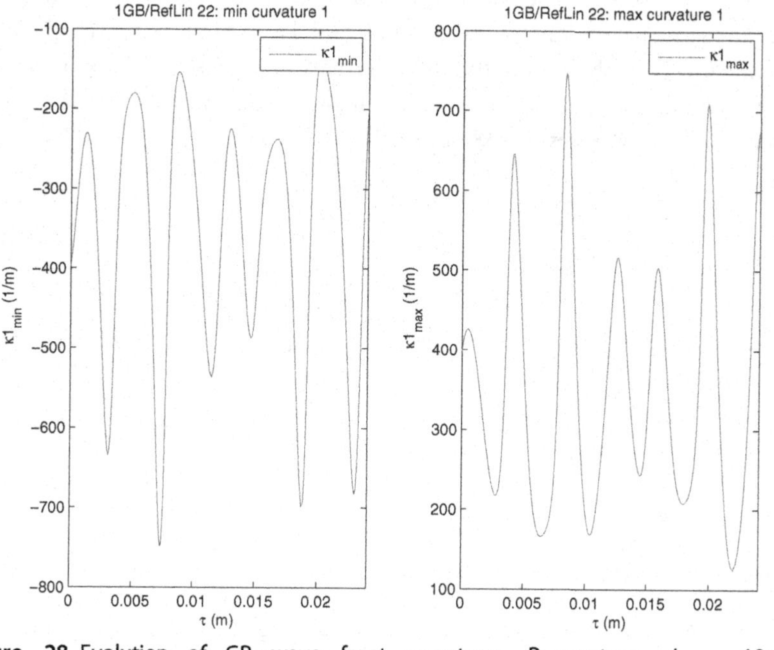

Figure 28 Evolution of GB wave front curvatures. Parameters: $\alpha l_0 = 10^{-2}$, $w1_{\min}(0) = 10^{-5}$, $w1_{\max}(0) = 2 \times 10^{-5}$, $R_{11}(0) = 0$, $R_{12}(0) = 397$, $R_{22}(0) = 0$, $I_{11}(0) = 1592$, $I_{12}(0) = 0$, $I_{22}(0) = 397$.

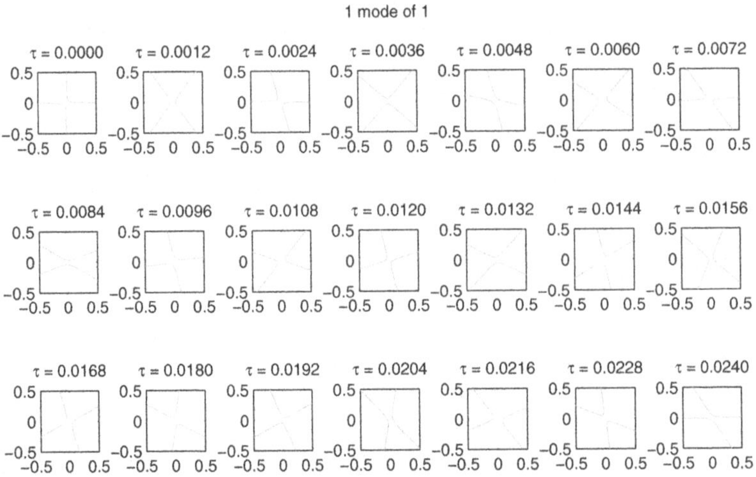

Figure 29 Evolution of wave front cross section. Parameters: $\alpha l_0 = 10^{-2}$, $w1_{\min}(0) = 10^{-5}$, $w1_{\max}(0) = 2 \times 10^{-5}$, $R_{11}(0) = 0$, $R_{12}(0) = 397$, $R_{22}(0) = 0$, $I_{11}(0) = 1592$, $I_{12}(0) = 0$, $I_{22}(0) = 397$.

In Figure 30, it can be observed that the GB cross sections of two interacting beams rotate clockwise only. Figure 34 demonstrates that the wave front cross section of the first mode rotates clockwise, but the wave front cross section of the second mode rotates counterclockwise. Figures 31–33 show the evolution of GB widths, GB spots and GB curvature calculated with the same parameters as in Figures 30 and 34. In Figure 35, it can be seen that the GB cross sections of two modes rotate clockwise only. In Figure 39, it can be observed that wave front cross sections of two modes rotate clockwise only. Figures 36–38 depict the evolution of GB widths, GB spots

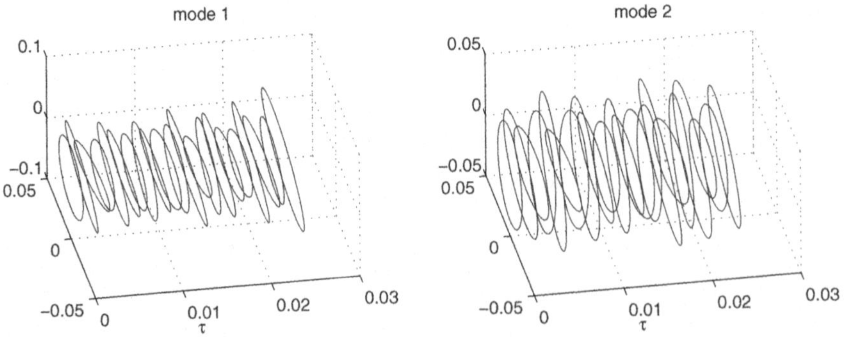

Figure 30 Evolution of GB cross sections. Parameters: $\alpha_1 l_{01} = 10^{-2}$, $\alpha_2 l_{02} = 10^{-1}$, $w1_{\min}(0) = 10^{-5}$, $w1_{\max}(0) = 2 \times 10^{-5}$, $w2_{\min}(0) = 10^{-5}$, $w2_{\max}(0) = 1.5 \times 10^{-5}$, ${}^1R_{11}(0) = 707$, ${}^1R_{12}(0) = 397$, ${}^1R_{22}(0) = 0$, ${}^1I_{11}(0) = 1592$, ${}^1I_{12}(0) = 0$, ${}^1I_{22}(0) = 397$, ${}^2R_{11}(0) = -707$, ${}^2R_{12}(0) = 397$, ${}^2R_{22}(0) = 0$, ${}^2I_{11}(0) = 1592$, ${}^2I_{12}(0) = 0$, ${}^2I_{22}(0) = 707$.

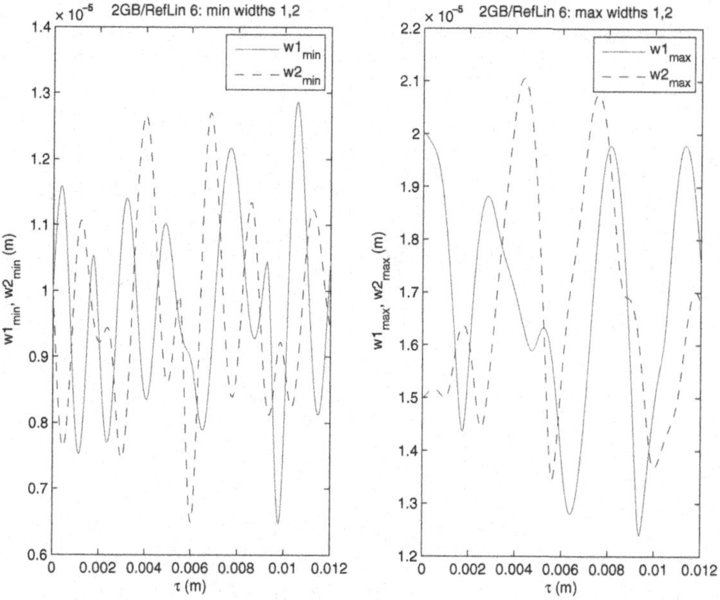

Figure 31 Evolution of GB widths. Parameters: $\alpha_1 l_{01} = 10^{-2}$, $\alpha_2 l_{02} = 10^{-1}$, $w1_{\min}(0) = 10^{-5}$, $w1_{\max}(0) = 2 \times 10^{-5}$, $w2_{\min}(0) = 10^{-5}$, $w2_{\max}(0) = 1.5 \times 10^{-5}$, ${}^1R_{11}(0) = 707$, ${}^1R_{12}(0) = 397$, ${}^1R_{22}(0) = 0$, ${}^1I_{11}(0) = 1592$, ${}^1I_{12}(0) = 0$, ${}^1I_{22}(0) = 397$, ${}^2R_{11}(0) = -707$, ${}^2R_{12}(0) = 397$, ${}^2R_{22}(0) = 0$, ${}^2I_{11}(0) = 1592$, ${}^2I_{12}(0) = 0$, ${}^2I_{22}(0) = 707$. (See the color plate.)

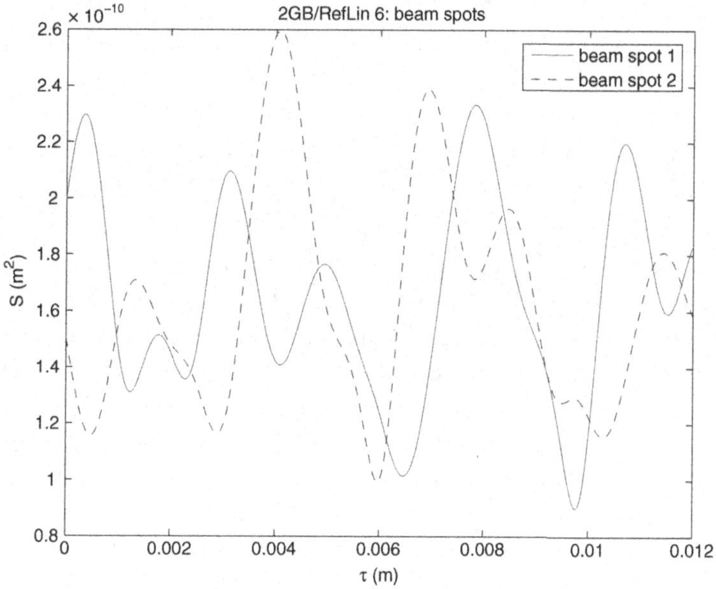

Figure 32 Evolution of GB spots. Parameters: $\alpha_1 l_{01} = 10^{-2}$, $\alpha_2 l_{02} = 10^{-1}$, $w1_{\min}(0) = 10^{-5}$, $w1_{\max}(0) = 2 \times 10^{-5}$, $w2_{\min}(0) = 10^{-5}$, $w2_{\max}(0) = 1.5 \times 10^{-5}$, ${}^1R_{11}(0) = 707$, ${}^1R_{12}(0) = 397$, ${}^1R_{22}(0) = 0$, ${}^1I_{11}(0) = 1592$, ${}^1I_{12}(0) = 0$, ${}^1I_{22}(0) = 397$, ${}^2R_{11}(0) = -707$, ${}^2R_{12}(0) = 397$, ${}^2R_{22}(0) = 0$, ${}^2I_{11}(0) = 1592$, ${}^2I_{12}(0) = 0$, ${}^2I_{22}(0) = 707$. (See the color plate.)

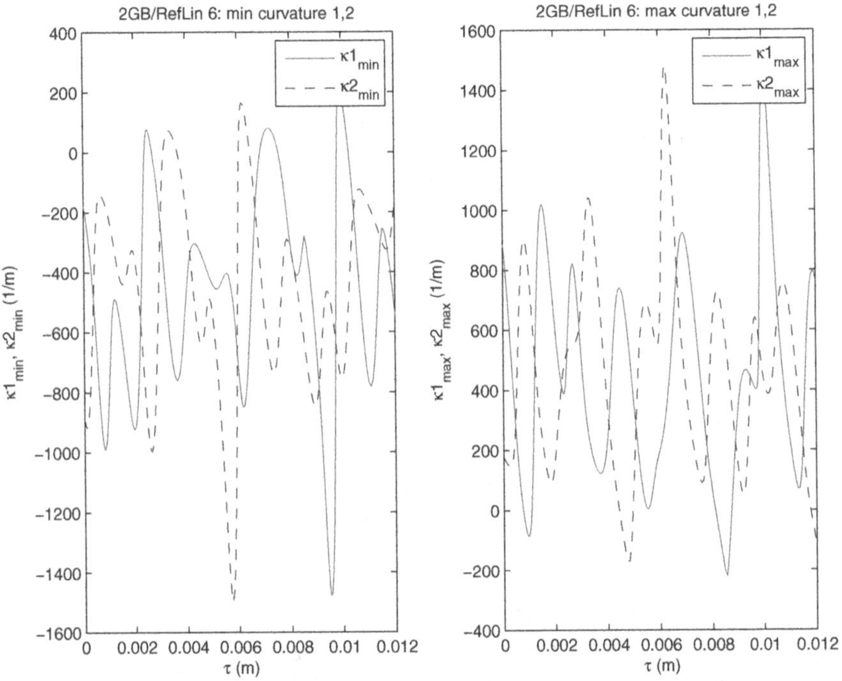

Figure 33 Evolution of GB wave front curvatures. Parameters: $\alpha_1 l_{01} = 10^{-2}$, $\alpha_2 l_{02} = 10^{-1}$, $w1_{\min}(0) = 10^{-5}$, $w1_{\max}(0) = 2 \times 10^{-5}$, $w2_{\min}(0) = 10^{-5}$, $w2_{\max}(0) = 1.5 \times 10^{-5}$, ${}^1R_{11}(0) = 707$, ${}^1R_{12}(0) = 397$, ${}^1R_{22}(0) = 0$, ${}^1I_{11}(0) = 1592$, ${}^1I_{12}(0) = 0$, ${}^1I_{22}(0) = 397$, ${}^2R_{11}(0) = -707$, ${}^2R_{12}(0) = 397$, ${}^2R_{22}(0) = 0$, ${}^2I_{11}(0) = 1592$, ${}^2I_{12}(0) = 0$, ${}^2I_{22}(0) = 707$. (See the color plate.)

and GB curvature calculated with the same parameters as in Figures 35 and 39. Figure 40 shows that GB cross sections of two modes rotate counterclockwise only. In Figure 44, it can be observed that the wave front cross sections of two modes rotate clockwise only. Figures 41–43 show the evolution of GB widths, GB spots and GB curvatures calculated with the same parameters as in Figures 40 and 44. In Figure 45, it can be seen that the GB cross section of the first mode rotates counterclockwise, but the second one rotates counterclockwise. In Figure 49, the wave front cross sections of the two modes rotate counterclockwise. Figures 46–48 represent the evolution of GB widths, GB spots and GB curvature calculated with the same parameters as in Figures 45 and 49. Figure 50 shows that the cross sections of the first mode rotate clockwise, but the cross section of the second mode rotates counterclockwise. In Figure 54, it can be observed that the wave front cross section the first mode rotates clockwise and the second

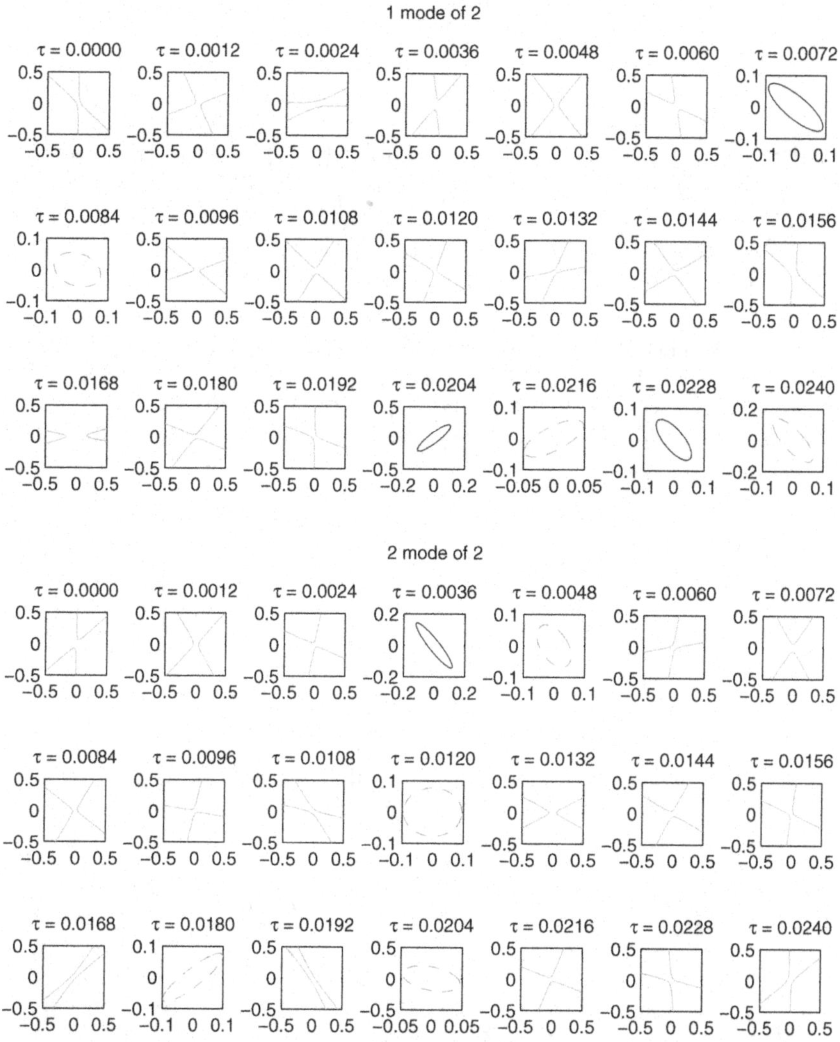

Figure 34 Evolution of wave front cross sections. Parameters: $\alpha_1 I_{01} = 10^{-2}$, $\alpha_2 I_{02} = 10^{-1}$, $w1_{\min}(0) = 10^{-5}$, $w1_{\max}(0) = 2 \times 10^{-5}$, $w2_{\min}(0) = 10^{-5}$, $w2_{\max}(0) = 1.5 \times 10^{-5}$, ${}^1R_{11}(0) = 707$, ${}^1R_{12}(0) = 397$, ${}^1R_{22}(0) = 0$, ${}^1I_{11}(0) = 1592$, ${}^1I_{12}(0) = 0$, ${}^1I_{22}(0) = 397$, ${}^2R_{11}(0) = -707$, ${}^2R_{12}(0) = 397$, ${}^2R_{22}(0) = 0$, ${}^2I_{11}(0) = 1592$, ${}^2I_{12}(0) = 0$, ${}^2I_{22}(0) = 707$.

one rotates counterclockwise; nevertheless, after a few diffraction lengths, they start to rotate in the opposite direction. Figures 51–53 show the evolution of GB widths, GB spots and GB curvature calculated with the same parameters as in Figures 50 and 54. In Figure 55 and 59, it can be observed that the evolution of two overlapping modes, where both GB cross

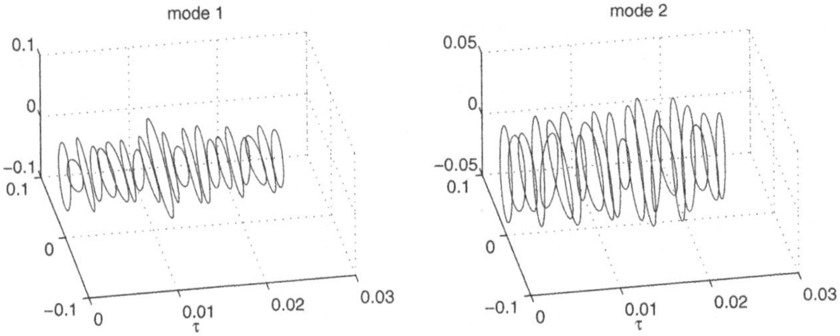

Figure 35 Evolution of GB cross sections. Parameters: $\alpha_1 l_{01} = 10^{-2}$, $\alpha_2 l_{02} = 10^{-1}$, $w1_{\min}(0) = 10^{-5}$, $w1_{\max}(0) = 2 \times 10^{-5}$, $w2_{\min}(0) = 10^{-5}$, $w2_{\max}(0) = 1.5 \times 10^{-5}$, ${}^1R_{11}(0) = 707$, ${}^1R_{12}(0) = -397$, ${}^1R_{22}(0) = 0$, ${}^1I_{11}(0) = 1592$, ${}^1I_{12}(0) = 0$, ${}^1I_{22}(0) = 397$, ${}^2R_{11}(0) = -707$, ${}^2R_{12}(0) = -397$, ${}^2R_{22}(0) = 0$, ${}^2I_{11}(0) = 1592$, ${}^2I_{12}(0) = 0$, ${}^2I_{22}(0) = 707$.

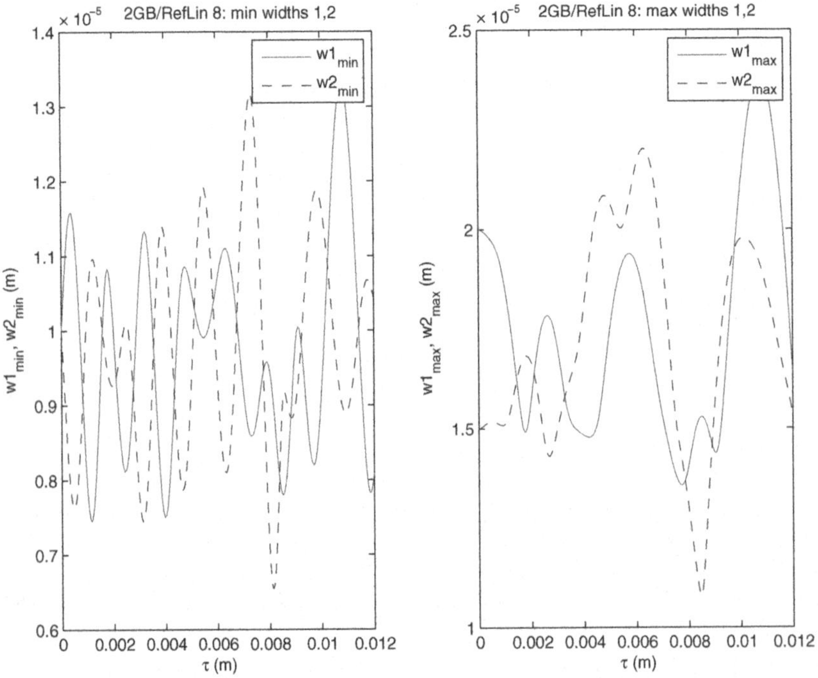

Figure 36 Evolution of GB widths. Parameters: $\alpha_1 l_{01} = 10^{-2}$, $\alpha_2 l_{02} = 10^{-1}$, $w1_{\min}(0) = 10^{-5}$, $w1_{\max}(0) = 2 \times 10^{-5}$, $w2_{\min}(0) = 10^{-5}$, $w2_{\max}(0) = 1.5 \times 10^{-5}$, ${}^1R_{11}(0) = 707$, ${}^1R_{12}(0) = 397$, ${}^1R_{22}(0) = 0$, ${}^1I_{11}(0) = 1592$, ${}^1I_{12}(0) = 0$, ${}^1I_{22}(0) = 397$, ${}^2R_{11}(0) = -707$, ${}^2R_{12}(0) = 397$, ${}^2R_{22}(0) = 0$, ${}^2I_{11}(0) = 1592$, ${}^2I_{12}(0) = 0$, ${}^2I_{22}(0) = 707$. (See the color plate.)

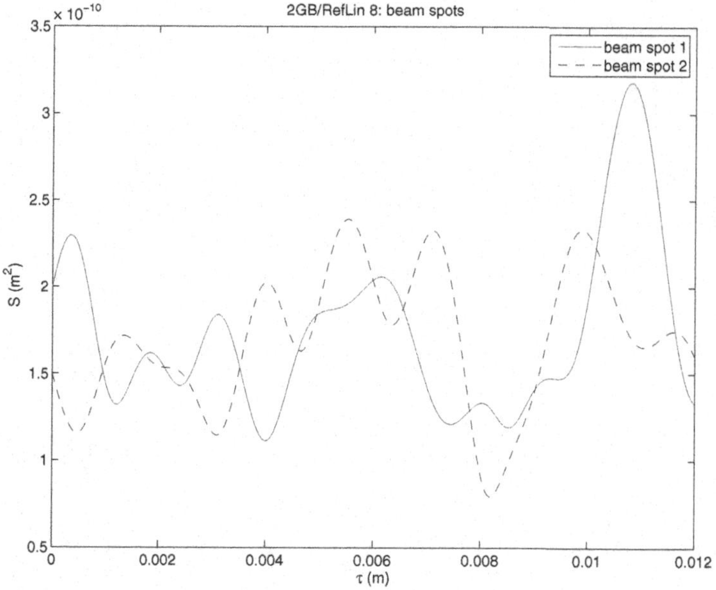

Figure 37 Evolution of GB spots. Parameters: $\alpha_1 l_{01} = 10^{-2}$, $\alpha_2 l_{02} = 10^{-1}$, $w1_{\min}(0) = 10^{-5}$, $w1_{\max}(0) = 2 \times 10^{-5}$, $w2_{\min}(0) = 10^{-5}$, $w2_{\max}(0) = 1.5 \times 10^{-5}$, ${}^1R_{11}(0) = 707$, ${}^1R_{12}(0) = 397$, ${}^1R_{22}(0) = 0$, ${}^1I_{11}(0) = 1592$, ${}^1I_{12}(0) = 0$, ${}^1I_{22}(0) = 397$, ${}^2R_{11}(0) = -707$, ${}^2R_{12}(0) = 397$, ${}^2R_{22}(0) = 0$, ${}^2I_{11}(0) = 1592$, ${}^2I_{12}(0) = 0$, ${}^2I_{22}(0) = 707$. (See the color plate.)

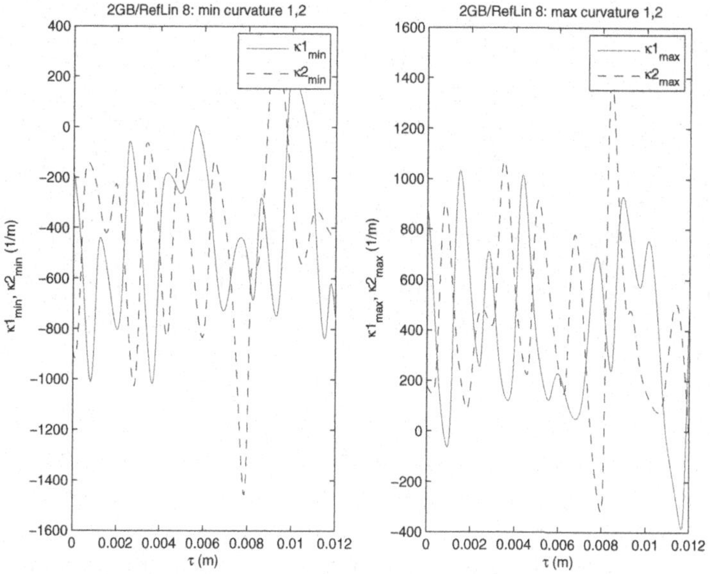

Figure 38 Evolution of GB wave front curvatures. Parameters: $\alpha_1 l_{01} = 10^{-2}$, $\alpha_2 l_{02} = 10^{-1}$, $w1_{\min}(0) = 10^{-5}$, $w1_{\max}(0) = 2 \times 10^{-5}$, $w2_{\min}(0) = 10^{-5}$, $w2_{\max}(0) = 1.5 \times 10^{-5}$, ${}^1R_{11}(0) = 707$, ${}^1R_{12}(0) = 397$, ${}^1R_{22}(0) = 0$, ${}^1I_{11}(0) = 1592$, ${}^1I_{12}(0) = 0$, ${}^1I_{22}(0) = 397$, ${}^2R_{11}(0) = -707$, ${}^2R_{12}(0) = 397$, ${}^2R_{22}(0) = 0$, ${}^2I_{11}(0) = 1592$, ${}^2I_{12}(0) = 0$, ${}^2I_{22}(0) = 707$. (See the color plate.)

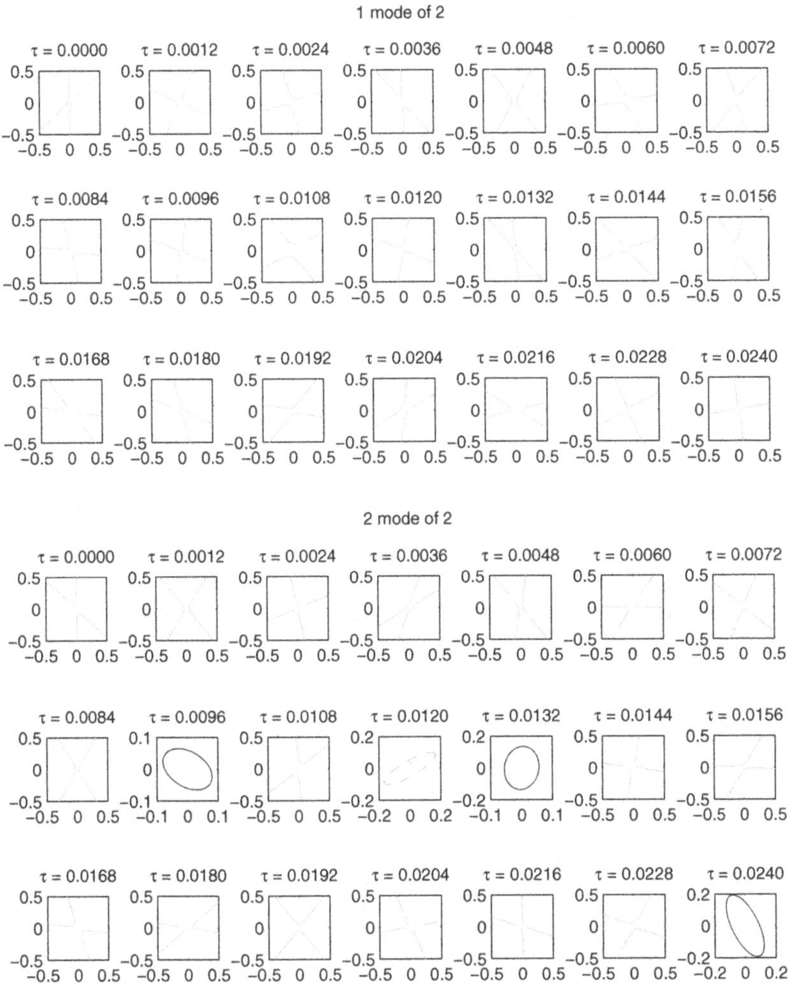

Figure 39 Evolution of wave front cross sections. Parameters: $\alpha_1 l_{01} = 10^{-2}$, $\alpha_2 l_{02} = 10^{-1}$, $w1_{\min}(0) = 10^{-5}$, $w1_{\max}(0) = 2 \times 10^{-5}$, $w2_{\min}(0) = 10^{-5}$, $w2_{\max}(0) = 1.5 \times 10^{-5}$, ${}^1R_{11}(0) = 707$, ${}^1R_{12}(0) = 397$, ${}^1R_{22}(0) = 0$, ${}^1I_{11}(0) = 1592$, ${}^1I_{12}(0) = 0$, ${}^1I_{22}(0) = 397$, ${}^2R_{11}(0) = -707$, ${}^2R_{12}(0) = 397$, ${}^2R_{22}(0) = 0$, ${}^2I_{11}(0) = 1592$, ${}^2I_{12}(0) = 0$, ${}^2I_{22}(0) = 707$.

sections and wave front cross sections evolve in the same manner. In Fig. 60–64 we observe also the evolution two overlapping modes, but with another parameter set. Figures 60 and 64 also show the evolution two overlapping modes. In Figure 65, GB cross sections rotate in the opposite direction. In Figure 69, it can be observed that the wave front cross section

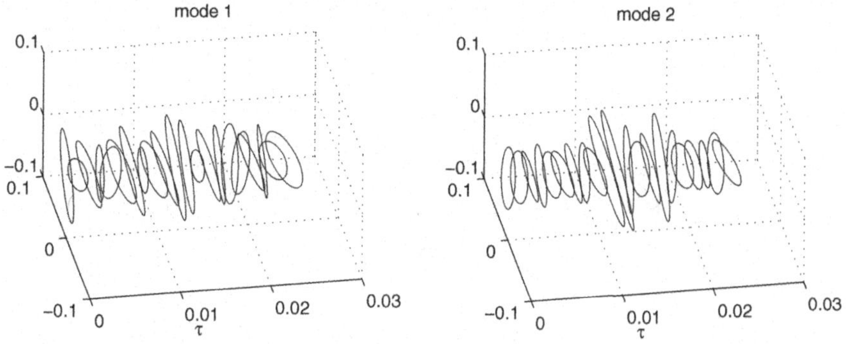

Figure 40 Evolution of GB cross sections. Parameters: $\alpha_1 I_{01} = 10^{-2}$, $\alpha_2 I_{02} = 10^{-1}$, $w1_{\min}(0) = 10^{-5}$, $w1_{\max}(0) = 2.4 \times 10^{-5}$, $w2_{\min}(0) = 10^{-5}$, $w2_{\max}(0) = 2.4 \times 10^{-5}$, $^1R_{11}(0) = 0$, $^1R_{12}(0) = -397$, $^1R_{22}(0) = 0$, $^1I_{11}(0) = 1592$, $^1I_{12}(0) = -397$, $^1I_{22}(0) = 397$, $^2R_{11}(0) = 0$, $^2R_{12}(0) = -397$, $^2R_{22}(0) = 0$, $^2I_{11}(0) = 1592$, $^2I_{12}(0) = 397$, $^2I_{22}(0) = 397$.

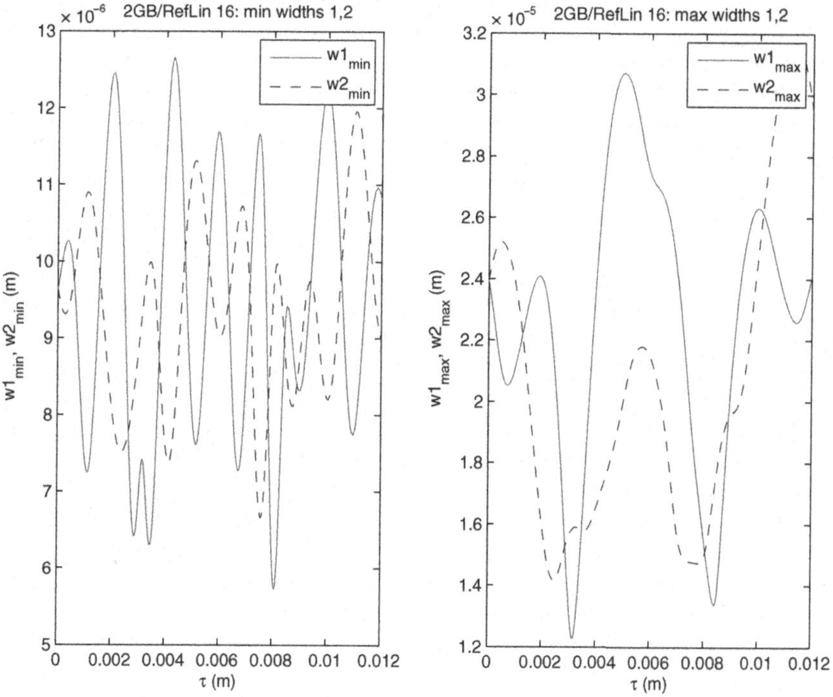

Figure 41 Evolution of GB widths. Parameters: $\alpha_1 I_{01} = 10^{-2}$, $\alpha_2 I_{02} = 10^{-1}$, $w1_{\min}(0) = 10^{-5}$, $w1_{\max}(0) = 2.4 \times 10^{-5}$, $w2_{\min}(0) = 10^{-5}$, $w2_{\max}(0) = 2.4 \times 10^{-5}$, $^1R_{11}(0) = 0$, $^1R_{12}(0) = -397$, $^1R_{22}(0) = 0$, $^1I_{11}(0) = 1592$, $^1I_{12}(0) = -397$, $^1I_{22}(0) = 397$, $^2R_{11}(0) = 0$, $^2R_{12}(0) = -397$, $^2R_{22}(0) = 0$, $^2I_{11}(0) = 1592$, $^2I_{12}(0) = 397$, $^2I_{22}(0) = 397$. (See the color plate.)

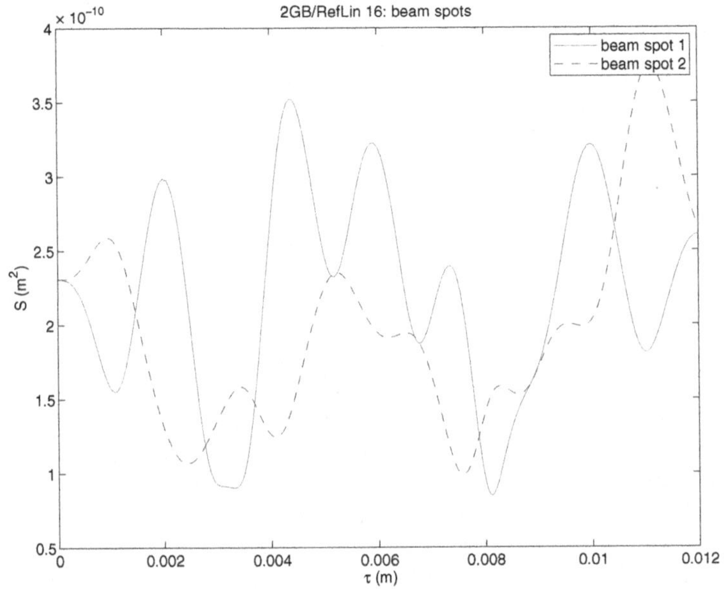

Figure 42 Evolution of GB spots. Parameters: $\alpha_1 l_{01} = 10^{-2}$, $\alpha_2 l_{02} = 10^{-1}$, $w1_{\min}(0) = 10^{-5}$, $w1_{\max}(0) = 2.4 \times 10^{-5}$, $w2_{\min}(0) = 10^{-5}$, $w2_{\max}(0) = 2.4 \times 10^{-5}$, ${}^1R_{11}(0) = 0$, ${}^1R_{12}(0) = -397$, ${}^1R_{22}(0) = 0$, ${}^1I_{11}(0) = 1592$, ${}^1I_{12}(0) = -397$, ${}^1I_{22}(0) = 397$, ${}^2R_{11}(0) = 0$, ${}^2R_{12}(0) = -397$, ${}^2R_{22}(0) = 0$, ${}^2I_{11}(0) = 1592$, ${}^2I_{12}(0) = 397$, ${}^2I_{22}(0) = 397$. (See the color plate.)

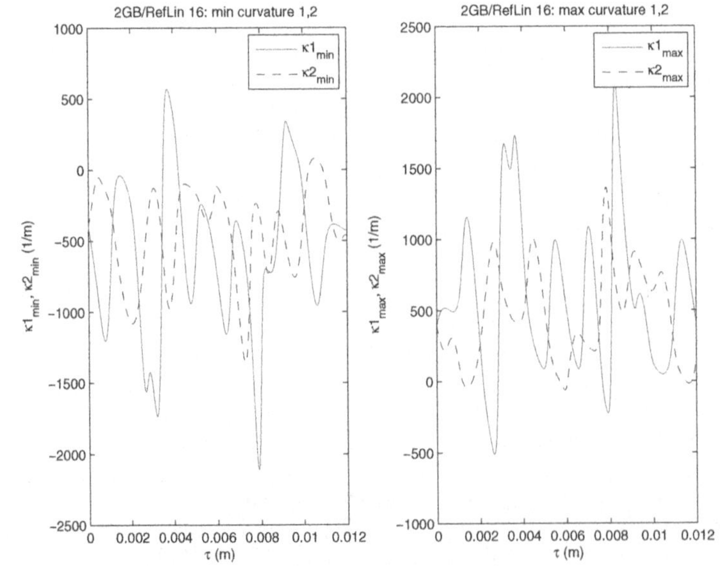

Figure 43 Evolution of GB wave front curvatures. Parameters: $\alpha_1 l_{01} = 10^{-2}$, $\alpha_2 l_{02} = 10^{-1}$, $w1_{\min}(0) = 10^{-5}$, $w1_{\max}(0) = 2.4 \times 10^{-5}$, $w2_{\min}(0) = 10^{-5}$, $w2_{\max}(0) = 2.4 \times 10^{-5}$, ${}^1R_{11}(0) = 0$, ${}^1R_{12}(0) = -397$, ${}^1R_{22}(0) = 0$, ${}^1I_{11}(0) = 1592$, ${}^1I_{12}(0) = -397$, ${}^1I_{22}(0) = 397$, ${}^2R_{11}(0) = 0$, ${}^2R_{12}(0) = -397$, ${}^2R_{22}(0) = 0$, ${}^2I_{11}(0) = 1592$, ${}^2I_{12}(0) = 397$, ${}^2I_{22}(0) = 397$. (See the color plate.)

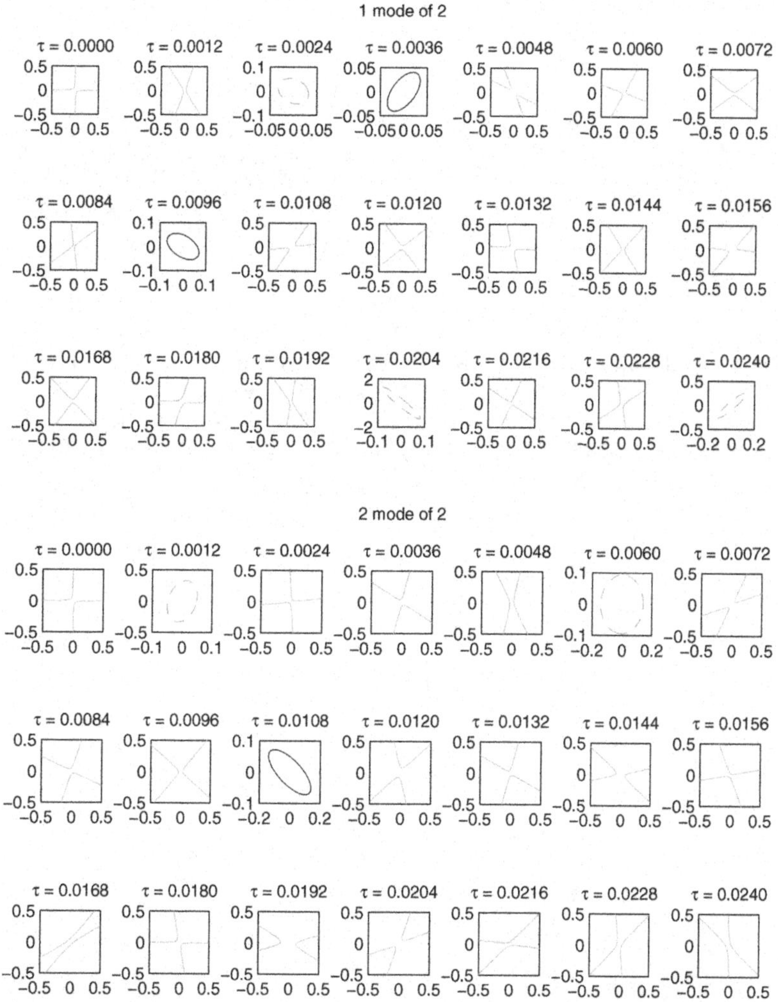

Figure 44 Evolution of wave front cross sections. Parameters: $\alpha_1 l_{01} = 10^{-2}$, $\alpha_2 l_{02} = 10^{-1}$, $w1_{\min}(0) = 10^{-5}$, $w1_{\max}(0) = 2.4 \times 10^{-5}$, $w2_{\min}(0) = 10^{-5}$, $w2_{\max}(0) = 2.4 \times 10^{-5}$, ${}^1R_{11}(0) = 0$, ${}^1R_{12}(0) = -397$, ${}^1R_{22}(0) = 0$, ${}^1I_{11}(0) = 1592$, ${}^1I_{12}(0) = -397$, ${}^1I_{22}(0) = 397$, ${}^2R_{11}(0) = 0$, ${}^2R_{12}(0) = -397$, ${}^2R_{22}(0) = 0$, ${}^2I_{11}(0) = 1592$, ${}^2I_{12}(0) = 397$, ${}^2I_{22}(0) = 397$.

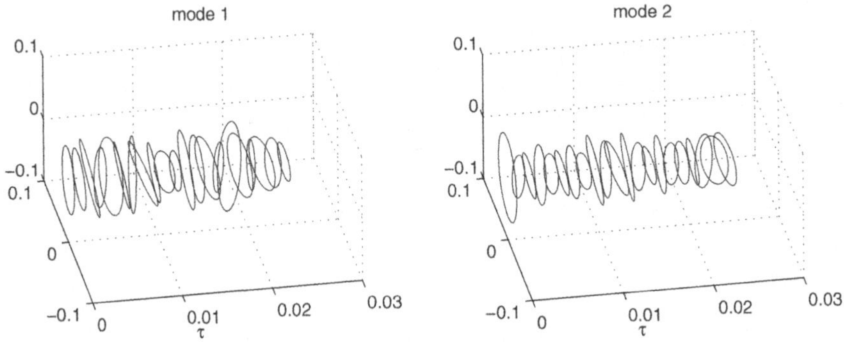

Figure 45 Evolution of GB cross sections. Parameters: $\alpha_1 l_{01} = 10^{-2}$, $\alpha_2 l_{02} = 10^{-1}$, $w1_{\min}(0) = 10^{-5}$, $w1_{\max}(0) = 2 \times 10^{-5}$, $w2_{\min}(0) = 1.5 \times 10^{-5}$, $w2_{\max}(0) = 2.5 \times 10^{-5}$, ${}^1R_{11}(0) = 0$, ${}^1R_{12}(0) = 397$, ${}^1R_{22}(0) = 0$, ${}^1I_{11}(0) = 1592$, ${}^1I_{12}(0) = 0$, ${}^1I_{22}(0) = 397$, ${}^2R_{11}(0) = -397$, ${}^2R_{12}(0) = -397$, ${}^2R_{22}(0) = 0$, ${}^2I_{11}(0) = 707$, ${}^2I_{12}(0) = 0$, ${}^2I_{22}(0) = 254$.

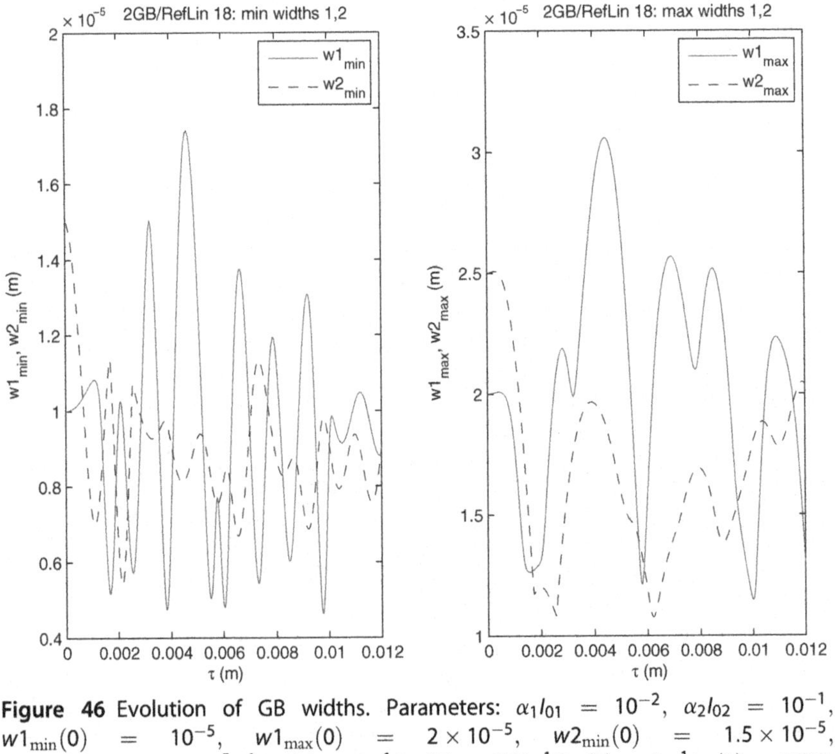

Figure 46 Evolution of GB widths. Parameters: $\alpha_1 l_{01} = 10^{-2}$, $\alpha_2 l_{02} = 10^{-1}$, $w1_{\min}(0) = 10^{-5}$, $w1_{\max}(0) = 2 \times 10^{-5}$, $w2_{\min}(0) = 1.5 \times 10^{-5}$, $w2_{\max}(0) = 2.5 \times 10^{-5}$, ${}^1R_{11}(0) = 0$, ${}^1R_{12}(0) = 397$, ${}^1R_{22}(0) = 0$, ${}^1I_{11}(0) = 1592$, ${}^1I_{12}(0) = 0$, ${}^1I_{22}(0) = 397$, ${}^2R_{11}(0) = -397$, ${}^2R_{12}(0) = -397$, ${}^2R_{22}(0) = 0$, ${}^2I_{11}(0) = 707$, ${}^2I_{12}(0) = 0$, ${}^2I_{22}(0) = 254$. (See the color plate.)

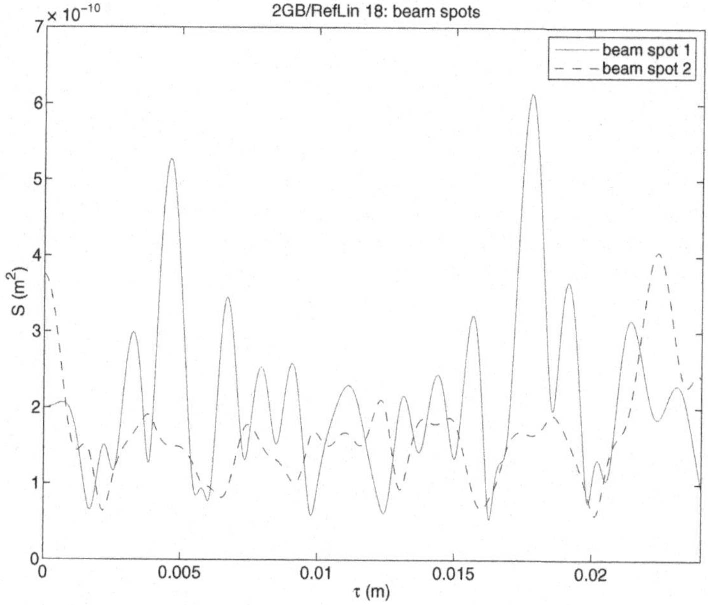

Figure 47 Evolution of GB spots. Parameters: $\alpha_1 l_{01} = 10^{-2}$, $\alpha_2 l_{02} = 10^{-1}$, $w1_{\min}(0) = 10^{-5}$, $w1_{\max}(0) = 2 \times 10^{-5}$, $w2_{\min}(0) = 1.5 \times 10^{-5}$, $w2_{\max}(0) = 2.5 \times 10^{-5}$, ${}^1R_{11}(0) = 0$, ${}^1R_{12}(0) = 397$, ${}^1R_{22}(0) = 0$, ${}^1I_{11}(0) = 1592$, ${}^1I_{12}(0) = 0$, ${}^1I_{22}(0) = 397$, ${}^2R_{11}(0) = -397$, ${}^2R_{12}(0) = -397$, ${}^2R_{22}(0) = 0$, ${}^2I_{11}(0) = 707$, ${}^2I_{12}(0) = 0$, ${}^2I_{22}(0) = 254$. (See the color plate.)

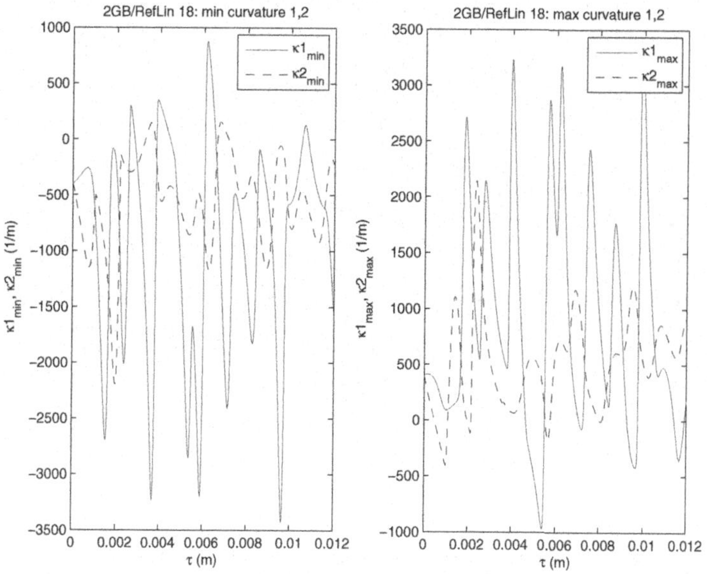

Figure 48 Evolution of GB wave front curvatures. Parameters: $\alpha_1 l_{01} = 10^{-2}$, $\alpha_2 l_{02} = 10^{-1}$, $w1_{\min}(0) = 10^{-5}$, $w1_{\max}(0) = 2 \times 10^{-5}$, $w2_{\min}(0) = 1.5 \times 10^{-5}$, $w2_{\max}(0) = 2.5 \times 10^{-5}$, ${}^1R_{11}(0) = 0$, ${}^1R_{12}(0) = 397$, ${}^1R_{22}(0) = 0$, ${}^1I_{11}(0) = 1592$, ${}^1I_{12}(0) = 0$, ${}^1I_{22}(0) = 397$, ${}^2R_{11}(0) = -397$, ${}^2R_{12}(0) = -397$, ${}^2R_{22}(0) = 0$, ${}^2I_{11}(0) = 707$, ${}^2I_{12}(0) = 0$, ${}^2I_{22}(0) = 254$. (See the color plate.)

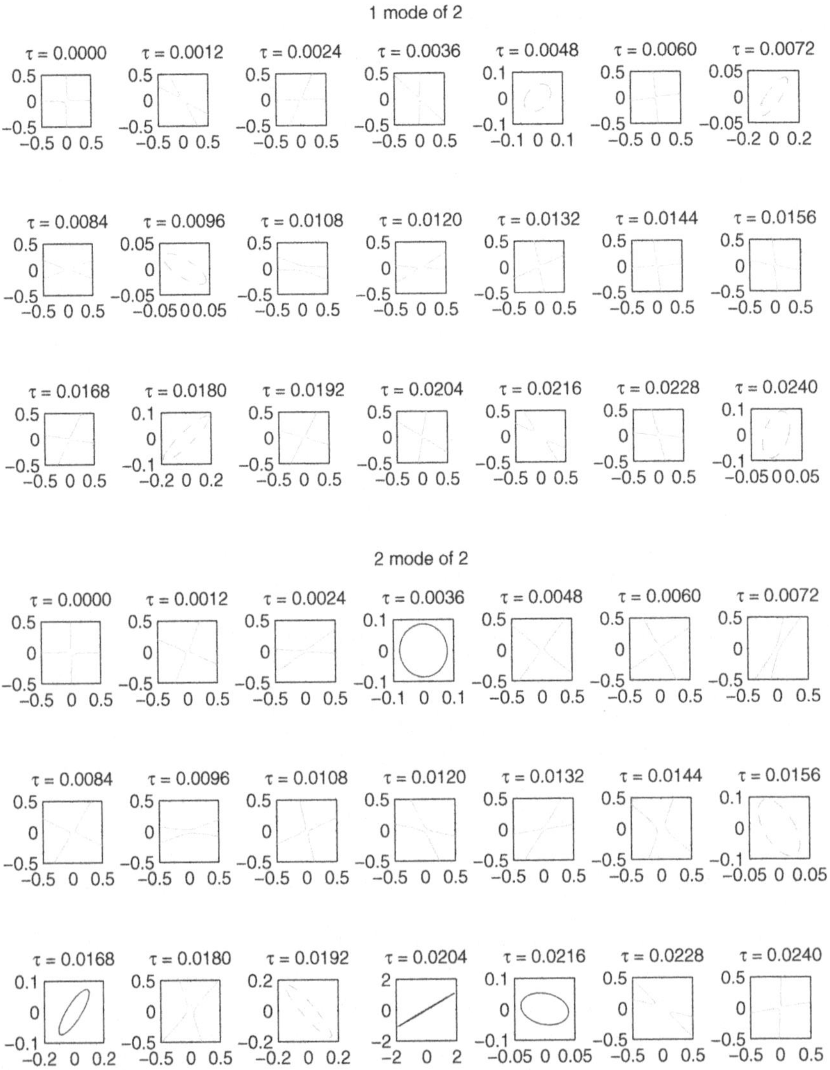

Figure 49 Evolution of wave front cross sections. Parameters: $\alpha_1 l_{01} = 10^{-2}$, $\alpha_2 l_{02} = 10^{-1}$, $w1_{\min}(0) = 10^{-5}$, $w1_{\max}(0) = 2 \times 10^{-5}$, $w2_{\min}(0) = 1.5 \times 10^{-5}$, $w2_{\max}(0) = 2.5 \times 10^{-5}$, ${}^1R_{11}(0) = 0$, ${}^1R_{12}(0) = 397$, ${}^1R_{22}(0) = 0$, ${}^1I_{11}(0) = 1592$, ${}^1I_{12}(0) = 0$, ${}^1I_{22}(0) = 397$, ${}^2R_{11}(0) = -397$, ${}^2R_{12}(0) = -397$, ${}^2R_{22}(0) = 0$, ${}^2I_{11}(0) = 707$, ${}^2I_{12}(0) = 0$, ${}^2I_{22}(0) = 254$.

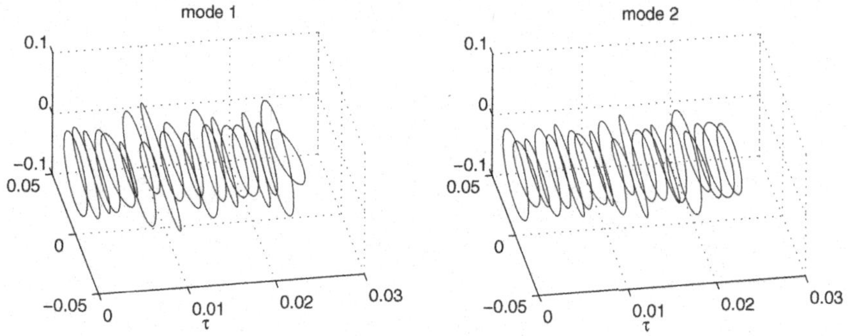

Figure 50 Evolution of GB cross sections. Parameters: $\alpha_1 I_{01} = 10^{-2}$, $\alpha_2 I_{02} = 10^{-1}$, $w1_{\min}(0) = 10^{-5}$, $w1_{\max}(0) = 2 \times 10^{-5}$, $w2_{\min}(0) = 1.1 \times 10^{-5}$, $w2_{\max}(0) = 2.1 \times 10^{-5}$, ${}^1R_{11}(0) = 0$, ${}^1R_{12}(0) \doteq 397$, ${}^1R_{22}(0) = 0$, ${}^1I_{11}(0) = 1592$, ${}^1I_{12}(0) = 0$, ${}^1I_{22}(0) = 397$, ${}^2R_{11}(0) = 0$, ${}^2R_{12}(0) = -397$, ${}^2R_{22}(0) = 0$, ${}^2I_{11}(0) = 1315$, ${}^2I_{12}(0) = 0$, ${}^2I_{22}(0) = 360$.

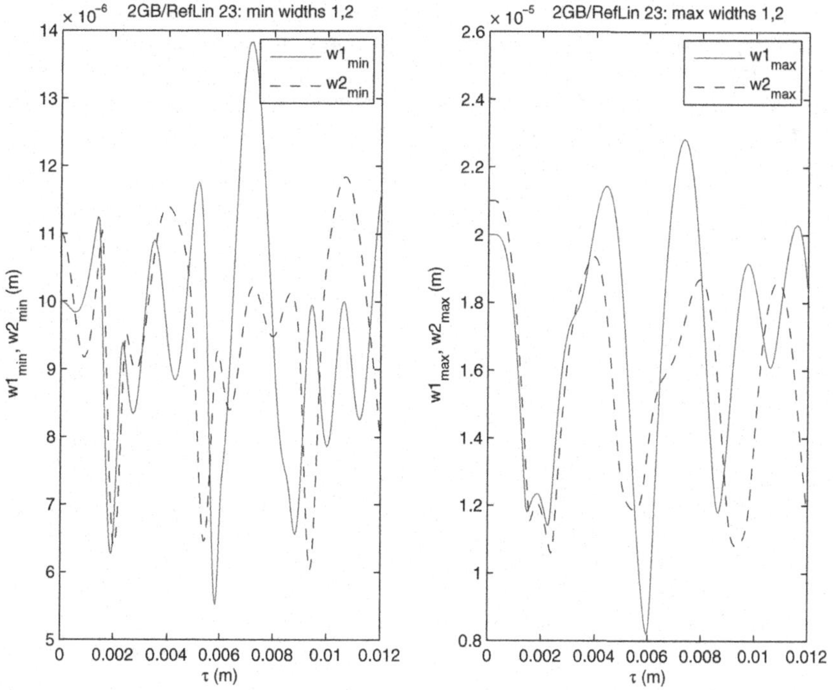

Figure 51 Evolution of GB widths. Parameters: $\alpha_1 I_{01} = 10^{-2}$, $\alpha_2 I_{02} = 10^{-1}$, $w1_{\min}(0) = 10^{-5}$, $w1_{\max}(0) = 2 \times 10^{-5}$, $w2_{\min}(0) = 1.1 \times 10^{-5}$, $w2_{\max}(0) = 2.1 \times 10^{-5}$, ${}^1R_{11}(0) = 0$, ${}^1R_{12}(0) = 397$, ${}^1R_{22}(0) = 0$, ${}^1I_{11}(0) = 1592$, ${}^1I_{12}(0) = 0$, ${}^1I_{22}(0) = 397$, ${}^2R_{11}(0) = 0$, ${}^2R_{12}(0) = -397$, ${}^2R_{22}(0) = 0$, ${}^2I_{11}(0) = 1315$, ${}^2I_{12}(0) = 0$, ${}^2I_{22}(0) = 360$. (See the color plate.)

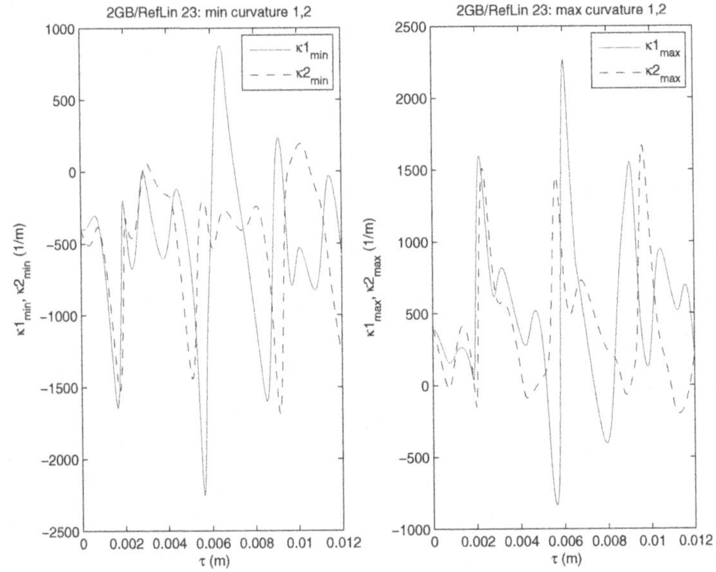

Figure 52 Evolution of GB wave front curvatures. Parameters: $\alpha_1 l_{01} = 10^{-2}$, $\alpha_2 l_{02} = 10^{-1}$, $w1_{\min}(0) = 10^{-5}$, $w1_{\max}(0) = 2 \times 10^{-5}$, $w2_{\min}(0) = 1.1 \times 10^{-5}$, $w2_{\max}(0) = 2.1 \times 10^{-5}$, ${}^1R_{11}(0) = 0$, ${}^1R_{12}(0) = 397$, ${}^1R_{22}(0) = 0$, ${}^1I_{11}(0) = 1592$, ${}^1I_{12}(0) = 0$, ${}^1I_{22}(0) = 397$, ${}^2R_{11}(0) = 0$, ${}^2R_{12}(0) = -397$, ${}^2R_{22}(0) = 0$, ${}^2I_{11}(0) = 1315$, ${}^2I_{12}(0) = 0$, ${}^2I_{22}(0) = 360$. (See the color plate.)

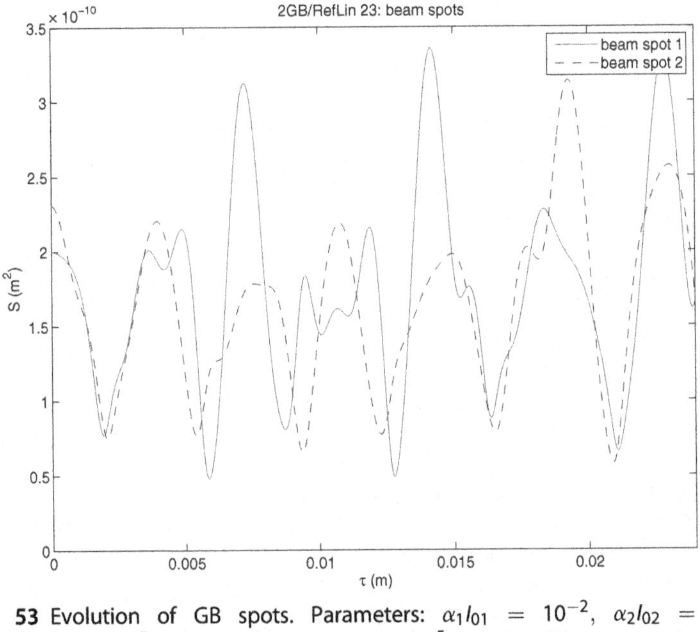

Figure 53 Evolution of GB spots. Parameters: $\alpha_1 l_{01} = 10^{-2}$, $\alpha_2 l_{02} = 10^{-1}$, $w1_{\min}(0) = 10^{-5}$, $w1_{\max}(0) = 2 \times 10^{-5}$, $w2_{\min}(0) = 1.1 \times 10^{-5}$, $w2_{\max}(0) = 2.1 \times 10^{-5}$, ${}^1R_{11}(0) = 0$, ${}^1R_{12}(0) = 397$, ${}^1R_{22}(0) = 0$, ${}^1I_{11}(0) = 1592$, ${}^1I_{12}(0) = 0$, ${}^1I_{22}(0) = 397$, ${}^2R_{11}(0) = 0$, ${}^2R_{12}(0) = -397$, ${}^2R_{22}(0) = 0$, ${}^2I_{11}(0) = 1315$, ${}^2I_{12}(0) = 0$, ${}^2I_{22}(0) = 360$. (See the color plate.)

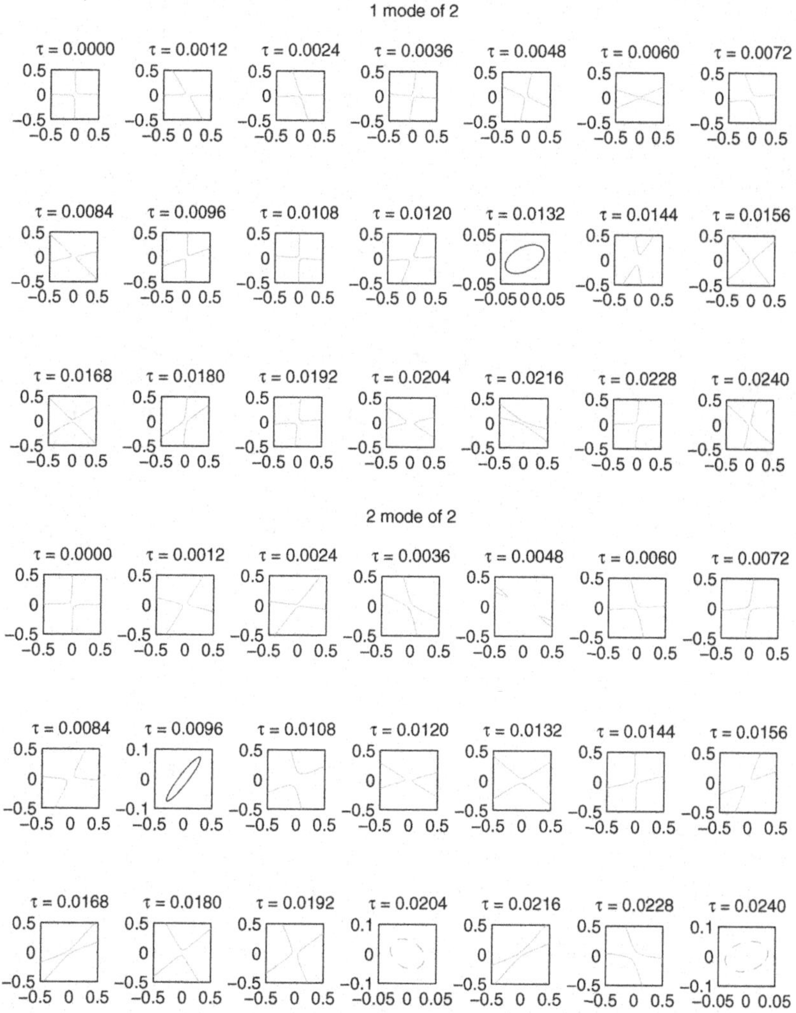

Figure 54 Evolution of wave front cross sections. Parameters: $\alpha_1 l_{01} = 10^{-2}$, $\alpha_2 l_{02} = 10^{-1}$, $w1_{\min}(0) = 10^{-5}$, $w1_{\max}(0) = 2 \times 10^{-5}$, $w2_{\min}(0) = 1.1 \times 10^{-5}$, $w2_{\max}(0) = 2.1 \times 10^{-5}$, ${}^1R_{11}(0) = 0$, ${}^1R_{12}(0) = 397$, ${}^1R_{22}(0) = 0$, ${}^1I_{11}(0) = 1592$, ${}^1I_{12}(0) = 0$, ${}^1I_{22}(0) = 397$, ${}^2R_{11}(0) = 0$, ${}^2R_{12}(0) = -397$, ${}^2R_{22}(0) = 0$, ${}^2I_{11}(0) = 1315$, ${}^2I_{12}(0) = 0$, ${}^2I_{22}(0) = 360$.

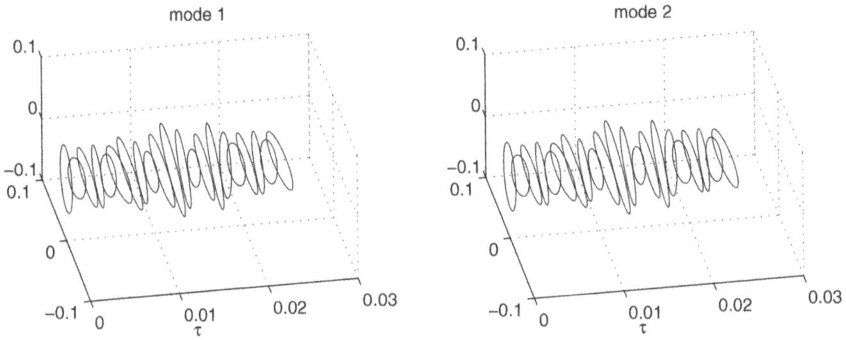

Figure 55 Evolution of GB cross sections. Parameters: $\alpha_1 l_{01} = 0.2$, $\alpha_2 l_{02} = 0.2$, $w1_{\min}(0) = 10^{-5}$, $w1_{\max}(0) = 2 \times 10^{-5}$, $w2_{\min}(0) = 10^{-5}$, $w2_{\max}(0) = 2 \times 10^{-5}$, ${}^1R_{11}(0) = 0$, ${}^1R_{12}(0) = -397$, ${}^1R_{22}(0) = 0$, ${}^1I_{11}(0) = 1592$, ${}^1I_{12}(0) = 0$, ${}^1I_{22}(0) = 397$, ${}^2R_{11}(0) = 0$, ${}^2R_{12}(0) = -397$, ${}^2R_{22}(0) = 0$, ${}^2I_{11}(0) = 1592$, ${}^2I_{12}(0) = 0$, ${}^2I_{22}(0) = 397$.

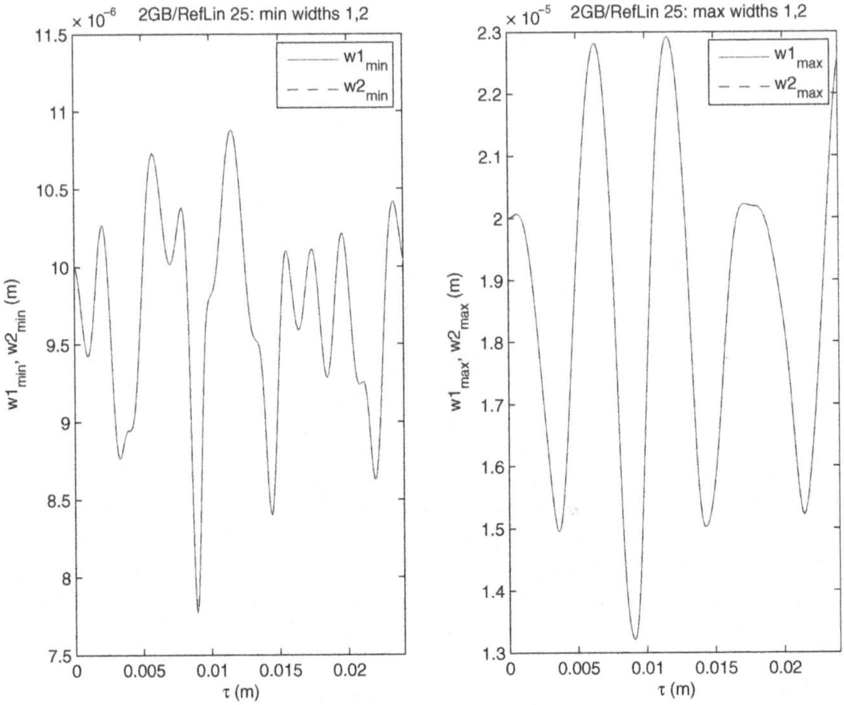

Figure 56 Evolution of GB widths. Parameters: $\alpha_1 l_{01} = 0.2$, $\alpha_2 l_{02} = 0.2$, $w1_{\min}(0) = 10^{-5}$, $w1_{\max}(0) = 2 \times 10^{-5}$, $w2_{\min}(0) = 10^{-5}$, $w2_{\max}(0) = 2 \times 10^{-5}$, ${}^1R_{11}(0) = 0$, ${}^1R_{12}(0) = -397$, ${}^1R_{22}(0) = 0$, ${}^1I_{11}(0) = 1592$, ${}^1I_{12}(0) = 0$, ${}^1I_{22}(0) = 397$, ${}^2R_{11}(0) = 0$, ${}^2R_{12}(0) = -397$, ${}^2R_{22}(0) = 0$, ${}^2I_{11}(0) = 1592$, ${}^2I_{12}(0) = 0$, ${}^2I_{22}(0) = 397$. (See the color plate.)

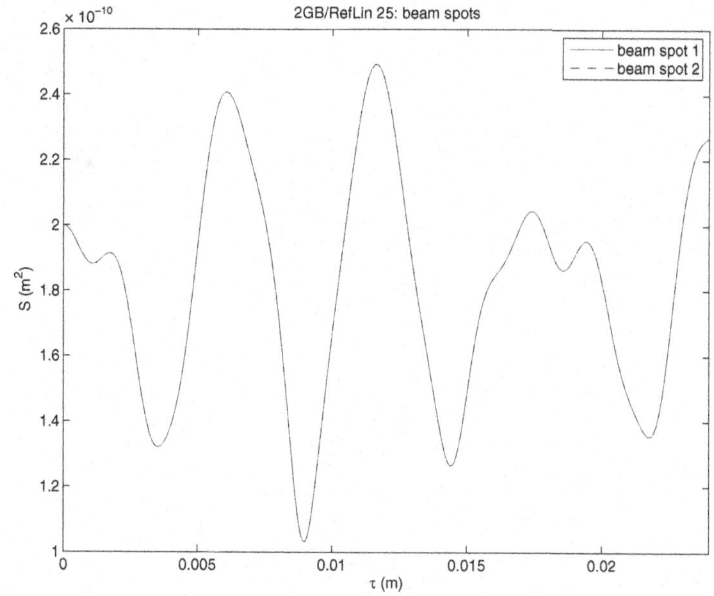

Figure 57 Evolution of GB spots. Parameters: $\alpha_1 l_{01} = 0.2$, $\alpha_2 l_{02} = 0.2$, $w1_{\min}(0) = 10^{-5}$, $w1_{\max}(0) = 2 \times 10^{-5}$, $w2_{\min}(0) = 10^{-5}$, $w2_{\max}(0) = 2 \times 10^{-5}$, ${}^1R_{11}(0) = 0$, ${}^1R_{12}(0) = -397$, ${}^1R_{22}(0) = 0$, ${}^1I_{11}(0) = 1592$, ${}^1I_{12}(0) = 0$, ${}^1I_{22}(0) = 397$, ${}^2R_{11}(0) = 0$, ${}^2R_{12}(0) = -397$, ${}^2R_{22}(0) = 0$, ${}^2I_{11}(0) = 1592$, ${}^2I_{12}(0) = 0$, ${}^2I_{22}(0) = 397$. (See the color plate.)

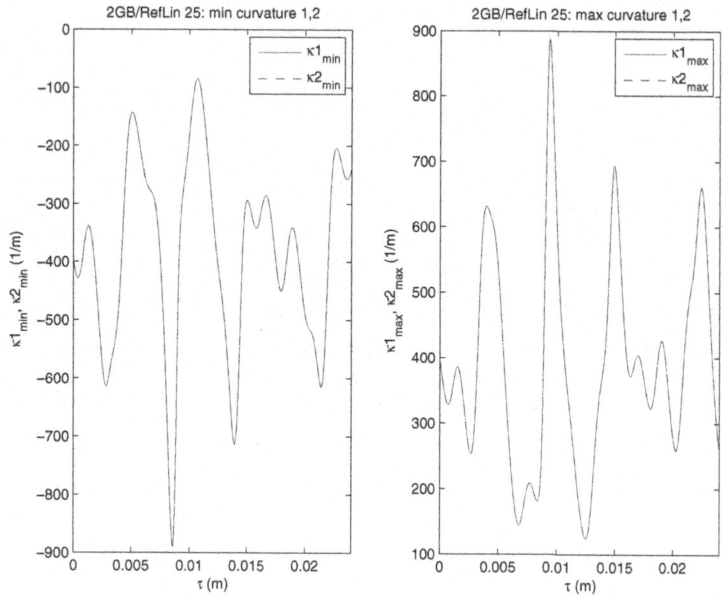

Figure 58 Evolution of GB wave front curvatures. Parameters: $\alpha_1 l_{01} = 0.2$, $\alpha_2 l_{02} = 0.2$, $w1_{\min}(0) = 10^{-5}$, $w1_{\max}(0) = 2 \times 10^{-5}$, $w2_{\min}(0) = 10^{-5}$, $w2_{\max}(0) = 2 \times 10^{-5}$, ${}^1R_{11}(0) = 0$, ${}^1R_{12}(0) = -397$, ${}^1R_{22}(0) = 0$, ${}^1I_{11}(0) = 1592$, ${}^1I_{12}(0) = 0$, ${}^1I_{22}(0) = 397$, ${}^2R_{11}(0) = 0$, ${}^2R_{12}(0) = -397$, ${}^2R_{22}(0) = 0$, ${}^2I_{11}(0) = 1592$, ${}^2I_{12}(0) = 0$, ${}^2I_{22}(0) = 397$. (See the color plate.)

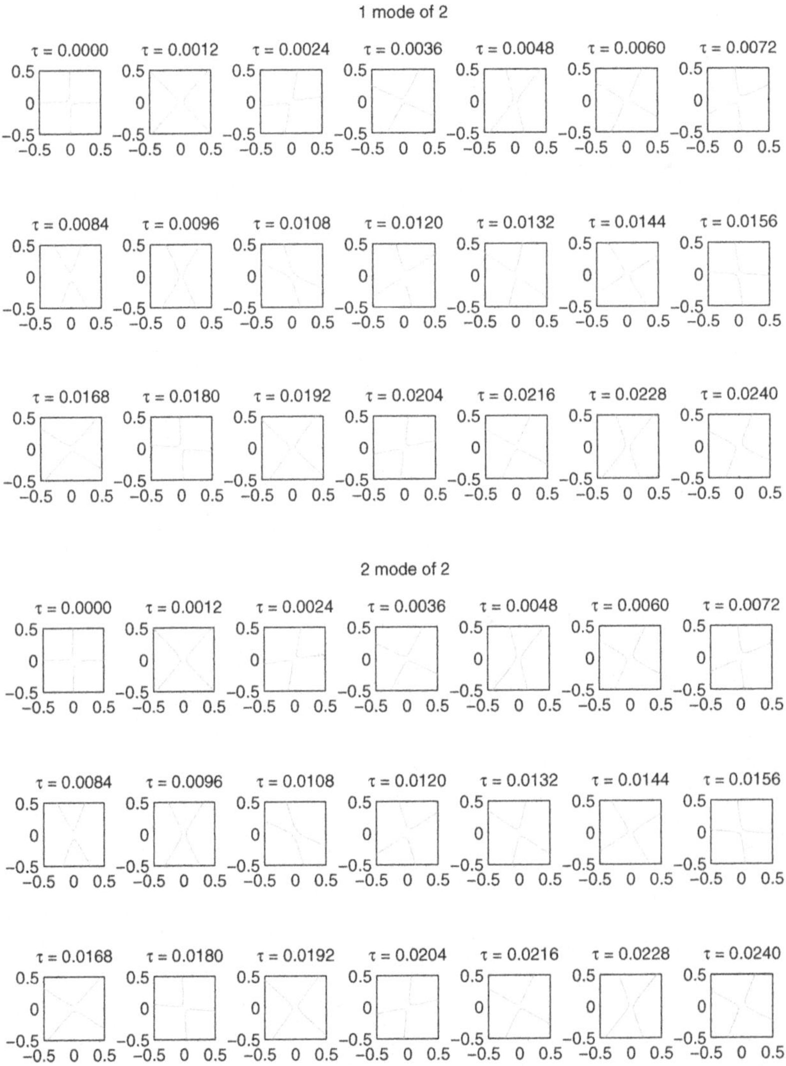

Figure 59 Evolution of wave front cross sections. Parameters: $\alpha_1 l_{01} = 0.2$, $\alpha_2 l_{02} = 0.2$, $w1_{\min}(0) = 10^{-5}$, $w1_{\max}(0) = 2 \times 10^{-5}$, $w2_{\min}(0) = 10^{-5}$, $w2_{\max}(0) = 2 \times 10^{-5}$, ${}^1R_{11}(0) = 0$, ${}^1R_{12}(0) = -397$, ${}^1R_{22}(0) = 0$, ${}^1I_{11}(0) = 1592$, ${}^1I_{12}(0) = 0$, ${}^1I_{22}(0) = 397$, ${}^2R_{11}(0) = 0$, ${}^2R_{12}(0) = -397$, ${}^2R_{22}(0) = 0$, ${}^2I_{11}(0) = 1592$, ${}^2I_{12}(0) = 0$, ${}^2I_{22}(0) = 397$.

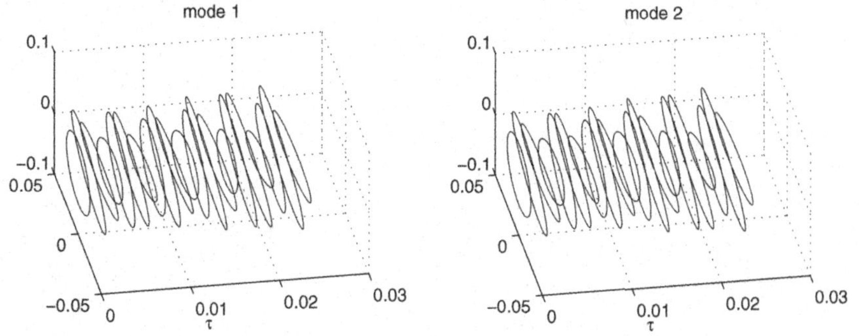

Figure 60 Evolution of GB cross sections. Parameters: $\alpha_1 l_{01} = 0.2$, $\alpha_2 l_{02} = 0.2$, $w1_{\min}(0) = 10^{-5}$, $w1_{\max}(0) = 2 \times 10^{-5}$, $w2_{\min}(0) = 10^{-5}$, $w2_{\max}(0) = 2 \times 10^{-5}$, ${}^1R_{11}(0) = 0$, ${}^1R_{12}(0) = 397$, ${}^1R_{22}(0) = 0$, ${}^1I_{11}(0) = 1592$, ${}^1I_{12}(0) = 0$, ${}^1I_{22}(0) = 397$, ${}^2R_{11}(0) = 0$, ${}^2R_{12}(0) = 397$, ${}^2R_{22}(0) = 0$, ${}^2I_{11}(0) = 1592$, ${}^2I_{12}(0) = 0$, ${}^2I_{22}(0) = 397$.

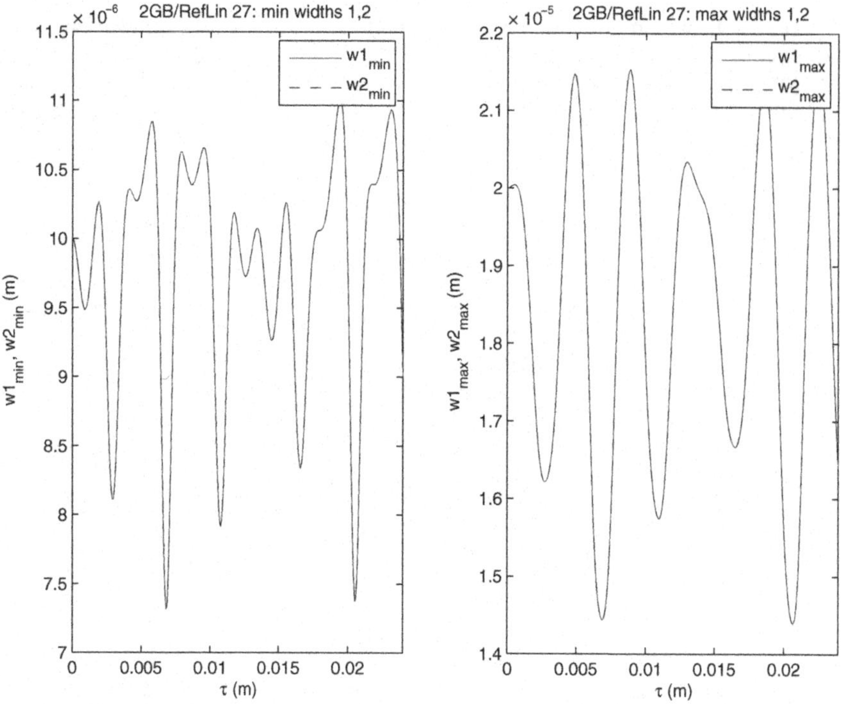

Figure 61 Evolution of GB widths. Parameters: $\alpha_1 l_{01} = 0.2$, $\alpha_2 l_{02} = 0.2$, $w1_{\min}(0) = 10^{-5}$, $w1_{\max}(0) = 2 \times 10^{-5}$, $w2_{\min}(0) = 10^{-5}$, $w2_{\max}(0) = 2 \times 10^{-5}$, ${}^1R_{11}(0) = 0$, ${}^1R_{12}(0) = 397$, ${}^1R_{22}(0) = 0$, ${}^1I_{11}(0) = 1592$, ${}^1I_{12}(0) = 0$, ${}^1I_{22}(0) = 397$, ${}^2R_{11}(0) = 0$, ${}^2R_{12}(0) = 397$, ${}^2R_{22}(0) = 0$, ${}^2I_{11}(0) = 1592$, ${}^2I_{12}(0) = 0$, ${}^2I_{22}(0) = 397$. (See the color plate.)

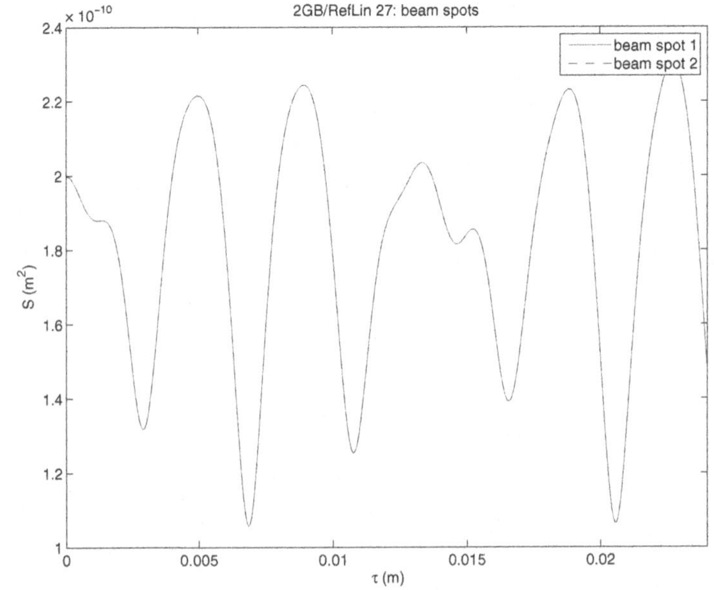

Figure 62 Evolution of GB spots. Parameters: $\alpha_1 l_{01} = 0.2$, $\alpha_2 l_{02} = 0.2$, $w1_{\min}(0) = 10^{-5}$, $w1_{\max}(0) = 2 \times 10^{-5}$, $w2_{\min}(0) = 10^{-5}$, $w2_{\max}(0) = 2 \times 10^{-5}$, ${}^1R_{11}(0) = 0$, ${}^1R_{12}(0) = 397$, ${}^1R_{22}(0) = 0$, ${}^1I_{11}(0) = 1592$, ${}^1I_{12}(0) = 0$, ${}^1I_{22}(0) = 397$, ${}^2R_{11}(0) = 0$, ${}^2R_{12}(0) = 397$, ${}^2R_{22}(0) = 0$, ${}^2I_{11}(0) = 1592$, ${}^2I_{12}(0) = 0$, ${}^2I_{22}(0) = 397$. (See the color plate.)

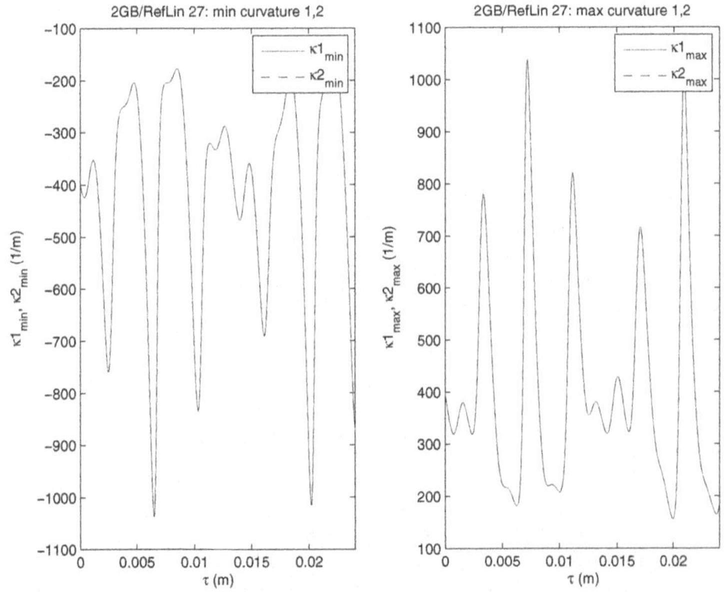

Figure 63 Evolution of GB wave front curvatures. Parameters: $\alpha_1 l_{01} = 0.2$, $\alpha_2 l_{02} = 0.2$, $w1_{\min}(0) = 10^{-5}$, $w1_{\max}(0) = 2 \times 10^{-5}$, $w2_{\min}(0) = 10^{-5}$, $w2_{\max}(0) = 2 \times 10^{-5}$, ${}^1R_{11}(0) = 0$, ${}^1R_{12}(0) = 397$, ${}^1R_{22}(0) = 0$, ${}^1I_{11}(0) = 1592$, ${}^1I_{12}(0) = 0$, ${}^1I_{22}(0) = 397$, ${}^2R_{11}(0) = 0$, ${}^2R_{12}(0) = 397$, ${}^2R_{22}(0) = 0$, ${}^2I_{11}(0) = 1592$, ${}^2I_{12}(0) = 0$, ${}^2I_{22}(0) = 397$. (See the color plate.)

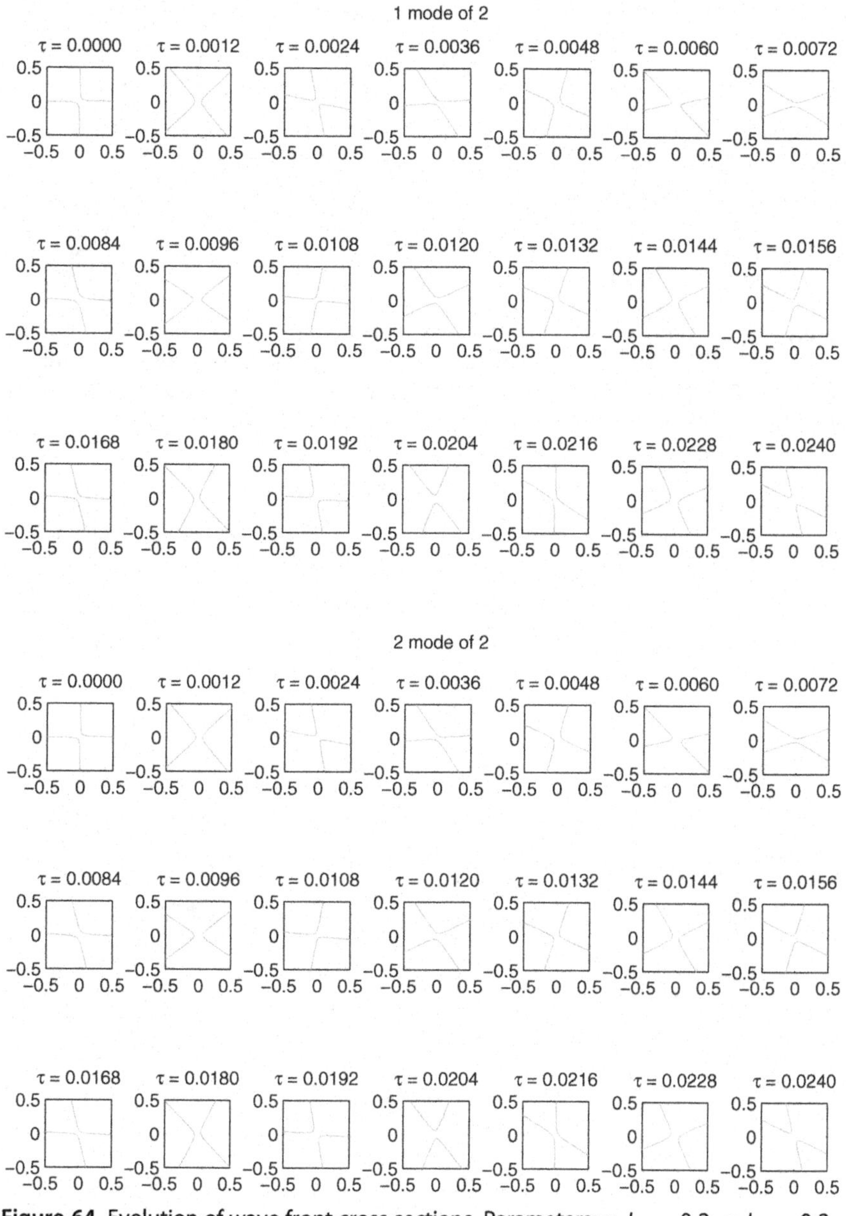

Figure 64 Evolution of wave front cross sections. Parameters: $\alpha_1 l_{01} = 0.2$, $\alpha_2 l_{02} = 0.2$, $w1_{\min}(0) = 10^{-5}$, $w1_{\max}(0) = 2 \times 10^{-5}$, $w2_{\min}(0) = 10^{-5}$, $w2_{\max}(0) = 2 \times 10^{-5}$, ${}^1R_{11}(0) = 0$, ${}^1R_{12}(0) = 397$, ${}^1R_{22}(0) = 0$, ${}^1I_{11}(0) = 1592$, ${}^1I_{12}(0) = 0$, ${}^1I_{22}(0) = 397$, ${}^2R_{11}(0) = 0$, ${}^2R_{12}(0) = 397$, ${}^2R_{22}(0) = 0$, ${}^2I_{11}(0) = 1592$, ${}^2I_{12}(0) = 0$, ${}^2I_{22}(0) = 397$.

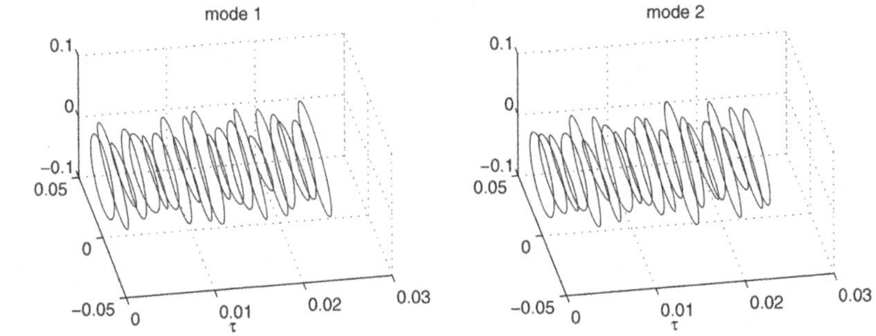

Figure 65 Evolution of GB cross sections. Parameters: $\alpha_1 l_{01} = 0.2$, $\alpha_2 l_{02} = 0.2$, $w1_{\min}(0) = 10^{-5}$, $w1_{\max}(0) = 2 \times 10^{-5}$, $w2_{\min}(0) = 10^{-5}$, $w2_{\max}(0) = 2 \times 10^{-5}$, ${}^1R_{11}(0) = 0$, ${}^1R_{12}(0) = 397$, ${}^1R_{22}(0) = 0$, ${}^1I_{11}(0) = 1592$, ${}^1I_{12}(0) = 0$, ${}^1I_{22}(0) = 397$, ${}^2R_{11}(0) = 0$, ${}^2R_{12}(0) = -198$, ${}^2R_{22}(0) = 0$, ${}^2I_{11}(0) = 1592$, ${}^2I_{12}(0) = 0$, ${}^2I_{22}(0) = 397$.

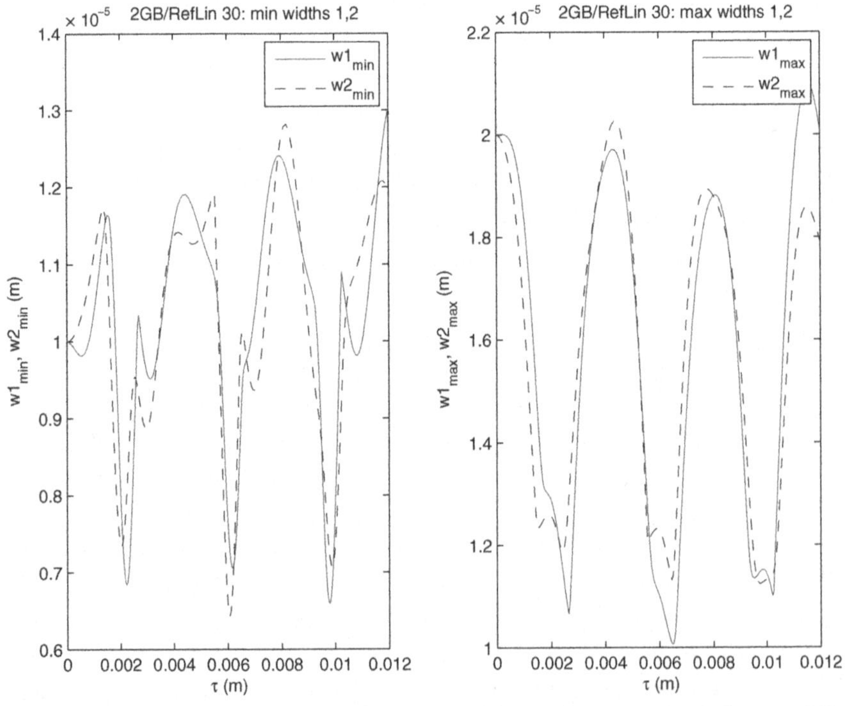

Figure 66 Evolution of GB widths. Parameters: $\alpha_1 l_{01} = 0.2$, $\alpha_2 l_{02} = 0.2$, $w1_{\min}(0) = 10^{-5}$, $w1_{\max}(0) = 2 \times 10^{-5}$, $w2_{\min}(0) = 10^{-5}$, $w2_{\max}(0) = 2 \times 10^{-5}$, ${}^1R_{11}(0) = 0$, ${}^1R_{12}(0) = 397$, ${}^1R_{22}(0) = 0$, ${}^1I_{11}(0) = 1592$, ${}^1I_{12}(0) = 0$, ${}^1I_{22}(0) = 397$, ${}^2R_{11}(0) = 0$, ${}^2R_{12}(0) = -198$, ${}^2R_{22}(0) = 0$, ${}^2I_{11}(0) = 1592$, ${}^2I_{12}(0) = 0$, ${}^2I_{22}(0) = 397$. (See the color plate.)

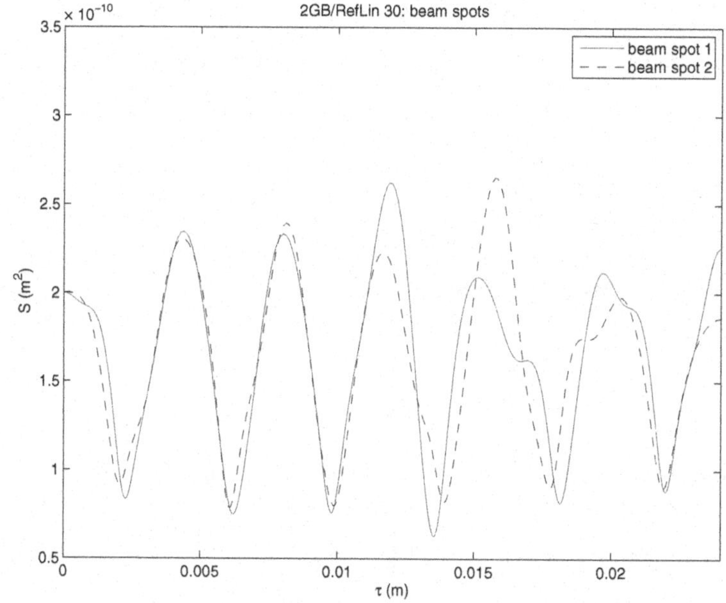

Figure 67 Evolution of GB spots. Parameters: $\alpha_1 l_{01} = 0.2$, $\alpha_2 l_{02} = 0.2$, $w1_{\min}(0) = 10^{-5}$, $w1_{\max}(0) = 2 \times 10^{-5}$, $w2_{\min}(0) = 10^{-5}$, $w2_{\max}(0) = 2 \times 10^{-5}$, ${}^1R_{11}(0) = 0$, ${}^1R_{12}(0) = 397$, ${}^1R_{22}(0) = 0$, ${}^1I_{11}(0) = 1592$, ${}^1I_{12}(0) = 0$, ${}^1I_{22}(0) = 397$, ${}^2R_{11}(0) = 0$, ${}^2R_{12}(0) = -198$, ${}^2R_{22}(0) = 0$, ${}^2I_{11}(0) = 1592$, ${}^2I_{12}(0) = 0$, ${}^2I_{22}(0) = 397$. (See the color plate.)

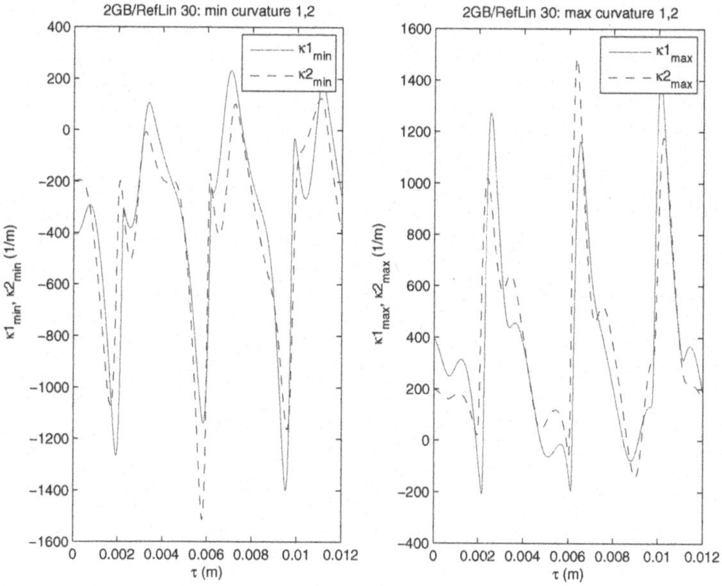

Figure 68 Evolution of GB wave front curvatures. Parameters: $\alpha_1 l_{01} = 0.2$, $\alpha_2 l_{02} = 0.2$, $w1_{\min}(0) = 10^{-5}$, $w1_{\max}(0) = 2 \times 10^{-5}$, $w2_{\min}(0) = 10^{-5}$, $w2_{\max}(0) = 2 \times 10^{-5}$, ${}^1R_{11}(0) = 0$, ${}^1R_{12}(0) = 397$, ${}^1R_{22}(0) = 0$, ${}^1I_{11}(0) = 1592$, ${}^1I_{12}(0) = 0$, ${}^1I_{22}(0) = 397$, ${}^2R_{11}(0) = 0$, ${}^2R_{12}(0) = -198$, ${}^2R_{22}(0) = 0$, ${}^2I_{11}(0) = 1592$, ${}^2I_{12}(0) = 0$, ${}^2I_{22}(0) = 397$. (See the color plate.)

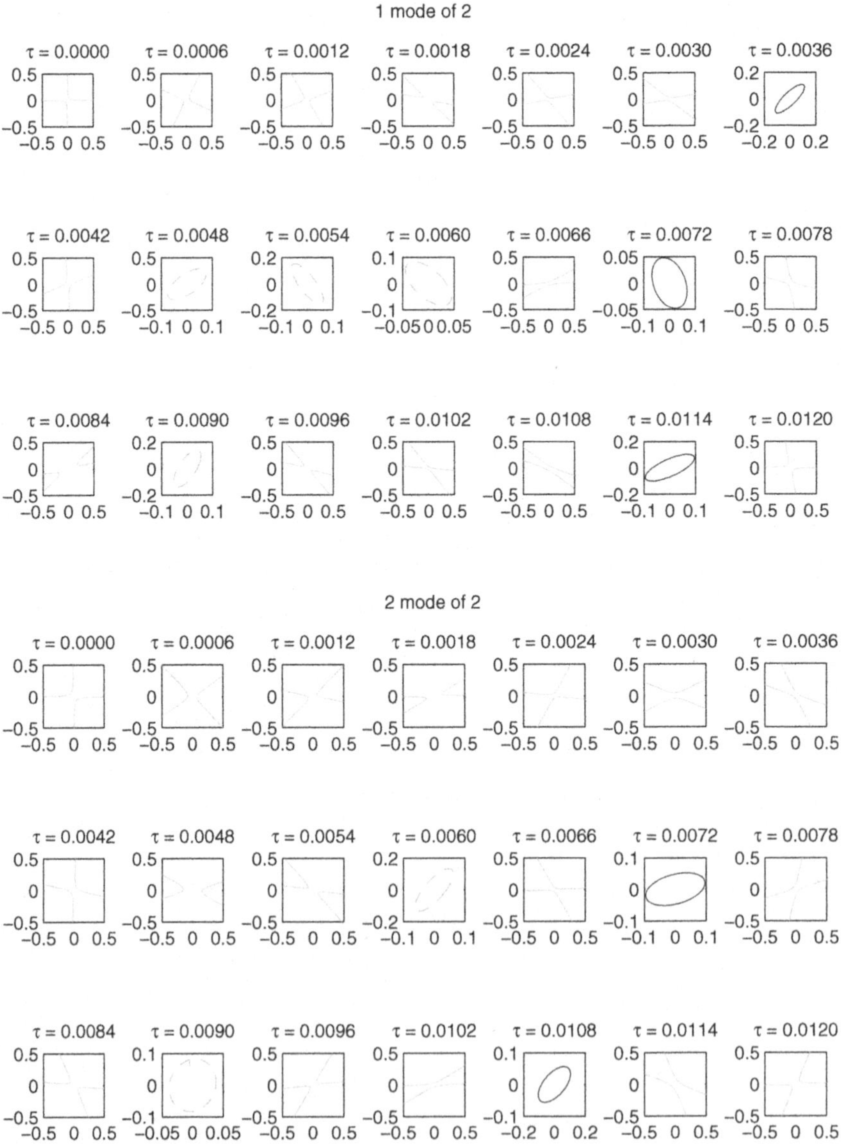

Figure 69 Evolution of wave front cross sections. Parameters: $\alpha_1 l_{01} = 0.2$, $\alpha_2 l_{02} = 0.2$, $w1_{\min}(0) = 10^{-5}$, $w1_{\max}(0) = 2 \times 10^{-5}$, $w2_{\min}(0) = 10^{-5}$, $w2_{\max}(0) = 2 \times 10^{-5}$, ${}^1R_{11}(0) = 0$, ${}^1R_{12}(0) = 397$, ${}^1R_{22}(0) = 0$, ${}^1I_{11}(0) = 1592$, ${}^1I_{12}(0) = 0$, ${}^1I_{22}(0) = 397$, ${}^2R_{11}(0) = 0$, ${}^2R_{12}(0) = -198$, ${}^2R_{22}(0) = 0$, ${}^2I_{11}(0) = 1592$, ${}^2I_{12}(0) = 0$, ${}^2I_{22}(0) = 397$.

changes the rotation direction after a distance of 12 diffraction lengths. Figures 66–68 depict the evolution of GB widths, GB spots and GB curvatures calculated with the same parameters as in Figures 65 and 69.

15. THREE- AND FOUR-ROTATING GBs

For the case of a three and four rotating GB, we present numerical simulations based on CGO equations for the evolution of a GB cross sections, evolution of GB widths, evolution of GB wave front curvatures, and evolution of the wave front cross sections in Figures 70–108.

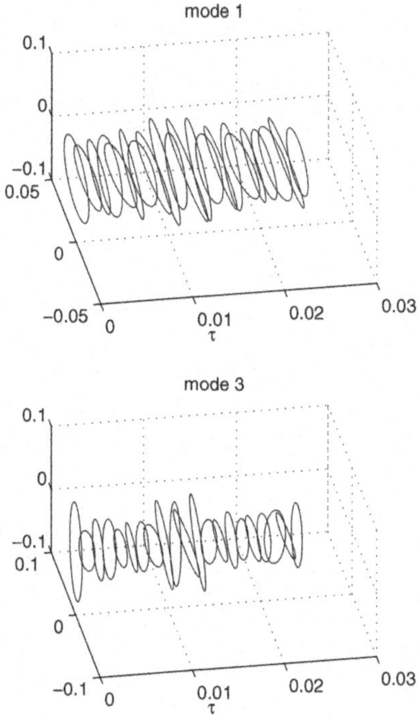

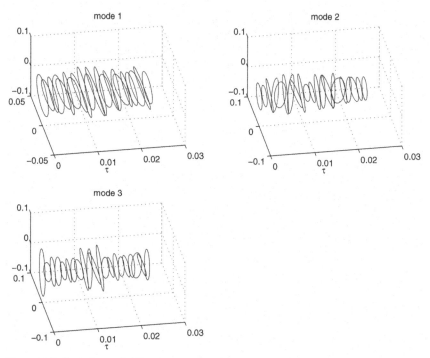

Figure 70 Evolution of GB cross sections. Parameters: $\alpha_1 l_{01} = 0.1$, $\alpha_2 l_{02} = 0.2$, $\alpha_3 l_{03} = 0.07$, $w1_{\min}(0) = 10^{-5}$, $w1_{\max}(0) = 2 \times 10^{-5}$, $w2_{\min}(0) = 10^{-5}$, $w2_{\max}(0) = 1.5 \times 10^{-5}$, $w3_{\min}(0) = 10^{-5}$, $w3_{\max}(0) = 3 \times 10^{-5}$, ${}^1R_{11}(0) = 0$, ${}^1R_{12}(0) = -397$, ${}^1R_{22}(0) = 0$, ${}^1I_{11}(0) = 1592$, ${}^1I_{12}(0) = 0$, ${}^1I_{22}(0) = 397$, ${}^2R_{11}(0) = 0$, ${}^2R_{12}(0) = 397$, ${}^2R_{22}(0) = 0$, ${}^2I_{11}(0) = 1592$, ${}^2I_{12}(0) = 0$, ${}^2I_{22}(0) = 707$, ${}^3R_{11}(0) = 0$, ${}^3R_{12}(0) = -397$, ${}^3R_{22}(0) = 0$, ${}^3I_{11}(0) = 1592$, ${}^3I_{12}(0) = 0$, ${}^3I_{22}(0) = 176$.

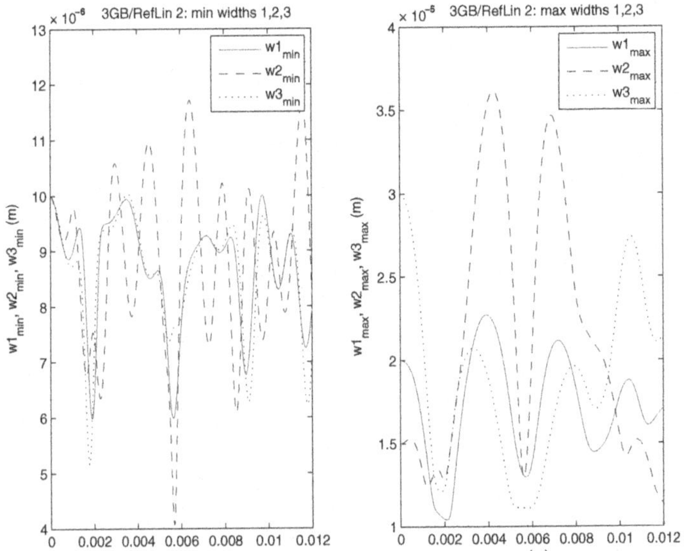

Figure 71 Evolution of GB widths. Parameters: $\alpha_1 l_{01} = 0.1$, $\alpha_2 l_{02} = 0.2$, $\alpha_3 l_{03} = 0.07$, $w1_{\min}(0) = 10^{-5}$, $w1_{\max}(0) = 2 \times 10^{-5}$, $w2_{\min}(0) = 10^{-5}$, $w2_{\max}(0) = 1.5 \times 10^{-5}$, $w3_{\min}(0) = 10^{-5}$, $w3_{\max}(0) = 3 \times 10^{-5}$, ${}^1R_{11}(0) = 0$, ${}^1R_{12}(0) = -397$, ${}^1R_{22}(0) = 0$, ${}^1I_{11}(0) = 1592$, ${}^1I_{12}(0) = 0$, ${}^1I_{22}(0) = 397$, ${}^2R_{11}(0) = 0$, ${}^2R_{12}(0) = 397$, ${}^2R_{22}(0) = 0$, ${}^2I_{11}(0) = 1592$, ${}^2I_{12}(0) = 0$, ${}^2I_{22}(0) = 707$, ${}^3R_{11}(0) = 0$, ${}^3R_{12}(0) = -397$, ${}^3R_{22}(0) = 0$, ${}^3I_{11}(0) = 1592$, ${}^3I_{12}(0) = 0$, ${}^3I_{22}(0) = 176$. (See the color plate.)

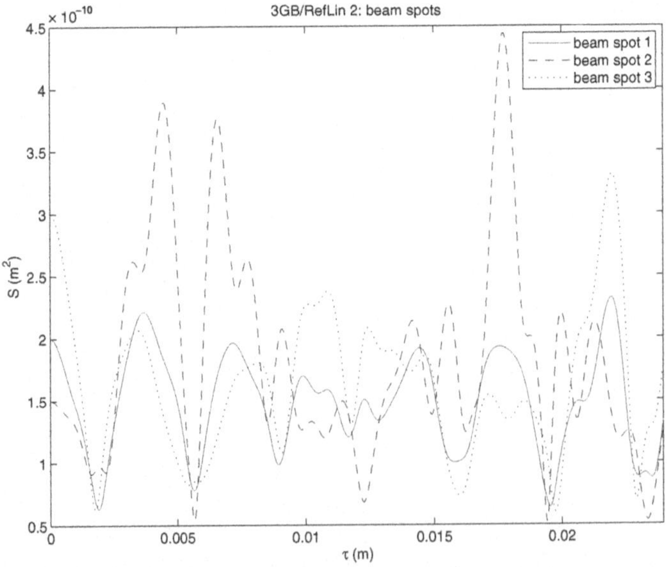

Figure 72 Evolution of GB spots. Parameters: $\alpha_1 l_{01} = 0.1$, $\alpha_2 l_{02} = 0.2$, $\alpha_3 l_{03} = 0.07$, $w1_{\min}(0) = 10^{-5}$, $w1_{\max}(0) = 2 \times 10^{-5}$, $w2_{\min}(0) = 10^{-5}$, $w2_{\max}(0) = 1.5 \times 10^{-5}$, $w3_{\min}(0) = 10^{-5}$, $w3_{\max}(0) = 3 \times 10^{-5}$, ${}^1R_{11}(0) = 0$, ${}^1R_{12}(0) = -397$, ${}^1R_{22}(0) = 0$, ${}^1I_{11}(0) = 1592$, ${}^1I_{12}(0) = 0$, ${}^1I_{22}(0) = 397$, ${}^2R_{11}(0) = 0$, ${}^2R_{12}(0) = 397$, ${}^2R_{22}(0) = 0$, ${}^2I_{11}(0) = 1592$, ${}^2I_{12}(0) = 0$, ${}^2I_{22}(0) = 707$, ${}^3R_{11}(0) = 0$, ${}^3R_{12}(0) = -397$, ${}^3R_{22}(0) = 0$, ${}^3I_{11}(0) = 1592$, ${}^3I_{12}(0) = 0$, ${}^3I_{22}(0) = 176$. (See the color plate.)

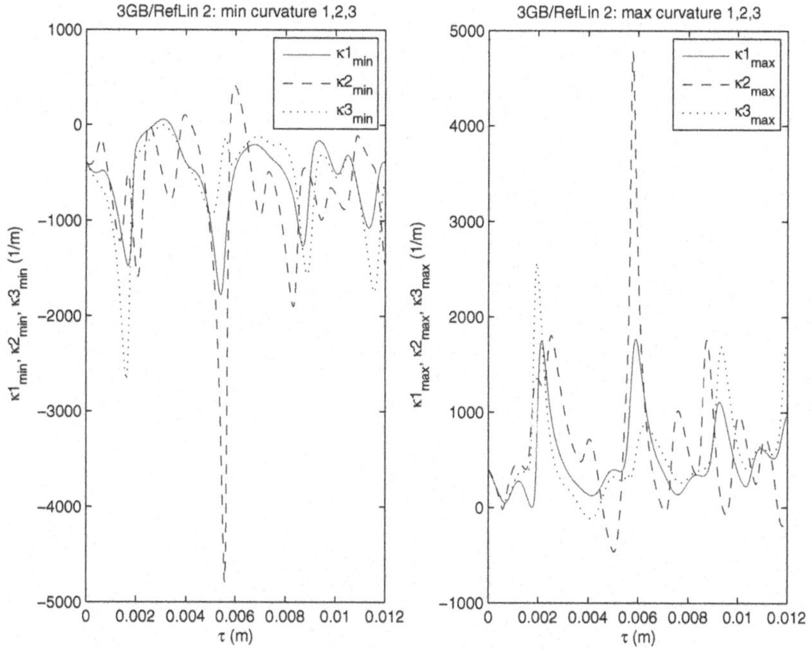

Figure 73 Evolution of GB wave front curvatures. Parameters: $\alpha_1 l_{01} = 0.1$, $\alpha_2 l_{02} = 0.2$, $\alpha_3 l_{03} = 0.07$, $w1_{\min}(0) = 10^{-5}$, $w1_{\max}(0) = 2 \times 10^{-5}$, $w2_{\min}(0) = 10^{-5}$, $w2_{\max}(0) = 1.5 \times 10^{-5}$, $w3_{\min}(0) = 10^{-5}$, $w3_{\max}(0) = 3 \times 10^{-5}$, $^{1}R_{11}(0) = 0$, $^{1}R_{12}(0) = -397$, $^{1}R_{22}(0) = 0$, $^{1}I_{11}(0) = 1592$, $^{1}I_{12}(0) = 0$, $^{1}I_{22}(0) = 397$, $^{2}R_{11}(0) = 0$, $^{2}R_{12}(0) = 397$, $^{2}R_{22}(0) = 0$, $^{2}I_{11}(0) = 1592$, $^{2}I_{12}(0) = 0$, $^{2}I_{22}(0) = 707$, $^{3}R_{11}(0) = 0$, $^{3}R_{12}(0) = -397$, $^{3}R_{22}(0) = 0$, $^{3}I_{11}(0) = 1592$, $^{3}I_{12}(0) = 0$, $^{3}I_{22}(0) = 176$. (See the color plate.)

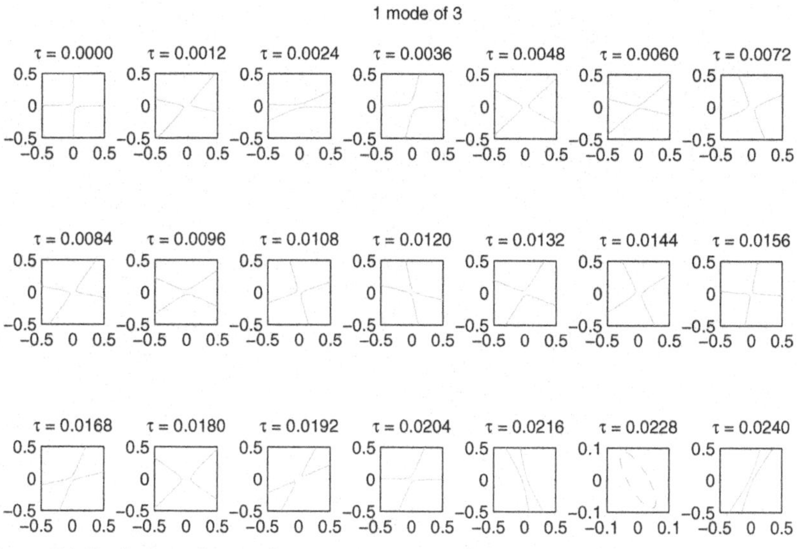

Figure 74 Evolution of wave front cross sections. Parameters: $\alpha_1 l_{01} = 0.1$, $\alpha_2 l_{02} = 0.2$, $\alpha_3 l_{03} = 0.07$, $w1_{\min}(0) = 10^{-5}$, $w1_{\max}(0) = 2 \times 10^{-5}$, $w2_{\min}(0) = 10^{-5}$, $w2_{\max}(0) = 1.5 \times 10^{-5}$, $w3_{\min}(0) = 10^{-5}$, $w3_{\max}(0) = 3 \times 10^{-5}$, $^{1}R_{11}(0) = 0$, $^{1}R_{12}(0) = -397$, $^{1}R_{22}(0) = 0$, $^{1}I_{11}(0) = 1592$, $^{1}I_{12}(0) = 0$, $^{1}I_{22}(0) = 397$, $^{2}R_{11}(0) = 0$, $^{2}R_{12}(0) = 397$, $^{2}R_{22}(0) = 0$, $^{2}I_{11}(0) = 1592$, $^{2}I_{12}(0) = 0$, $^{2}I_{22}(0) = 707$, $^{3}R_{11}(0) = 0$, $^{3}R_{12}(0) = -397$, $^{3}R_{22}(0) = 0$, $^{3}I_{11}(0) = 1592$, $^{3}I_{12}(0) = 0$, $^{3}I_{22}(0) = 176$.

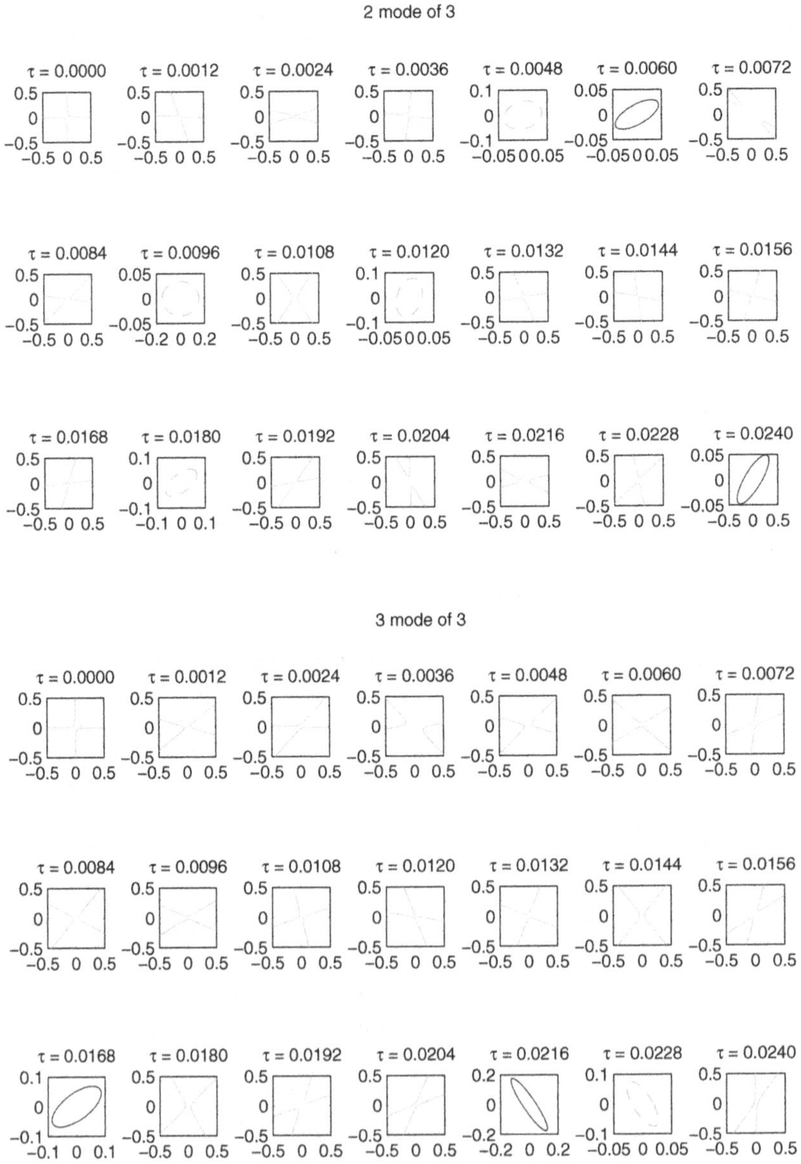

Figure 74 *(continued)*

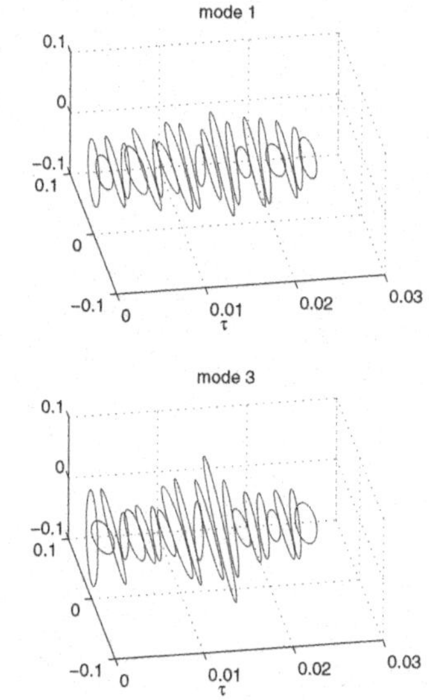

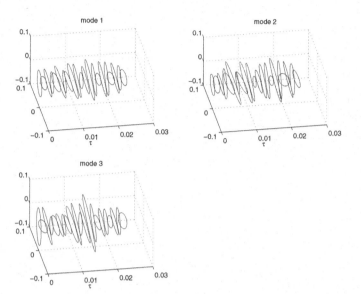

Figure 75 Evolution of GB cross sections. Parameters: $\alpha_1 l_{01} = 0.1$, $\alpha_2 l_{02} = 0.2$, $\alpha_3 l_{03} = 0.07$, $w1_{\min}(0) = 10^{-5}$, $w1_{\max}(0) = 2 \times 10^{-5}$, $w2_{\min}(0) = 10^{-5}$, $w2_{\max}(0) = 1.5 \times 10^{-5}$, $w3_{\min}(0) = 10^{-5}$, $w3_{\max}(0) = 3 \times 10^{-5}$, ${}^1R_{11}(0) = 0$, ${}^1R_{12}(0) = -397$, ${}^1R_{22}(0) = 0$, ${}^1I_{11}(0) = 1592$, ${}^1I_{12}(0) = 0$, ${}^1I_{22}(0) = 397$, ${}^2R_{11}(0) = 0$, ${}^2R_{12}(0) = -795$, ${}^2R_{22}(0) = 0$, ${}^2I_{11}(0) = 1592$, ${}^2I_{12}(0) = 0$, ${}^2I_{22}(0) = 707$, ${}^3R_{11}(0) = 0$, ${}^3R_{12}(0) = -397$, ${}^3R_{22}(0) = 0$, ${}^3I_{11}(0) = 1592$, ${}^3I_{12}(0) = 0$, ${}^3I_{22}(0) = 176$.

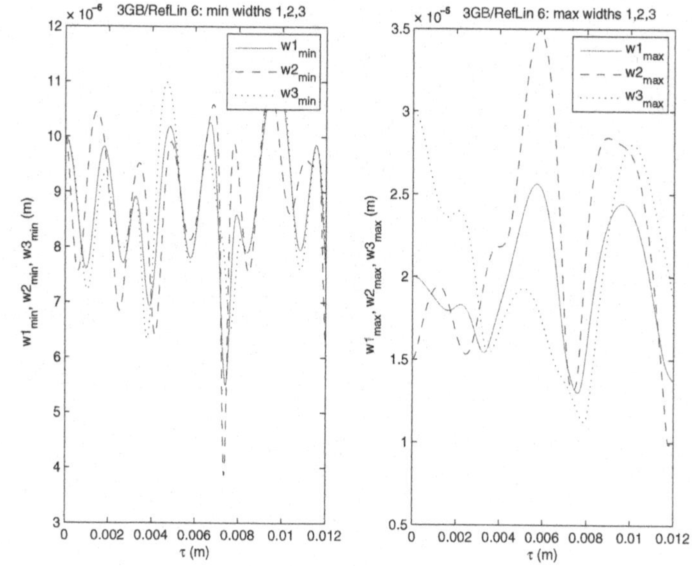

Figure 76 Evolution of GB widths. Parameters: $\alpha_1 l_{01} = 0.1$, $\alpha_2 l_{02} = 0.2$, $\alpha_3 l_{03} = 0.07$, $w1_{\min}(0) = 10^{-5}$, $w1_{\max}(0) = 2 \times 10^{-5}$, $w2_{\min}(0) = 10^{-5}$, $w2_{\max}(0) = 1.5 \times 10^{-5}$, $w3_{\min}(0) = 10^{-5}$, $w3_{\max}(0) = 3 \times 10^{-5}$, ${}^1R_{11}(0) = 0$, ${}^1R_{12}(0) = -397$, ${}^1R_{22}(0) = 0$, ${}^1I_{11}(0) = 1592$, ${}^1I_{12}(0) = 0$, ${}^1I_{22}(0) = 397$, ${}^2R_{11}(0) = 0$, ${}^2R_{12}(0) = -795$, ${}^2R_{22}(0) = 0$, ${}^2I_{11}(0) = 1592$, ${}^2I_{12}(0) = 0$, ${}^2I_{22}(0) = 707$, ${}^3R_{11}(0) = 0$, ${}^3R_{12}(0) = -397$, ${}^3R_{22}(0) = 0$, ${}^3I_{11}(0) = 1592$, ${}^3I_{12}(0) = 0$, ${}^3I_{22}(0) = 176$. (See the color plate.)

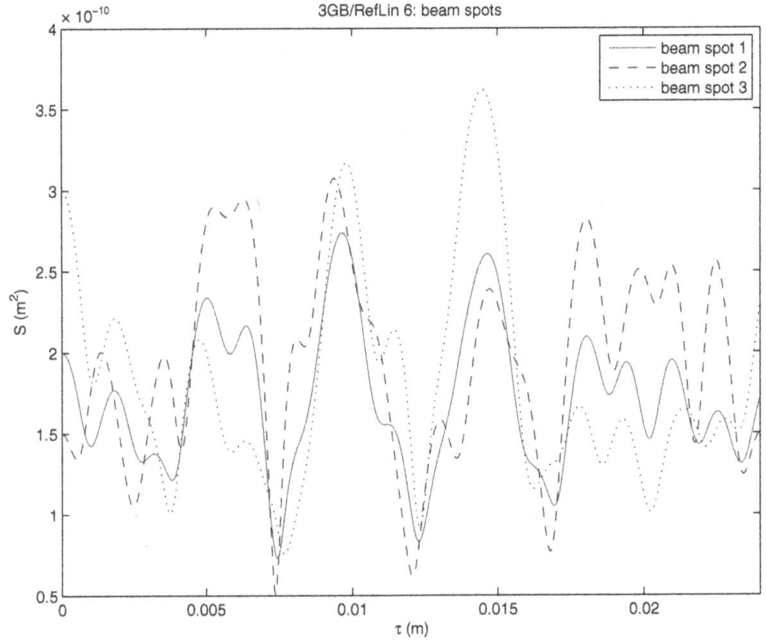

Figure 77 Evolution of GB spots. Parameters: $\alpha_1 l_{01} = 0.1$, $\alpha_2 l_{02} = 0.2$, $\alpha_3 l_{03} = 0.07$, $w1_{\min}(0) = 10^{-5}$, $w1_{\max}(0) = 2 \times 10^{-5}$, $w2_{\min}(0) = 10^{-5}$, $w2_{\max}(0) = 1.5 \times 10^{-5}$, $w3_{\min}(0) = 10^{-5}$, $w3_{\max}(0) = 3 \times 10^{-5}$, ${}^1R_{11}(0) = 0$, ${}^1R_{12}(0) = -397$, ${}^1R_{22}(0) = 0$, ${}^1I_{11}(0) = 1592$, ${}^1I_{12}(0) = 0$, ${}^1I_{22}(0) = 397$, ${}^2R_{11}(0) = 0$, ${}^2R_{12}(0) = -795$, ${}^2R_{22}(0) = 0$, ${}^2I_{11}(0) = 1592$, ${}^2I_{12}(0) = 0$, ${}^2I_{22}(0) = 707$, ${}^3R_{11}(0) = 0$, ${}^3R_{12}(0) = -397$, ${}^3R_{22}(0) = 0$, ${}^3I_{11}(0) = 1592$, ${}^3I_{12}(0) = 0$, ${}^3I_{22}(0) = 176$. (See the color plate.)

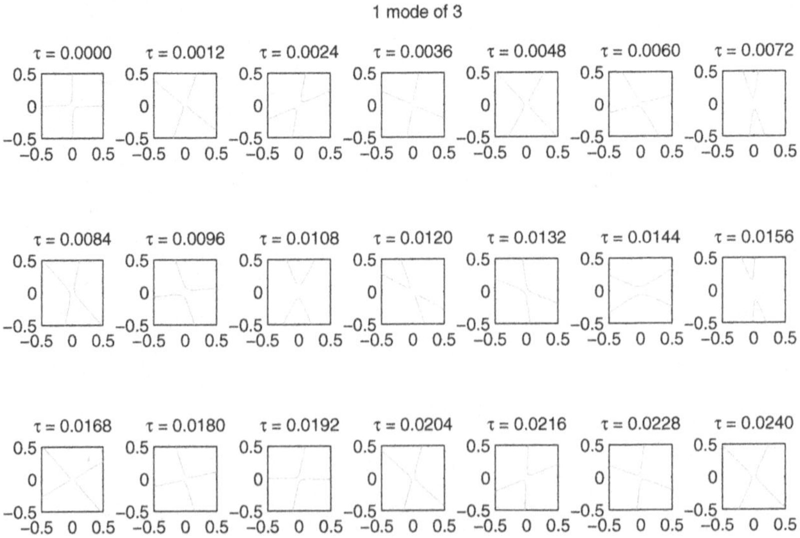

Figure 78 Evolution of wave front cross sections. Parameters: $\alpha_1 l_{01} = 0.1$, $\alpha_2 l_{02} = 0.2$, $\alpha_3 l_{03} = 0.07$, $w1_{\min}(0) = 10^{-5}$, $w1_{\max}(0) = 2 \times 10^{-5}$, $w2_{\min}(0) = 10^{-5}$, $w2_{\max}(0) = 1.5 \times 10^{-5}$, $w3_{\min}(0) = 10^{-5}$, $w3_{\max}(0) = 3 \times 10^{-5}$, ${}^1R_{11}(0) = 0$, ${}^1R_{12}(0) = -397$, ${}^1R_{22}(0) = 0$, ${}^1I_{11}(0) = 1592$, ${}^1I_{12}(0) = 0$, ${}^1I_{22}(0) = 397$, ${}^2R_{11}(0) = 0$, ${}^2R_{12}(0) = -795$, ${}^2R_{22}(0) = 0$, ${}^2I_{11}(0) = 1592$, ${}^2I_{12}(0) = 0$, ${}^2I_{22}(0) = 707$, ${}^3R_{11}(0) = 0$, ${}^3R_{12}(0) = -397$, ${}^3R_{22}(0) = 0$, ${}^3I_{11}(0) = 1592$, ${}^3I_{12}(0) = 0$, ${}^3I_{22}(0) = 176$.

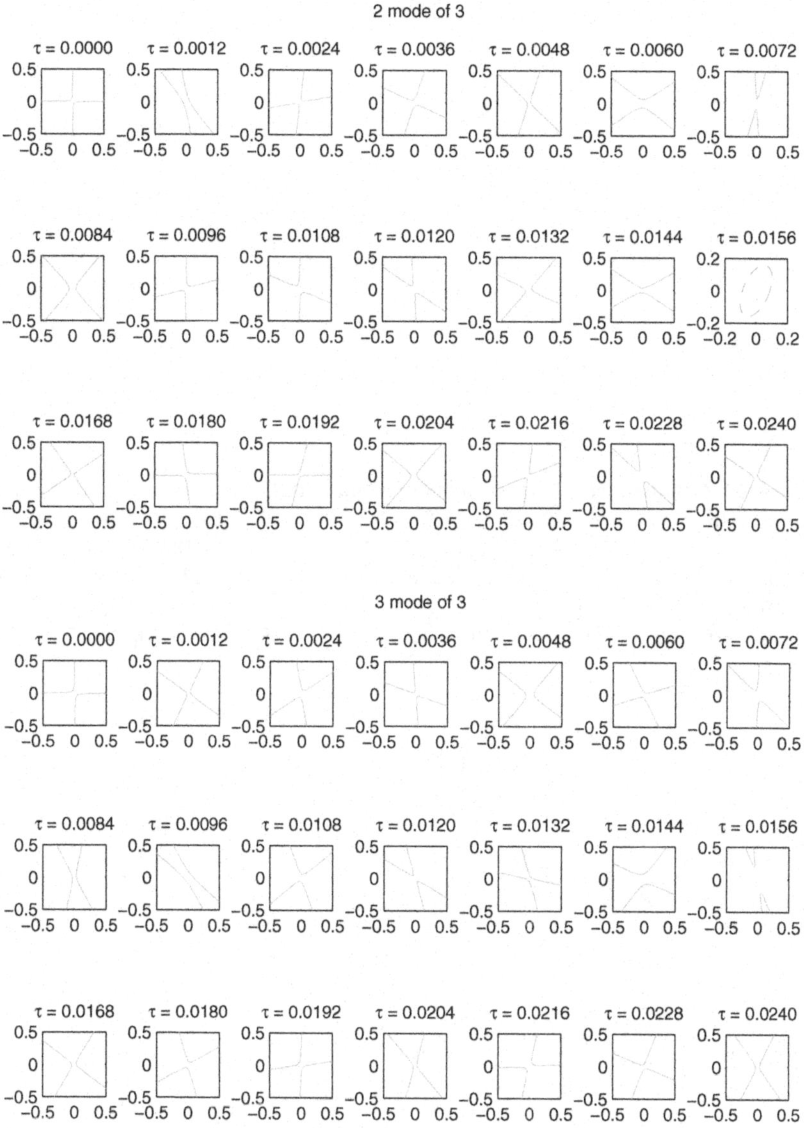

Figure 78 (*continued*)

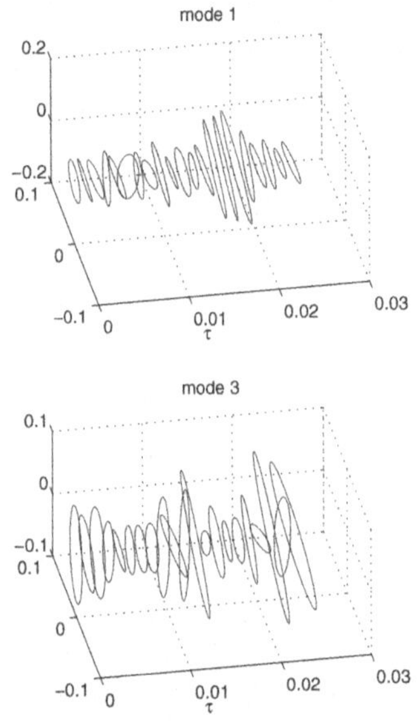

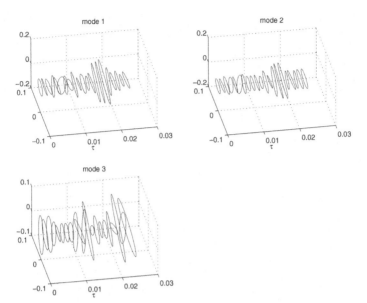

Figure 79 Evolution of GB cross sections. Parameters: $\alpha_1 l_{01} = 0.1$, $\alpha_2 l_{02} = 0.2$, $\alpha_3 l_{03} = 0.07$, $w1_{\min}(0) = 10^{-5}$, $w1_{\max}(0) = 2 \times 10^{-5}$, $w2_{\min}(0) = 10^{-5}$, $w2_{\max}(0) = 1.5 \times 10^{-5}$, $w3_{\min}(0) = 10^{-5}$, $w3_{\max}(0) = 3 \times 10^{-5}$, ${}^1R_{11}(0) = -707$, ${}^1R_{12}(0) = -397$, ${}^1R_{22}(0) = -400$, ${}^1l_{11}(0) = 1592$, ${}^1l_{12}(0) = 0$, ${}^1l_{22}(0) = 397$, ${}^2R_{11}(0) = -707$, ${}^2R_{12}(0) = 397$, ${}^2R_{22}(0) = -400$, ${}^2l_{11}(0) = 1592$, ${}^2l_{12}(0) = 0$, ${}^2l_{22}(0) = 707$, ${}^3R_{11}(0) = 707$, ${}^3R_{12}(0) = 397$, ${}^3R_{22}(0) = 400$, ${}^3l_{11}(0) = 1592$, ${}^3l_{12}(0) = 0$, ${}^3l_{22}(0) = 176$.

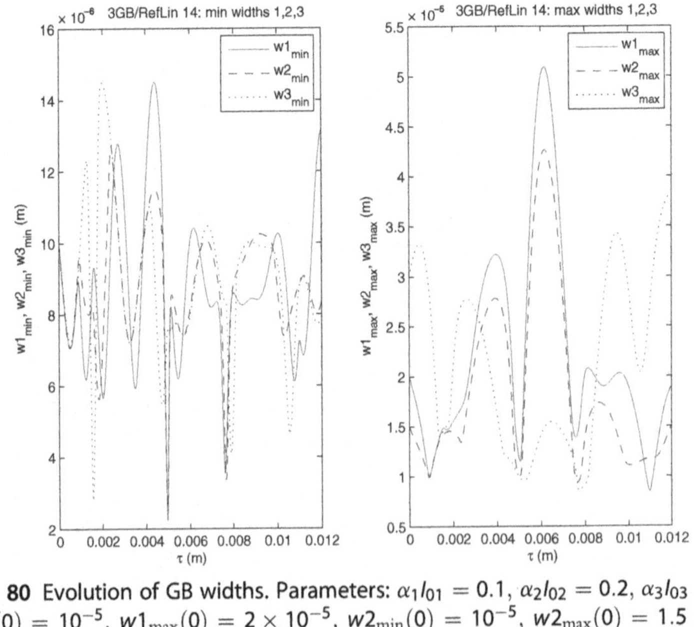

Figure 80 Evolution of GB widths. Parameters: $\alpha_1 l_{01} = 0.1$, $\alpha_2 l_{02} = 0.2$, $\alpha_3 l_{03} = 0.07$, $w1_{\min}(0) = 10^{-5}$, $w1_{\max}(0) = 2 \times 10^{-5}$, $w2_{\min}(0) = 10^{-5}$, $w2_{\max}(0) = 1.5 \times 10^{-5}$, $w3_{\min}(0) = 10^{-5}$, $w3_{\max}(0) = 3 \times 10^{-5}$, ${}^1R_{11}(0) = -707$, ${}^1R_{12}(0) = -397$, ${}^1R_{22}(0) = -400$, ${}^1l_{11}(0) = 1592$, ${}^1l_{12}(0) = 0$, ${}^1l_{22}(0) = 397$, ${}^2R_{11}(0) = -707$, ${}^2R_{12}(0) = 397$, ${}^2R_{22}(0) = -400$, ${}^2l_{11}(0) = 1592$, ${}^2l_{12}(0) = 0$, ${}^2l_{22}(0) = 707$, ${}^3R_{11}(0) = 707$, ${}^3R_{12}(0) = 397$, ${}^3R_{22}(0) = 400$, ${}^3l_{11}(0) = 1592$, ${}^3l_{12}(0) = 0$, ${}^3l_{22}(0) = 176$. (See the color plate.)

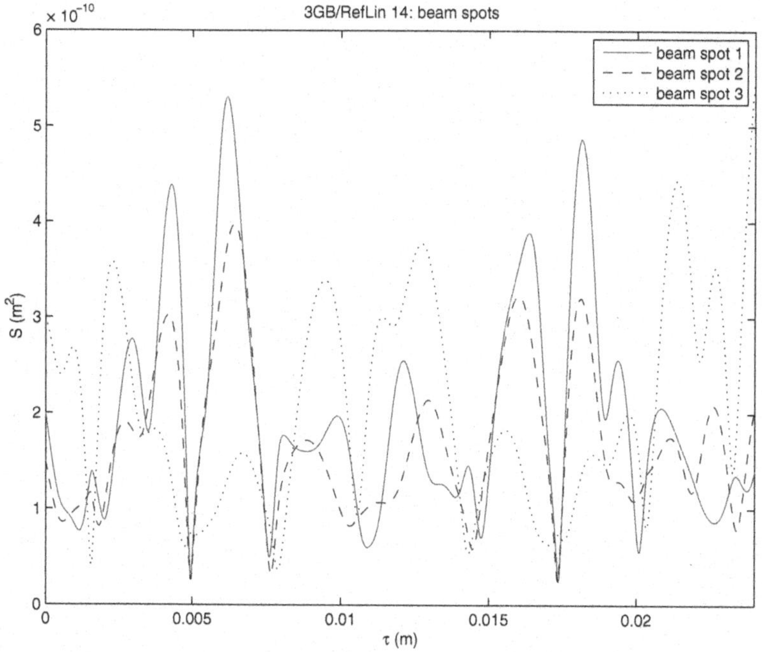

Figure 81 Evolution of GB spots. Parameters: $\alpha_1 l_{01} = 0.1$, $\alpha_2 l_{02} = 0.2$, $\alpha_3 l_{03} = 0.07$, $w1_{\min}(0) = 10^{-5}$, $w1_{\max}(0) = 2 \times 10^{-5}$, $w2_{\min}(0) = 10^{-5}$, $w2_{\max}(0) = 1.5 \times 10^{-5}$, $w3_{\min}(0) = 10^{-5}$, $w3_{\max}(0) = 3 \times 10^{-5}$, ${}^1R_{11}(0) = -707$, ${}^1R_{12}(0) = -397$, ${}^1R_{22}(0) = -400$, ${}^1I_{11}(0) = 1592$, ${}^1I_{12}(0) = 0$, ${}^1I_{22}(0) = 397$, ${}^2R_{11}(0) = -707$, ${}^2R_{12}(0) = 397$, ${}^2R_{22}(0) = -400$, ${}^2I_{11}(0) = 1592$, ${}^2I_{12}(0) = 0$, ${}^2I_{22}(0) = 707$, ${}^3R_{11}(0) = 707$, ${}^3R_{12}(0) = 397$, ${}^3R_{22}(0) = 400$, ${}^3I_{11}(0) = 1592$, ${}^3I_{12}(0) = 0$, ${}^3I_{22}(0) = 176$. (See the color plate.)

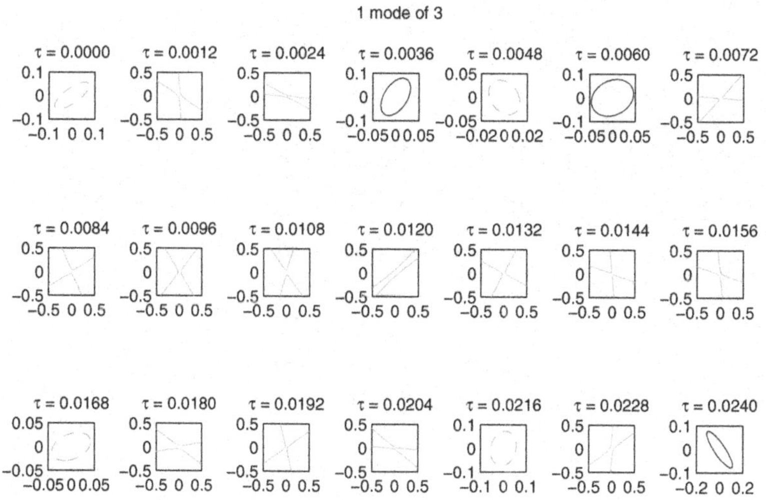

Figure 82 Evolution of wave front cross sections. Parameters: $\alpha_1 l_{01} = 0.1$, $\alpha_2 l_{02} = 0.2$, $\alpha_3 l_{03} = 0.07$, $w1_{\min}(0) = 10^{-5}$, $w1_{\max}(0) = 2 \times 10^{-5}$, $w2_{\min}(0) = 10^{-5}$, $w2_{\max}(0) = 1.5 \times 10^{-5}$, $w3_{\min}(0) = 10^{-5}$, $w3_{\max}(0) = 3 \times 10^{-5}$, ${}^1R_{11}(0) = -707$, ${}^1R_{12}(0) = -397$, ${}^1R_{22}(0) = -400$, ${}^1I_{11}(0) = 1592$, ${}^1I_{12}(0) = 0$, ${}^1I_{22}(0) = 397$, ${}^2R_{11}(0) = -707$, ${}^2R_{12}(0) = 397$, ${}^2R_{22}(0) = -400$, ${}^2I_{11}(0) = 1592$, ${}^2I_{12}(0) = 0$, ${}^2I_{22}(0) = 707$, ${}^3R_{11}(0) = 707$, ${}^3R_{12}(0) = 397$, ${}^3R_{22}(0) = 400$, ${}^3I_{11}(0) = 1592$, ${}^3I_{12}(0) = 0$, ${}^3I_{22}(0) = 176$.

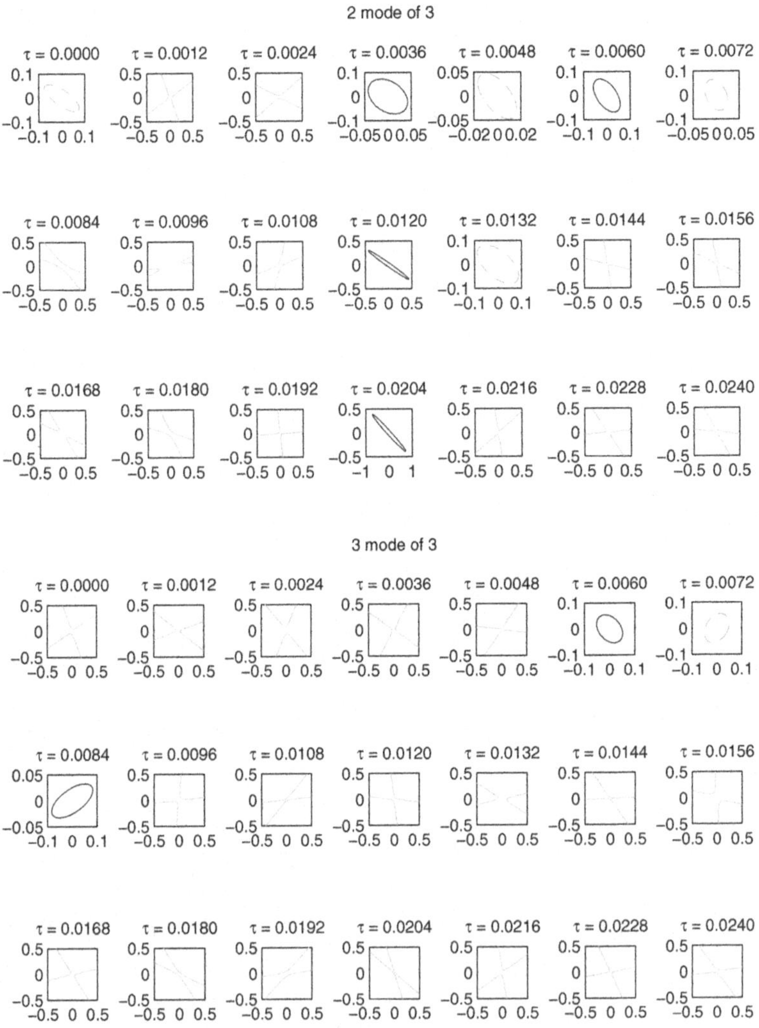

Figure 82 (*continued*)

In Figure 70, it can be observed that the cross sections of two modes rotate counterclockwise, whereas the GB cross section of the third mode rotates clockwise. A similar situation is shown in Figure 74, where wave front cross sections of two modes evolve in the same manner, but the third rotates in the opposite direction. Figures 71–73 show the evolution of GB widths, GB spots and GB curvatures calculated with the same parameters as in Figures 71 and 74. Figures 79–82 show analogues behavior as in the

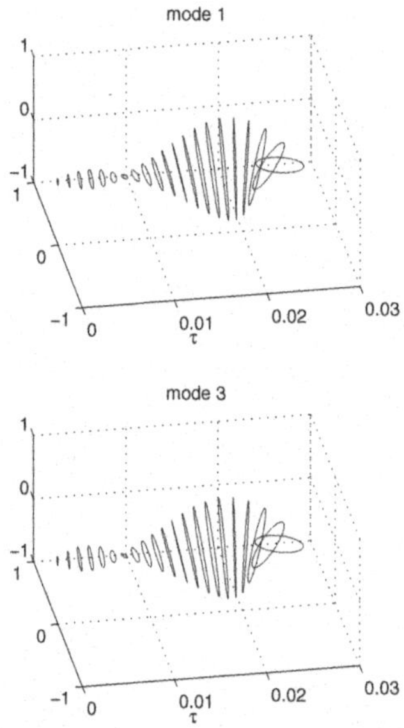

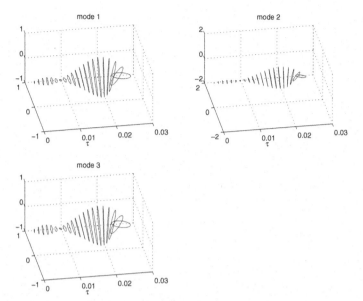

Figure 83 Evolution of GB cross sections. Parameters: $\alpha_1 l_{01} = 0.1$, $\alpha_2 l_{02} = 0.2$, $\alpha_3 l_{03} = 0.07$, $w1_{\min}(0) = 10^{-5}$, $w1_{\max}(0) = 2 \times 10^{-5}$, $w2_{\min}(0) = 10^{-5}$, $w2_{\max}(0) = 1.5 \times 10^{-5}$, $w3_{\min}(0) = 10^{-5}$, $w3_{\max}(0) = 3 \times 10^{-5}$, ${}^1R_{11}(0) = -707$, ${}^1R_{12}(0) = 397$, ${}^1R_{22}(0) = 400$, ${}^1I_{11}(0) = 1592$, ${}^1I_{12}(0) = 0$, ${}^1I_{22}(0) = 397$, ${}^2R_{11}(0) = 707$, ${}^2R_{12}(0) = 397$, ${}^2R_{22}(0) = 400$, ${}^2I_{11}(0) = 1592$, ${}^2I_{12}(0) = 0$, ${}^2I_{22}(0) = 707$, ${}^3R_{11}(0) = -707$, ${}^3R_{12}(0) = 397$, ${}^3R_{22}(0) = 400$, ${}^3I_{11}(0) = 1592$, ${}^3I_{12}(0) = 0$, ${}^3I_{22}(0) = 176$.

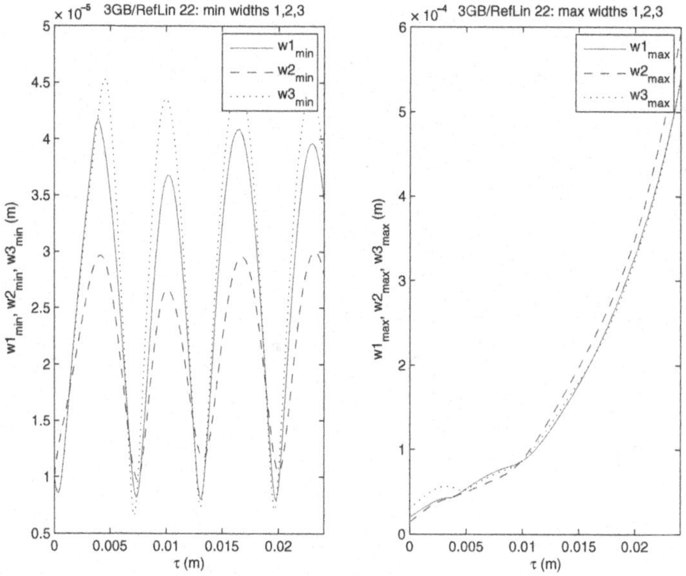

Figure 84 Evolution of GB widths. Parameters: $\alpha_1 l_{01} = 0.1$, $\alpha_2 l_{02} = 0.2$, $\alpha_3 l_{03} = 0.07$, $w1_{\min}(0) = 10^{-5}$, $w1_{\max}(0) = 2 \times 10^{-5}$, $w2_{\min}(0) = 10^{-5}$, $w2_{\max}(0) = 1.5 \times 10^{-5}$, $w3_{\min}(0) = 10^{-5}$, $w3_{\max}(0) = 3 \times 10^{-5}$, ${}^1R_{11}(0) = -707$, ${}^1R_{12}(0) = 397$, ${}^1R_{22}(0) = 400$, ${}^1I_{11}(0) = 1592$, ${}^1I_{12}(0) = 0$, ${}^1I_{22}(0) = 397$, ${}^2R_{11}(0) = 707$, ${}^2R_{12}(0) = 397$, ${}^2R_{22}(0) = 400$, ${}^2I_{11}(0) = 1592$, ${}^2I_{12}(0) = 0$, ${}^2I_{22}(0) = 707$, ${}^3R_{11}(0) = -707$, ${}^3R_{12}(0) = 397$, ${}^3R_{22}(0) = 400$, ${}^3I_{11}(0) = 1592$, ${}^3I_{12}(0) = 0$, ${}^3I_{22}(0) = 176$. (See the color plate.)

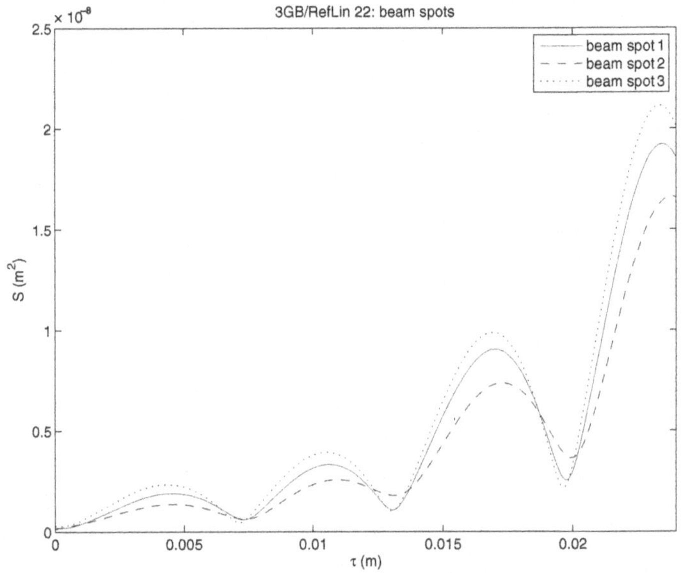

Figure 85 Evolution of GB spots. Parameters: $\alpha_1 l_{01} = 0.1$, $\alpha_2 l_{02} = 0.2$, $\alpha_3 l_{03} = 0.07$, $w1_{\min}(0) = 10^{-5}$, $w1_{\max}(0) = 2 \times 10^{-5}$, $w2_{\min}(0) = 10^{-5}$, $w2_{\max}(0) = 1.5 \times 10^{-5}$, $w3_{\min}(0) = 10^{-5}$, $w3_{\max}(0) = 3 \times 10^{-5}$, ${}^1R_{11}(0) = -707$, ${}^1R_{12}(0) = 397$, ${}^1R_{22}(0) = 400$, ${}^1I_{11}(0) = 1592$, ${}^1I_{12}(0) = 0$, ${}^1I_{22}(0) = 397$, ${}^2R_{11}(0) = 707$, ${}^2R_{12}(0) = 397$, ${}^2R_{22}(0) = 400$, ${}^2I_{11}(0) = 1592$, ${}^2I_{12}(0) = 0$, ${}^2I_{22}(0) = 707$, ${}^3R_{11}(0) = -707$, ${}^3R_{12}(0) = 397$, ${}^3R_{22}(0) = 400$, ${}^3I_{11}(0) = 1592$, ${}^3I_{12}(0) = 0$, ${}^3I_{22}(0) = 176$. (See the color plate.)

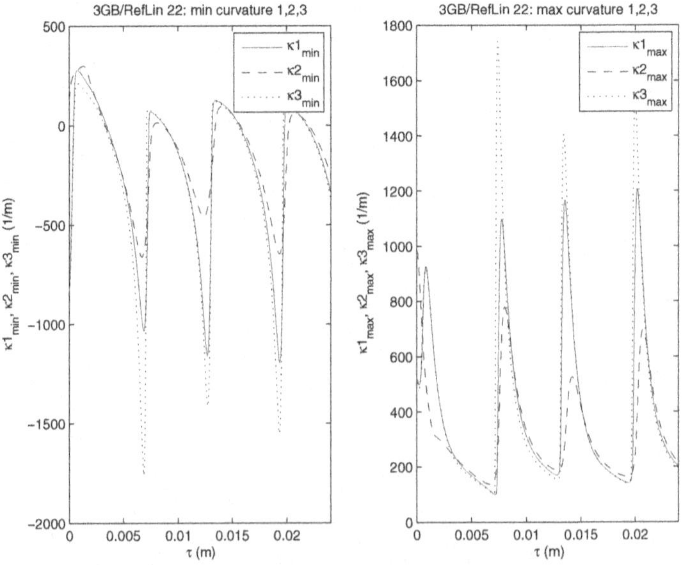

Figure 86 Evolution of GB wave front curvatures. Parameters: $\alpha_1 l_{01} = 0.1$, $\alpha_2 l_{02} = 0.2$, $\alpha_3 l_{03} = 0.07$, $w1_{\min}(0) = 10^{-5}$, $w1_{\max}(0) = 2 \times 10^{-5}$, $w2_{\min}(0) = 10^{-5}$, $w2_{\max}(0) = 1.5 \times 10^{-5}$, $w3_{\min}(0) = 10^{-5}$, $w3_{\max}(0) = 3 \times 10^{-5}$, ${}^1R_{11}(0) = -707$, ${}^1R_{12}(0) = 397$, ${}^1R_{22}(0) = 400$, ${}^1I_{11}(0) = 1592$, ${}^1I_{12}(0) = 0$, ${}^1I_{22}(0) = 397$, ${}^2R_{11}(0) = 707$, ${}^2R_{12}(0) = 397$, ${}^2R_{22}(0) = 400$, ${}^2I_{11}(0) = 1592$, ${}^2I_{12}(0) = 0$, ${}^2I_{22}(0) = 707$, ${}^3R_{11}(0) = -707$, ${}^3R_{12}(0) = 397$, ${}^3R_{22}(0) = 400$, ${}^3I_{11}(0) = 1592$, ${}^3I_{12}(0) = 0$, ${}^3I_{22}(0) = 176$. (See the color plate.)

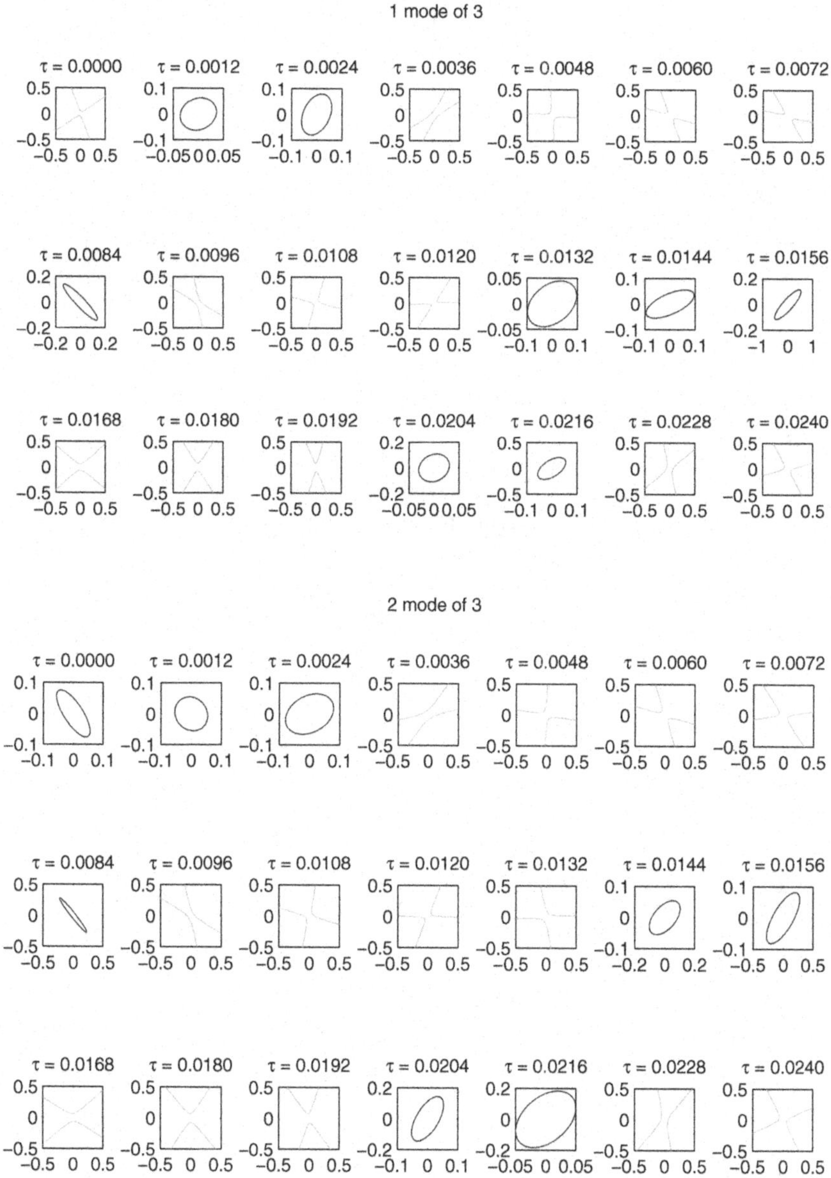

Figure 87 Evolution of GB wave front curvatures. Parameters: $\alpha_1 l_{01} = 0.1$, $\alpha_2 l_{02} = 0.2$, $\alpha_3 l_{03} = 0.07$, $w1_{\min}(0) = 10^{-5}$, $w1_{\max}(0) = 2 \times 10^{-5}$, $w2_{\min}(0) = 10^{-5}$, $w2_{\max}(0) = 1.5 \times 10^{-5}$, $w3_{\min}(0) = 10^{-5}$, $w3_{\max}(0) = 3 \times 10^{-5}$, ${}^1R_{11}(0) = -707$, ${}^1R_{12}(0) = 397$, ${}^1R_{22}(0) = 400$, ${}^1l_{11}(0) = 1592$, ${}^1l_{12}(0) = 0$, ${}^1l_{22}(0) = 397$, ${}^2R_{11}(0) = 707$, ${}^2R_{12}(0) = 397$, ${}^2R_{22}(0) = 400$, ${}^2l_{11}(0) = 1592$, ${}^2l_{12}(0) = 0$, ${}^2l_{22}(0) = 707$, ${}^3R_{11}(0) = -707$, ${}^3R_{12}(0) = 397$, ${}^3R_{22}(0) = 400$, ${}^3l_{11}(0) = 1592$, ${}^3l_{12}(0) = 0$, ${}^3l_{22}(0) = 176$.

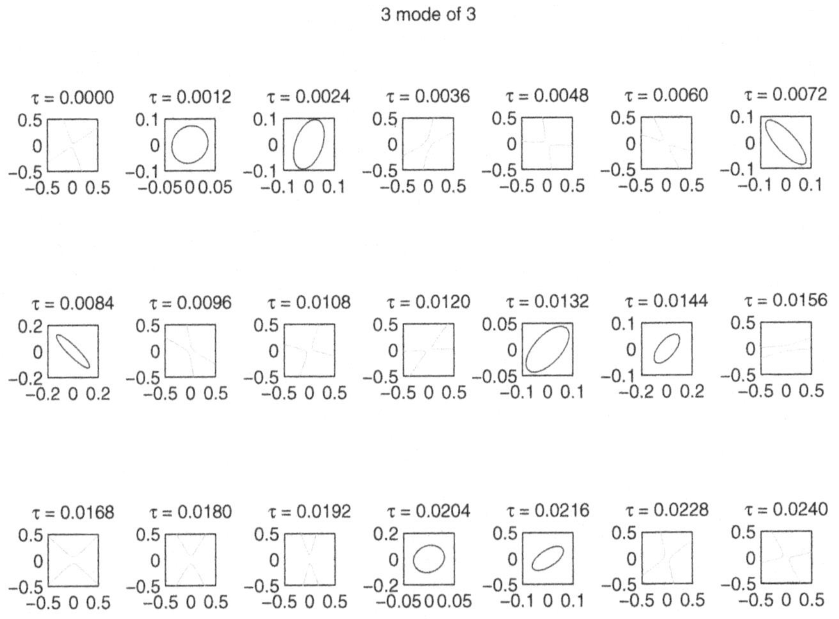

Figure 87 (*continued*)

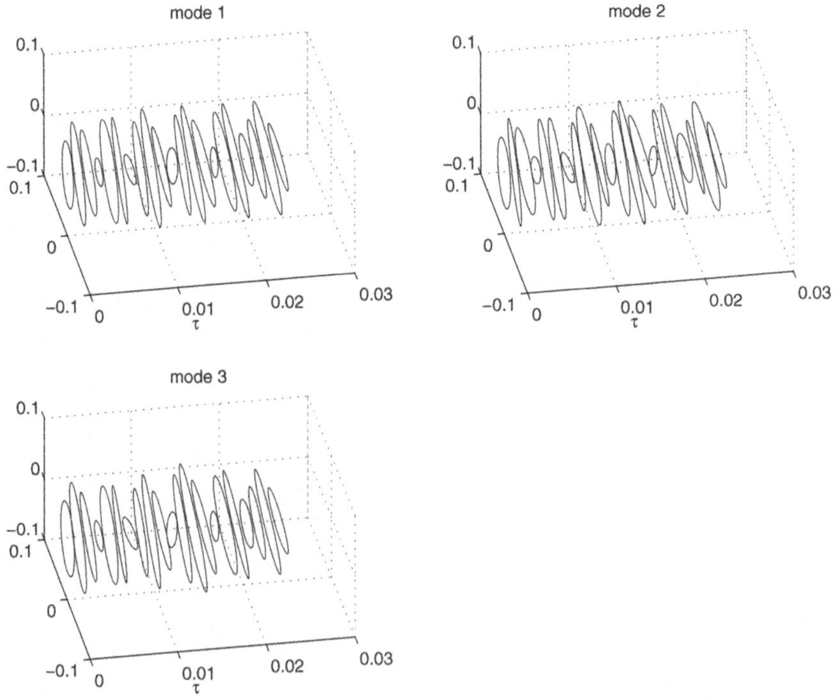

Figure 88 Evolution of GB cross sections. Parameters: $\alpha_1 l_{01} = 0.1$, $\alpha_2 l_{02} = 0.2$, $\alpha_3 l_{03} = 0.07$, $w1_{\min}(0) = 10^{-5}$, $w1_{\max}(0) = 2 \times 10^{-5}$, $w2_{\min}(0) = 1.1 \times 10^{-5}$, $w2_{\max}(0) = 2.1 \times 10^{-5}$, $w3_{\min}(0) = 1.2 \times 10^{-5}$, $w3_{\max}(0) = 2.2 \times 10^{-5}$, ${}^1R_{11}(0) = -707$, ${}^1R_{12}(0) = 397$, ${}^1R_{22}(0) = 400$, ${}^1I_{11}(0) = 1592$, ${}^1I_{12}(0) = 0$, ${}^1I_{22}(0) = 397$, ${}^2R_{11}(0) = 707$, ${}^2R_{12}(0) = 397$, ${}^2R_{22}(0) = 400$, ${}^2I_{11}(0) = 1315$, ${}^2I_{12}(0) = 0$, ${}^2I_{22}(0) = 360$, ${}^3R_{11}(0) = -707$, ${}^3R_{12}(0) = 397$, ${}^3R_{22}(0) = 400$, ${}^3I_{11}(0) = 1105$, ${}^3I_{12}(0) = 0$, ${}^3I_{22}(0) = 328$.

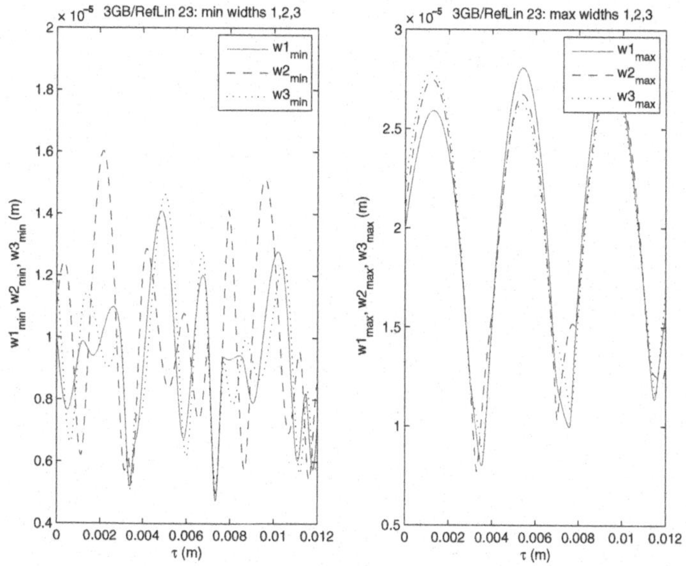

Figure 89 Evolution of GB widths. Parameters: $\alpha_1 l_{01} = 0.1$, $\alpha_2 l_{02} = 0.2$, $\alpha_3 l_{03} = 0.07$, $w1_{\min}(0) = 10^{-5}$, $w1_{\max}(0) = 2 \times 10^{-5}$, $w2_{\min}(0) = 1.1 \times 10^{-5}$, $w2_{\max}(0) = 2.1 \times 10^{-5}$, $w3_{\min}(0) = 1.2 \times 10^{-5}$, $w3_{\max}(0) = 2.2 \times 10^{-5}$, ${}^1R_{11}(0) = -707$, ${}^1R_{12}(0) = 397$, ${}^1R_{22}(0) = 400$, ${}^1I_{11}(0) = 1592$, ${}^1I_{12}(0) = 0$, ${}^1I_{22}(0) = 397$, ${}^2R_{11}(0) = 707$, ${}^2R_{12}(0) = 397$, ${}^2R_{22}(0) = 400$, ${}^2I_{11}(0) = 1315$, ${}^2I_{12}(0) = 0$, ${}^2I_{22}(0) = 360$, ${}^3R_{11}(0) = -707$, ${}^3R_{12}(0) = 397$, ${}^3R_{22}(0) = 400$, ${}^3I_{11}(0) = 1105$, ${}^3I_{12}(0) = 0$, ${}^3I_{22}(0) = 328$. (See the color plate.)

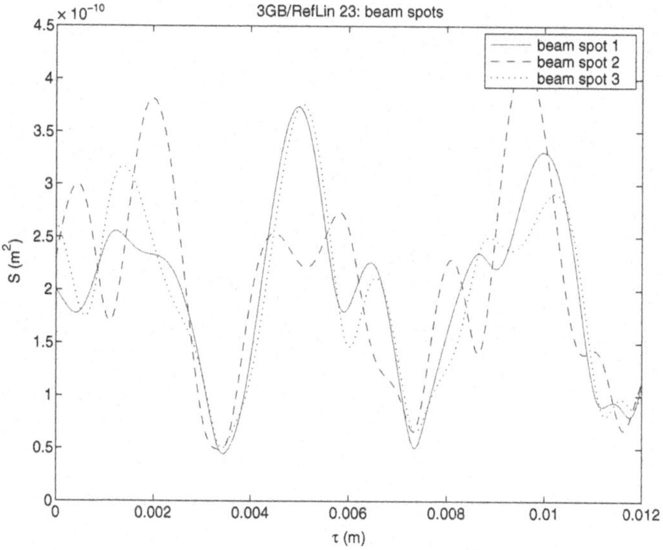

Figure 90 Evolution of GB spots. Parameters: $\alpha_1 l_{01} = 0.1$, $\alpha_2 l_{02} = 0.2$, $\alpha_3 l_{03} = 0.07$, $w1_{\min}(0) = 10^{-5}$, $w1_{\max}(0) = 2 \times 10^{-5}$, $w2_{\min}(0) = 1.1 \times 10^{-5}$, $w2_{\max}(0) = 2.1 \times 10^{-5}$, $w3_{\min}(0) = 1.2 \times 10^{-5}$, $w3_{\max}(0) = 2.2 \times 10^{-5}$, ${}^1R_{11}(0) = -707$, ${}^1R_{12}(0) = 397$, ${}^1R_{22}(0) = 400$, ${}^1I_{11}(0) = 1592$, ${}^1I_{12}(0) = 0$, ${}^1I_{22}(0) = 397$, ${}^2R_{11}(0) = 707$, ${}^2R_{12}(0) = 397$, ${}^2R_{22}(0) = 400$, ${}^2I_{11}(0) = 1315$, ${}^2I_{12}(0) = 0$, ${}^2I_{22}(0) = 360$, ${}^3R_{11}(0) = -707$, ${}^3R_{12}(0) = 397$, ${}^3R_{22}(0) = 400$, ${}^3I_{11}(0) = 1105$, ${}^3I_{12}(0) = 0$, ${}^3I_{22}(0) = 328$. (See the color plate.)

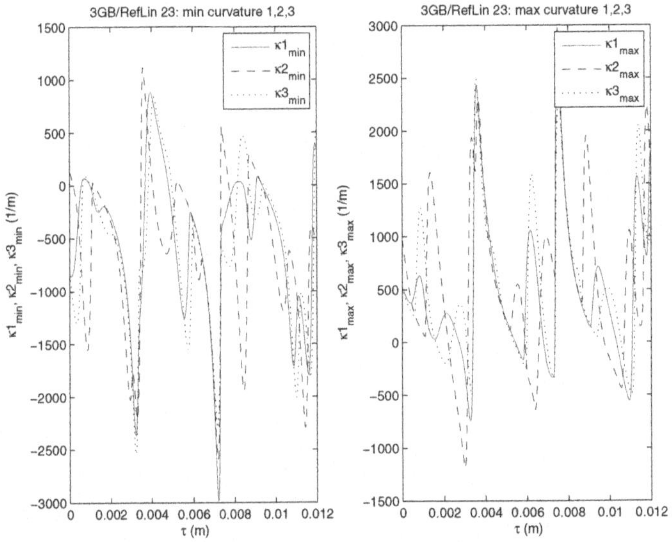

Figure 91 Evolution of GB wave front curvatures. Parameters: $\alpha_1 l_{01} = 0.1$, $\alpha_2 l_{02} = 0.2$, $\alpha_3 l_{03} = 0.07$, $w1_{\min}(0) = 10^{-5}$, $w1_{\max}(0) = 2 \times 10^{-5}$, $w2_{\min}(0) = 1.1 \times 10^{-5}$, $w2_{\max}(0) = 2.1 \times 10^{-5}$, $w3_{\min}(0) = 1.2 \times 10^{-5}$, $w3_{\max}(0) = 2.2 \times 10^{-5}$, ${}^1R_{11}(0) = -707$, ${}^1R_{12}(0) = 397$, ${}^1R_{22}(0) = 400$, ${}^1I_{11}(0) = 1592$, ${}^1I_{12}(0) = 0$, ${}^1I_{22}(0) = 397$, ${}^2R_{11}(0) = 707$, ${}^2R_{12}(0) = 397$, ${}^2R_{22}(0) = 400$, ${}^2I_{11}(0) = 1315$, ${}^2I_{12}(0) = 0$, ${}^2I_{22}(0) = 360$, ${}^3R_{11}(0) = -707$, ${}^3R_{12}(0) = 397$, ${}^3R_{22}(0) = 400$, ${}^3I_{11}(0) = 1105$, ${}^3I_{12}(0) = 0$, ${}^3I_{22}(0) = 328$. (See the color plate.)

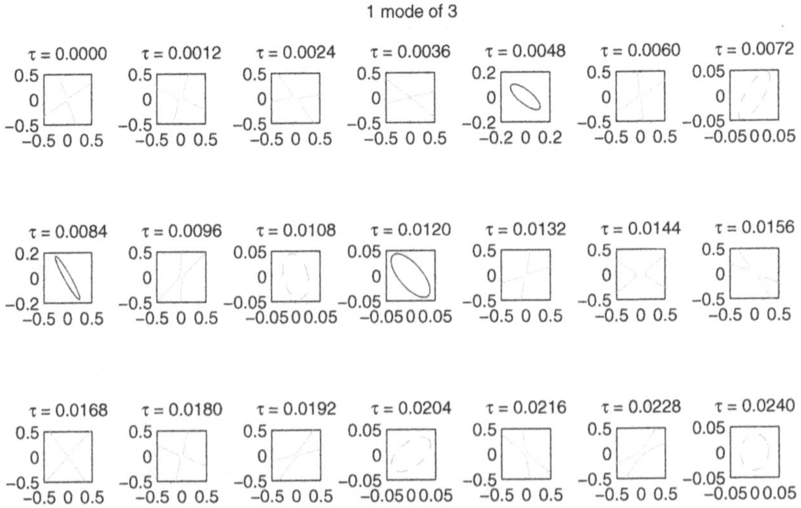

Figure 92 Evolution of wave front cross sections. Parameters: $\alpha_1 l_{01} = 0.1$, $\alpha_2 l_{02} = 0.2$, $\alpha_3 l_{03} = 0.07$, $w1_{\min}(0) = 10^{-5}$, $w1_{\max}(0) = 2 \times 10^{-5}$, $w2_{\min}(0) = 1.1 \times 10^{-5}$, $w2_{\max}(0) = 2.1 \times 10^{-5}$, $w3_{\min}(0) = 1.2 \times 10^{-5}$, $w3_{\max}(0) = 2.2 \times 10^{-5}$, ${}^1R_{11}(0) = -707$, ${}^1R_{12}(0) = 397$, ${}^1R_{22}(0) = 400$, ${}^1I_{11}(0) = 1592$, ${}^1I_{12}(0) = 0$, ${}^1I_{22}(0) = 397$, ${}^2R_{11}(0) = 707$, ${}^2R_{12}(0) = 397$, ${}^2R_{22}(0) = 400$, ${}^2I_{11}(0) = 1315$, ${}^2I_{12}(0) = 0$, ${}^2I_{22}(0) = 360$, ${}^3R_{11}(0) = -707$, ${}^3R_{12}(0) = 397$, ${}^3R_{22}(0) = 400$, ${}^3I_{11}(0) = 1105$, ${}^3I_{12}(0) = 0$, ${}^3I_{22}(0) = 328$.

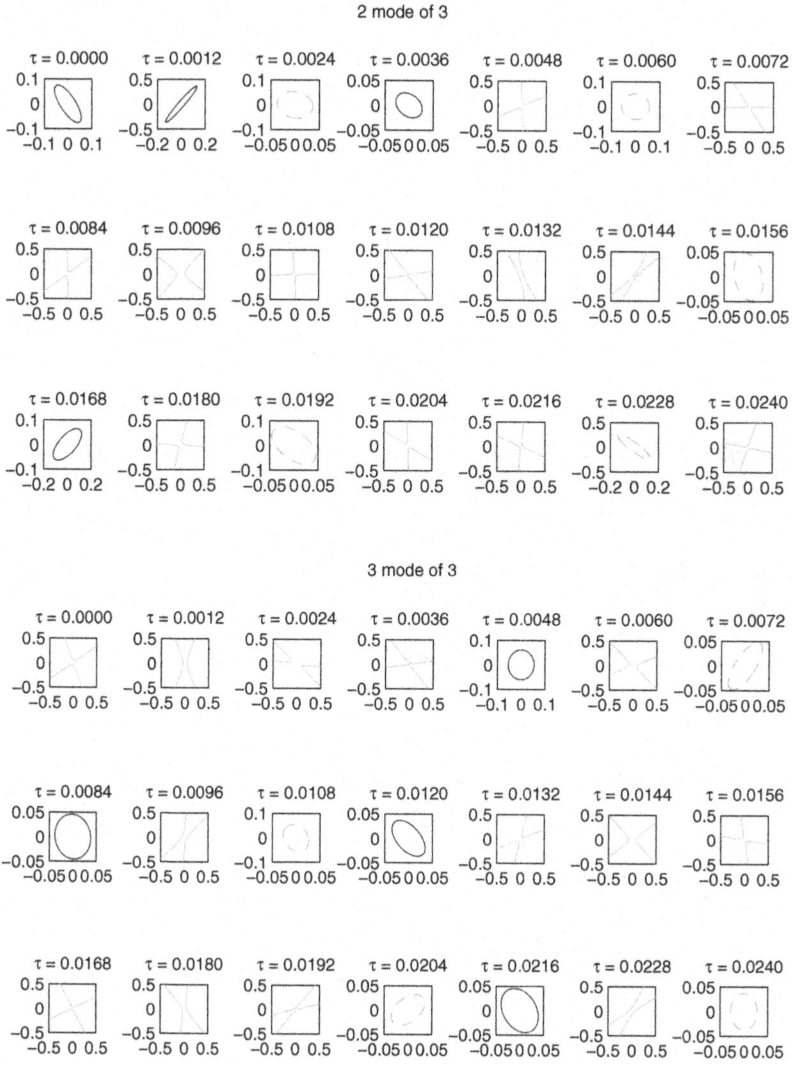

Figure 92 *(continued)*

previous case: beams evolve maintaining their initial directions of rotation. In Figure 83, modes mimic one another and all the GB cross sections rotate clockwise. In Figure 87 we can see wave fronts changing from hyperbolic to elliptic and vice versa. Figures 84–86 depict the evolution of GB widths, GB spots and GB curvatures calculated with the same parameters as in Figures 83 and 87. In Figure 88, there is a rather complicated evolution where the GB

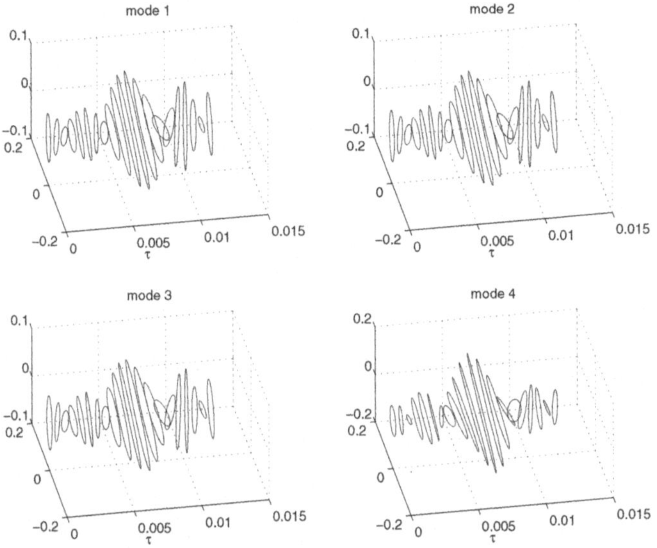

Figure 93 Evolution of GB cross sections. Parameters: $\alpha_1 l_{01} = 0.1$, $\alpha_2 l_{02} = 0.2$, $\alpha_3 l_{03} = 0.07$, $\alpha_4 l_{04} = 0.007$, $w1_{\min}(0) = 10^{-5}$, $w1_{\max}(0) = 2 \times 10^{-5}$, $w2_{\min}(0) = 1.1 \times 10^{-5}$, $w2_{\max}(0) = 2.1 \times 10^{-5}$, $w3_{\min}(0) = 1.2 \times 10^{-5}$, $w3_{\max}(0) = 2.2 \times 10^{-5}$, $w4_{\min}(0) = 1.3 \times 10^{-5}$, $w4_{\max}(0) = 2.3 \times 10^{-5}$, ${}^1R_{11}(0) = 0$, ${}^1R_{12}(0) = -397$, ${}^1R_{22}(0) = 0$, ${}^1I_{11}(0) = 1592$, ${}^1I_{12}(0) = 0$, ${}^1I_{22}(0) = 397$, ${}^2R_{11}(0) = 0$, ${}^2R_{12}(0) = -397$, ${}^2R_{22}(0) = 0$, ${}^2I_{11}(0) = 1315$, ${}^2I_{12}(0) = 0$, ${}^2I_{22}(0) = 360$, ${}^3R_{11}(0) = 0$, ${}^3R_{12}(0) = -397$, ${}^3R_{22}(0) = 0$, ${}^3I_{11}(0) = 1105$, ${}^3I_{12}(0) = 0$, ${}^3I_{22}(0) = 328$, ${}^4R_{11}(0) = 0$, ${}^4R_{12}(0) = 397$, ${}^4R_{22}(0) = 0$, ${}^4I_{11}(0) = 941$, ${}^4I_{12}(0) = 0$, ${}^4I_{22}(0) = 300$.

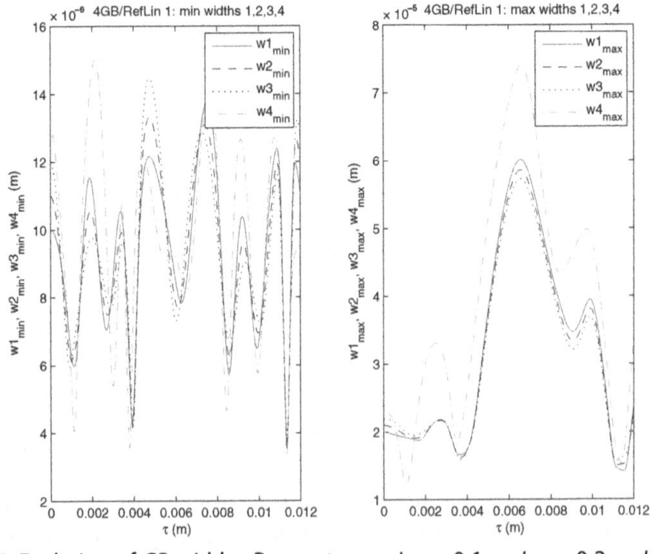

Figure 94 Evolution of GB widths. Parameters: $\alpha_1 l_{01} = 0.1$, $\alpha_2 l_{02} = 0.2$, $\alpha_3 l_{03} = 0.07$, $\alpha_4 l_{04} = 0.007$, $w1_{\min}(0) = 10^{-5}$, $w1_{\max}(0) = 2 \times 10^{-5}$, $w2_{\min}(0) = 1.1 \times 10^{-5}$, $w2_{\max}(0) = 2.1 \times 10^{-5}$, $w3_{\min}(0) = 1.2 \times 10^{-5}$, $w3_{\max}(0) = 2.2 \times 10^{-5}$, $w4_{\min}(0) = 1.3 \times 10^{-5}$, $w4_{\max}(0) = 2.3 \times 10^{-5}$, ${}^1R_{11}(0) = 0$, ${}^1R_{12}(0) = -397$, ${}^1R_{22}(0) = 0$, ${}^1I_{11}(0) = 1592$, ${}^1I_{12}(0) = 0$, ${}^1I_{22}(0) = 397$, ${}^2R_{11}(0) = 0$, ${}^2R_{12}(0) = -397$, ${}^2R_{22}(0) = 0$, ${}^2I_{11}(0) = 1315$, ${}^2I_{12}(0) = 0$, ${}^2I_{22}(0) = 360$, ${}^3R_{11}(0) = 0$, ${}^3R_{12}(0) = -397$, ${}^3R_{22}(0) = 0$, ${}^3I_{11}(0) = 1105$, ${}^3I_{12}(0) = 0$, ${}^3I_{22}(0) = 328$, ${}^4R_{11}(0) = 0$, ${}^4R_{12}(0) = 397$, ${}^4R_{22}(0) = 0$, ${}^4I_{11}(0) = 941$, ${}^4I_{12}(0) = 0$, ${}^4I_{22}(0) = 300$. (See the color plate.)

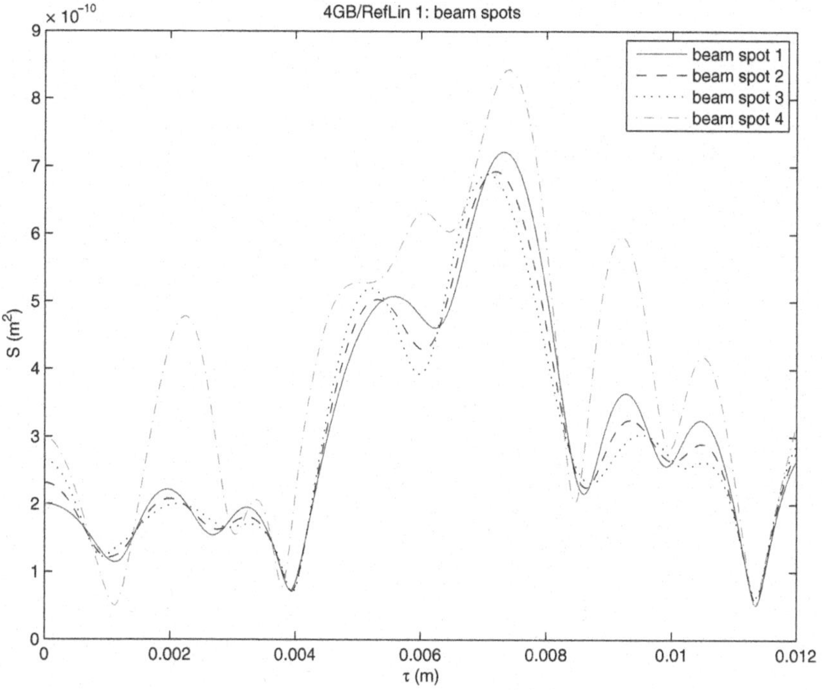

Figure 95 Evolution of GB spots. Parameters: $\alpha_1 l_{01} = 0.1$, $\alpha_2 l_{02} = 0.2$, $\alpha_3 l_{03} = 0.07$, $\alpha_4 l_{04} = 0.007$, $w1_{\min}(0) = 10^{-5}$, $w1_{\max}(0) = 2 \times 10^{-5}$, $w2_{\min}(0) = 1.1 \times 10^{-5}$, $w2_{\max}(0) = 2.1 \times 10^{-5}$, $w3_{\min}(0) = 1.2 \times 10^{-5}$, $w3_{\max}(0) = 2.2 \times 10^{-5}$, $w4_{\min}(0) = 1.3 \times 10^{-5}$, $w4_{\max}(0) = 2.3 \times 10^{-5}$, ${}^1R_{11}(0) = 0$, ${}^1R_{12}(0) = -397$, ${}^1R_{22}(0) = 0$, ${}^1I_{11}(0) = 1592$, ${}^1I_{12}(0) = 0$, ${}^1I_{22}(0) = 397$, ${}^2R_{11}(0) = 0$, ${}^2R_{12}(0) = -397$, ${}^2R_{22}(0) = 0$, ${}^2I_{11}(0) = 1315$, ${}^2I_{12}(0) = 0$, ${}^2I_{22}(0) = 360$, ${}^3R_{11}(0) = 0$, ${}^3R_{12}(0) = -397$, ${}^3R_{22}(0) = 0$, ${}^3I_{11}(0) = 1105$, ${}^3I_{12}(0) = 0$, ${}^3I_{22}(0) = 328$, ${}^4R_{11}(0) = 0$, ${}^4R_{12}(0) = 397$, ${}^4R_{22}(0) = 0$, ${}^4I_{11}(0) = 941$, ${}^4I_{12}(0) = 0$, ${}^4I_{22}(0) = 300$. (See the color plate.)

cross sections change orientation with respect to one another. In Figure 92, it can be observed that at the beginning, three modes rotate counter-clockwise and one of the four modes rotates clockwise. After a distance of about 10 diffraction lengths, the fourth mode starts to imitate the evolutions of the rest. Figures 89–91 represent the evolution of GB widths, GB spots and GB curvatures calculated with the same parameters as in Figures 88 and 92. Figures 93–98 show the results for another parameter set. In Figure 96, there is a rather complicated evolution of the wave front cross sections of four modes. In Figure 97, it can be observed that four modes rotate and evolve in the same manner. In Figure 98, after a distance of a few diffraction lengths, the wave front cross sections of each mode start to imitate the

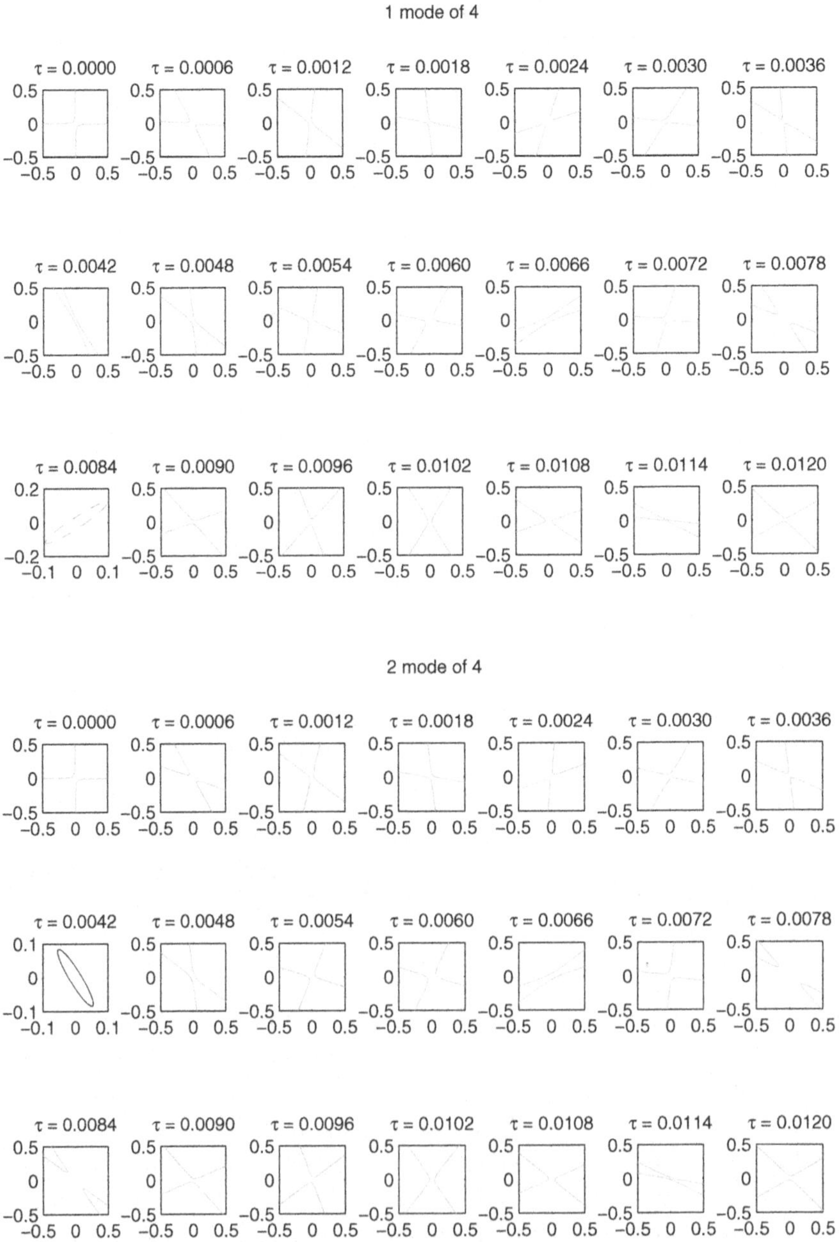

Figure 96 Evolution of wave front cross sections. Parameters: $\alpha_1 l_{01} = 0.1$, $\alpha_2 l_{02} = 0.2$, $\alpha_3 l_{03} = 0.07$, $\alpha_4 l_{04} = 0.007$, $w1_{\min}(0) = 10^{-5}$, $w1_{\max}(0) = 2 \times 10^{-5}$, $w2_{\min}(0) = 1.1 \times 10^{-5}$, $w2_{\max}(0) = 2.1 \times 10^{-5}$, $w3_{\min}(0) = 1.2 \times 10^{-5}$, $w3_{\max}(0) = 2.2 \times 10^{-5}$, $w4_{\min}(0) = 1.3 \times 10^{-5}$, $w4_{\max}(0) = 2.3 \times 10^{-5}$, ${}^1R_{11}(0) = 0$, ${}^1R_{12}(0) = -397$, ${}^1R_{22}(0) = 0$, ${}^1I_{11}(0) = 1592$, ${}^1I_{12}(0) = 0$, ${}^1I_{22}(0) = 397$, ${}^2R_{11}(0) = 0$, ${}^2R_{12}(0) = -397$, ${}^2R_{22}(0) = 0$, ${}^2I_{11}(0) = 1315$, ${}^2I_{12}(0) = 0$, ${}^2I_{22}(0) = 360$, ${}^3R_{11}(0) = 0$, ${}^3R_{12}(0) = -397$, ${}^3R_{22}(0) = 0$, ${}^3I_{11}(0) = 1105$, ${}^3I_{12}(0) = 0$, ${}^3I_{22}(0) = 328$, ${}^4R_{11}(0) = 0$, ${}^4R_{12}(0) = 397$, ${}^4R_{22}(0) = 0$, ${}^4I_{11}(0) = 941$, ${}^4I_{12}(0) = 0$, ${}^4I_{22}(0) = 300$.

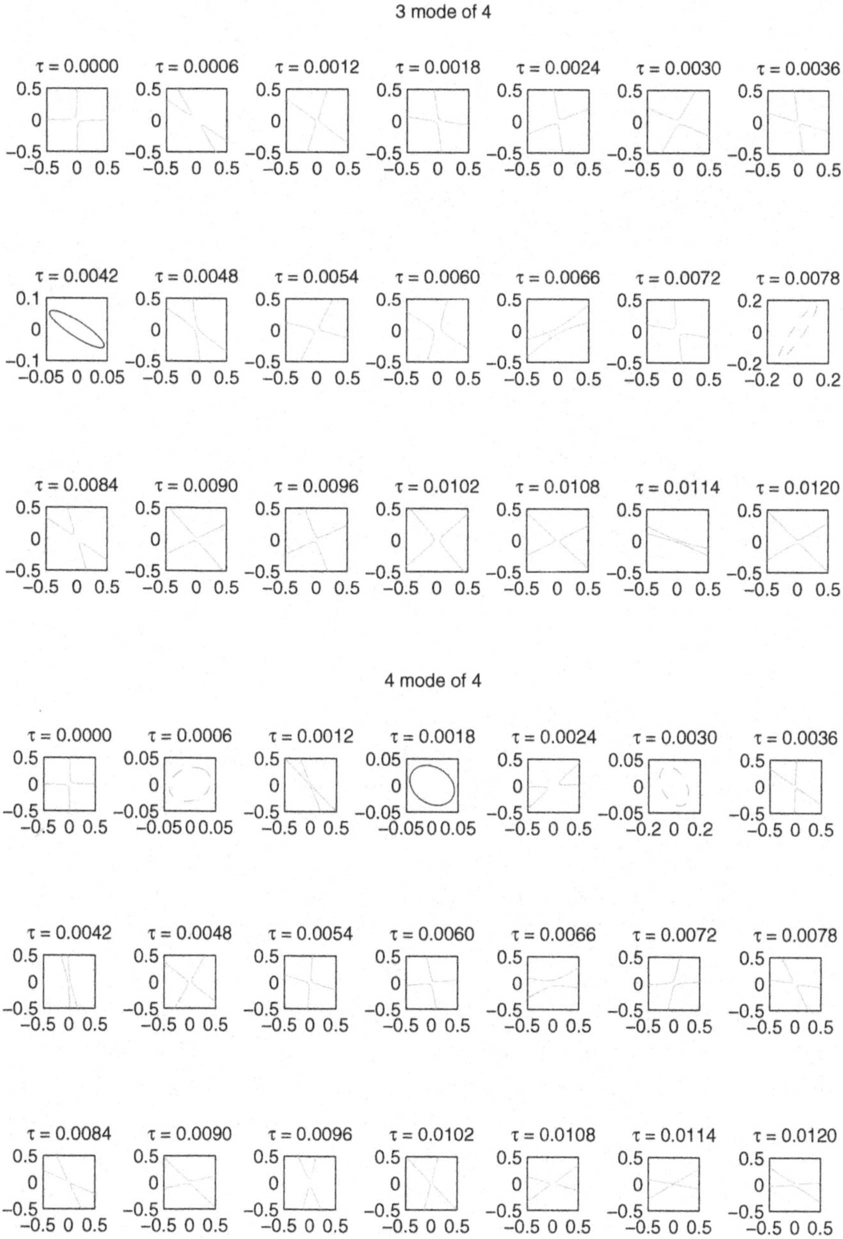

Figure 96 *(continued)*

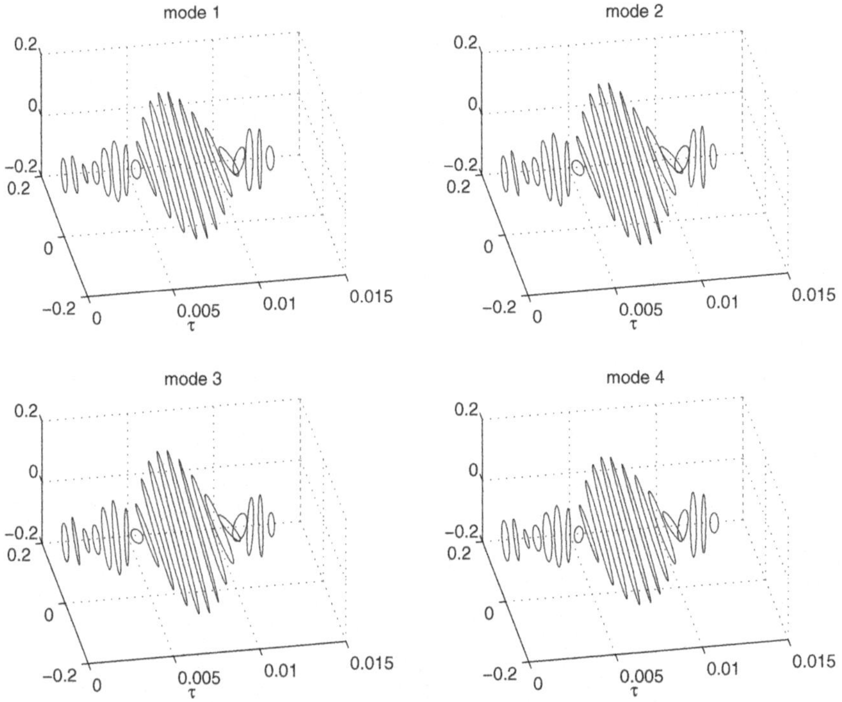

Figure 97 Evolution of GB cross sections. Parameters: $\alpha_1 l_{01} = 0.1$, $\alpha_2 l_{02} = 0.2$, $\alpha_3 l_{03} = 0.07$, $\alpha_4 l_{04} = 0.007$, $w1_{\min}(0) = 10^{-5}$, $w1_{\max}(0) = 2 \times 10^{-5}$, $w2_{\min}(0) = 1.1 \times 10^{-5}$, $w2_{\max}(0) = 2.1 \times 10^{-5}$, $w3_{\min}(0) = 1.2 \times 10^{-5}$, $w3_{\max}(0) = 2.2 \times 10^{-5}$, $w4_{\min}(0) = 1.3 \times 10^{-5}$, $w4_{\max}(0) = 2.3 \times 10^{-5}$, ${}^1R_{11}(0) = 0$, ${}^1R_{12}(0) = 397$, ${}^1R_{22}(0) = 0$, ${}^1I_{11}(0) = 1592$, ${}^1I_{12}(0) = 0$, ${}^1I_{22}(0) = 397$, ${}^2R_{11}(0) = 0$, ${}^2R_{12}(0) = 795$, ${}^2R_{22}(0) = 0$, ${}^2I_{11}(0) = 1315$, ${}^2I_{12}(0) = 0$, ${}^2I_{22}(0) = 360$, ${}^3R_{11}(0) = 0$, ${}^3R_{12}(0) = 795$, ${}^3R_{22}(0) = 0$, ${}^3I_{11}(0) = 1105$, ${}^3I_{12}(0) = 0$, ${}^3I_{22}(0) = 328$, ${}^4R_{11}(0) = 0$, ${}^4R_{12}(0) = 397$, ${}^4R_{22}(0) = 0$, ${}^4I_{11}(0) = 941$, ${}^4I_{12}(0) = 0$, ${}^4I_{22}(0) = 300$.

evolution of one another. Figure 99 illustrates that the evolution of the four modes splits into the development of two seemingly independent pairs of modes. In Fig. 100 GB spots are shown for this case. In Figure 101, there are very complicated dynamics of wave front cross sections, which are strongly coupled. In Figure 102, the four modes rotate in a similar manner. In Fig. 103 we can notice GB spots. In Figure 104, it can be seen that wave front cross sections of all four modes are heavily coupled. In Figure 105, four modes evolve separately, but they all rotate counterclockwise. The GB widths and GB spots are shown in Figures 106 and 107. In Figure 108, we can see that the wave front cross sections of four rotating modes are heavily coupled as well.

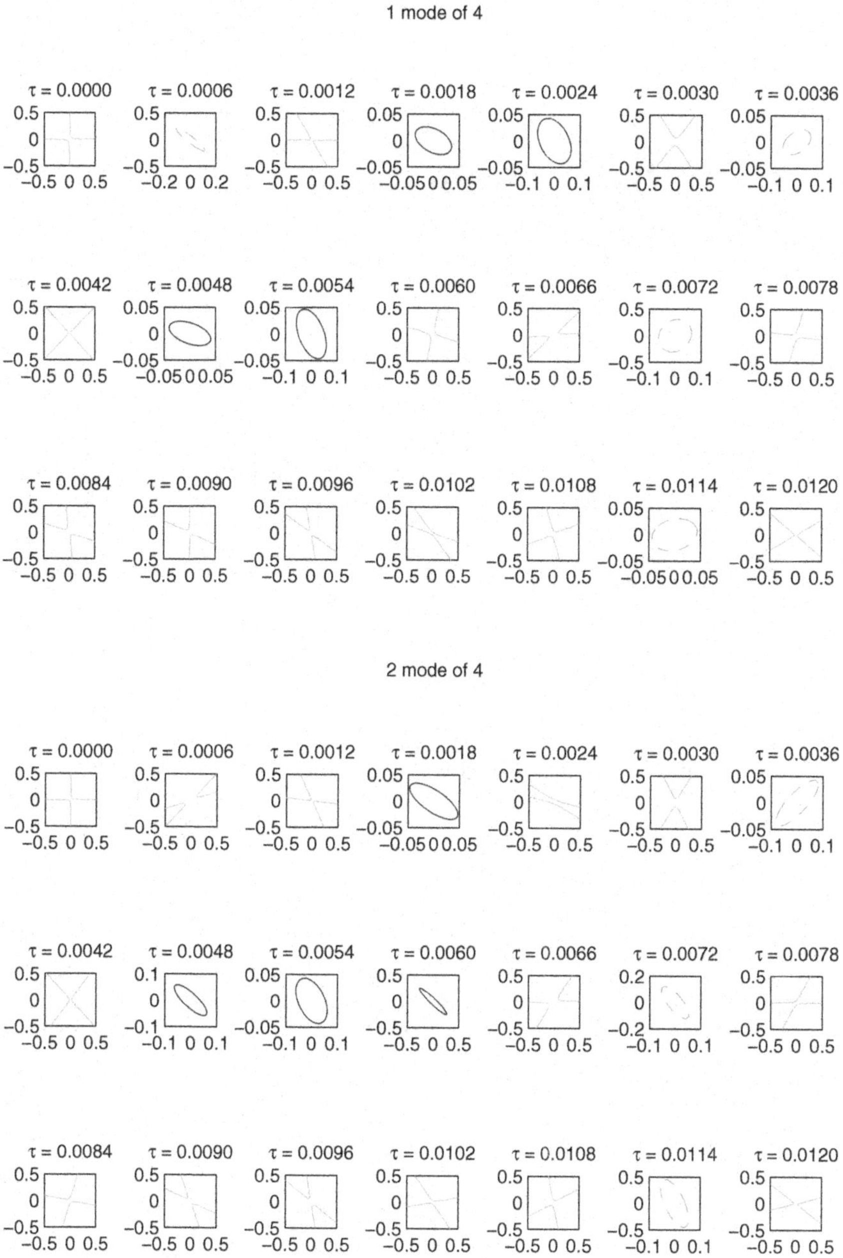

Figure 98 Evolution of wave front cross sections. Parameters: $\alpha_1 l_{01} = 0.1$, $\alpha_2 l_{02} = 0.2$, $\alpha_3 l_{03} = 0.07$, $\alpha_4 l_{04} = 0.007$, $w1_{\min}(0) = 10^{-5}$, $w1_{\max}(0) = 2 \times 10^{-5}$, $w2_{\min}(0) = 1.1 \times 10^{-5}$, $w2_{\max}(0) = 2.1 \times 10^{-5}$, $w3_{\min}(0) = 1.2 \times 10^{-5}$, $w3_{\max}(0) = 2.2 \times 10^{-5}$, $w4_{\min}(0) = 1.3 \times 10^{-5}$, $w4_{\max}(0) = 2.3 \times 10^{-5}$, ${}^1R_{11}(0) = 0$, ${}^1R_{12}(0) = 397$, ${}^1R_{22}(0) = 0$, ${}^1I_{11}(0) = 1592$, ${}^1I_{12}(0) = 0$, ${}^1I_{22}(0) = 397$, ${}^2R_{11}(0) = 0$, ${}^2R_{12}(0) = 795$, ${}^2R_{22}(0) = 0$, ${}^2I_{11}(0) = 1315$, ${}^2I_{12}(0) = 0$, ${}^2I_{22}(0) = 360$, ${}^3R_{11}(0) = 0$, ${}^3R_{12}(0) = 795$, ${}^3R_{22}(0) = 0$, ${}^3I_{11}(0) = 1105$, ${}^3I_{12}(0) = 0$, ${}^3I_{22}(0) = 328$, ${}^4R_{11}(0) = 0$, ${}^4R_{12}(0) = 397$, ${}^4R_{22}(0) = 0$, ${}^4I_{11}(0) = 941$, ${}^4I_{12}(0) = 0$, ${}^4I_{22}(0) = 300$.

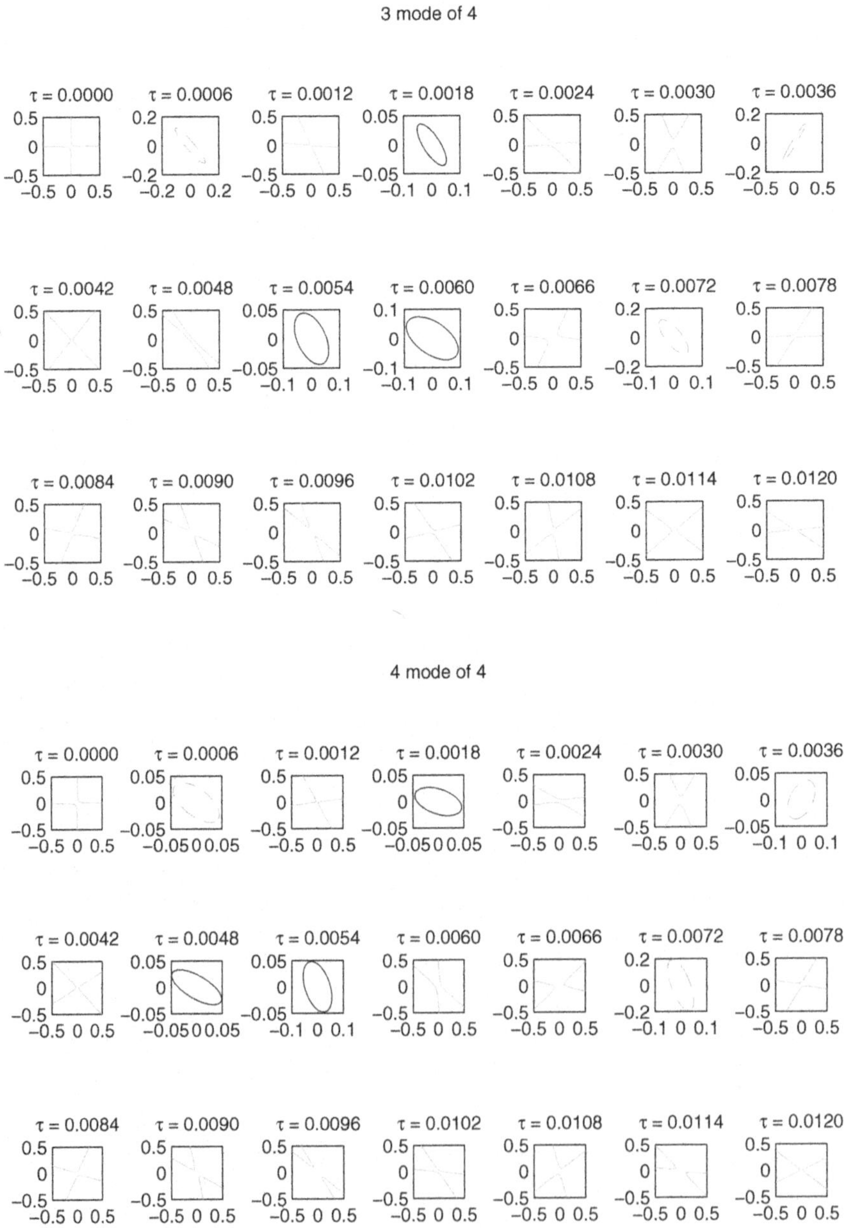

Figure 98 *(continued)*

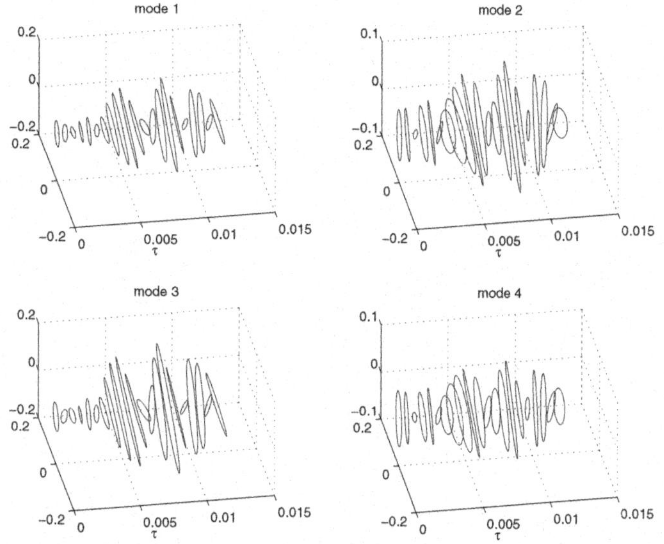

Figure 99 Evolution of GB cross sections. Parameters: $\alpha_1 l_{01} = 0.1$, $\alpha_2 l_{02} = 0.2$, $\alpha_3 l_{03} = 0.07$, $\alpha_4 l_{04} = 0.007$, $w1_{\min}(0) = 10^{-5}$, $w1_{\max}(0) = 2 \times 10^{-5}$, $w2_{\min}(0) = 1.1 \times 10^{-5}$, $w2_{\max}(0) = 2.1 \times 10^{-5}$, $w3_{\min}(0) = 1.2 \times 10^{-5}$, $w3_{\max}(0) = 2.2 \times 10^{-5}$, $w4_{\min}(0) = 1.3 \times 10^{-5}$, $w4_{\max}(0) = 2.3 \times 10^{-5}$, ${}^1R_{11}(0) = 0$, ${}^1R_{12}(0) = -397$, ${}^1R_{22}(0) = 0$, ${}^1I_{11}(0) = 1592$, ${}^1I_{12}(0) = 0$, ${}^1I_{22}(0) = 397$, ${}^2R_{11}(0) = 0$, ${}^2R_{12}(0) = 795$, ${}^2R_{22}(0) = 0$, ${}^2I_{11}(0) = 1315$, ${}^2I_{12}(0) = 0$, ${}^2I_{22}(0) = 360$, ${}^3R_{11}(0) = 0$, ${}^3R_{12}(0) = -795$, ${}^3R_{22}(0) = 0$, ${}^3I_{11}(0) = 1105$, ${}^3I_{12}(0) = 0$, ${}^3I_{22}(0) = 328$, ${}^4R_{11}(0) = 0$, ${}^4R_{12}(0) = 397$, ${}^4R_{22}(0) = 0$, ${}^4I_{11}(0) = 941$, ${}^4I_{12}(0) = 0$, ${}^4I_{22}(0) = 300$.

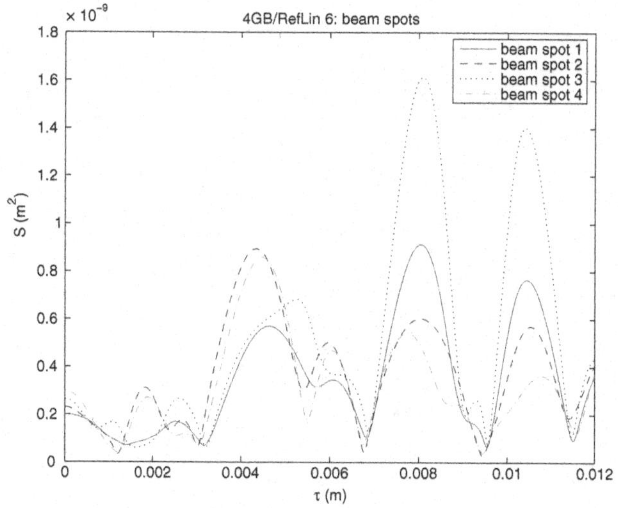

Figure 100 Evolution of GB spots. Parameters: $\alpha_1 l_{01} = 0.1$, $\alpha_2 l_{02} = 0.2$, $\alpha_3 l_{03} = 0.07$, $\alpha_4 l_{04} = 0.007$, $w1_{\min}(0) = 10^{-5}$, $w1_{\max}(0) = 2 \times 10^{-5}$, $w2_{\min}(0) = 1.1 \times 10^{-5}$, $w2_{\max}(0) = 2.1 \times 10^{-5}$, $w3_{\min}(0) = 1.2 \times 10^{-5}$, $w3_{\max}(0) = 2.2 \times 10^{-5}$, $w4_{\min}(0) = 1.3 \times 10^{-5}$, $w4_{\max}(0) = 2.3 \times 10^{-5}$, ${}^1R_{11}(0) = 0$, ${}^1R_{12}(0) = -397$, ${}^1R_{22}(0) = 0$, ${}^1I_{11}(0) = 1592$, ${}^1I_{12}(0) = 0$, ${}^1I_{22}(0) = 397$, ${}^2R_{11}(0) = 0$, ${}^2R_{12}(0) = 795$, ${}^2R_{22}(0) = 0$, ${}^2I_{11}(0) = 1315$, ${}^2I_{12}(0) = 0$, ${}^2I_{22}(0) = 360$, ${}^3R_{11}(0) = 0$, ${}^3R_{12}(0) = -795$, ${}^3R_{22}(0) = 0$, ${}^3I_{11}(0) = 1105$, ${}^3I_{12}(0) = 0$, ${}^3I_{22}(0) = 328$, ${}^4R_{11}(0) = 0$, ${}^4R_{12}(0) = 397$, ${}^4R_{22}(0) = 0$, ${}^4I_{11}(0) = 941$, ${}^4I_{12}(0) = 0$, ${}^4I_{22}(0) = 300$. (See the color plate.)

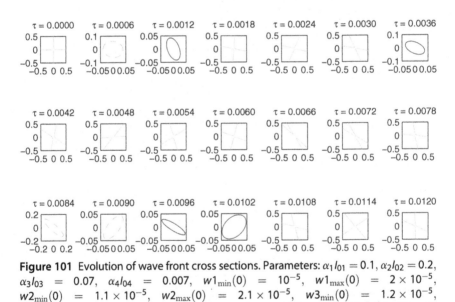

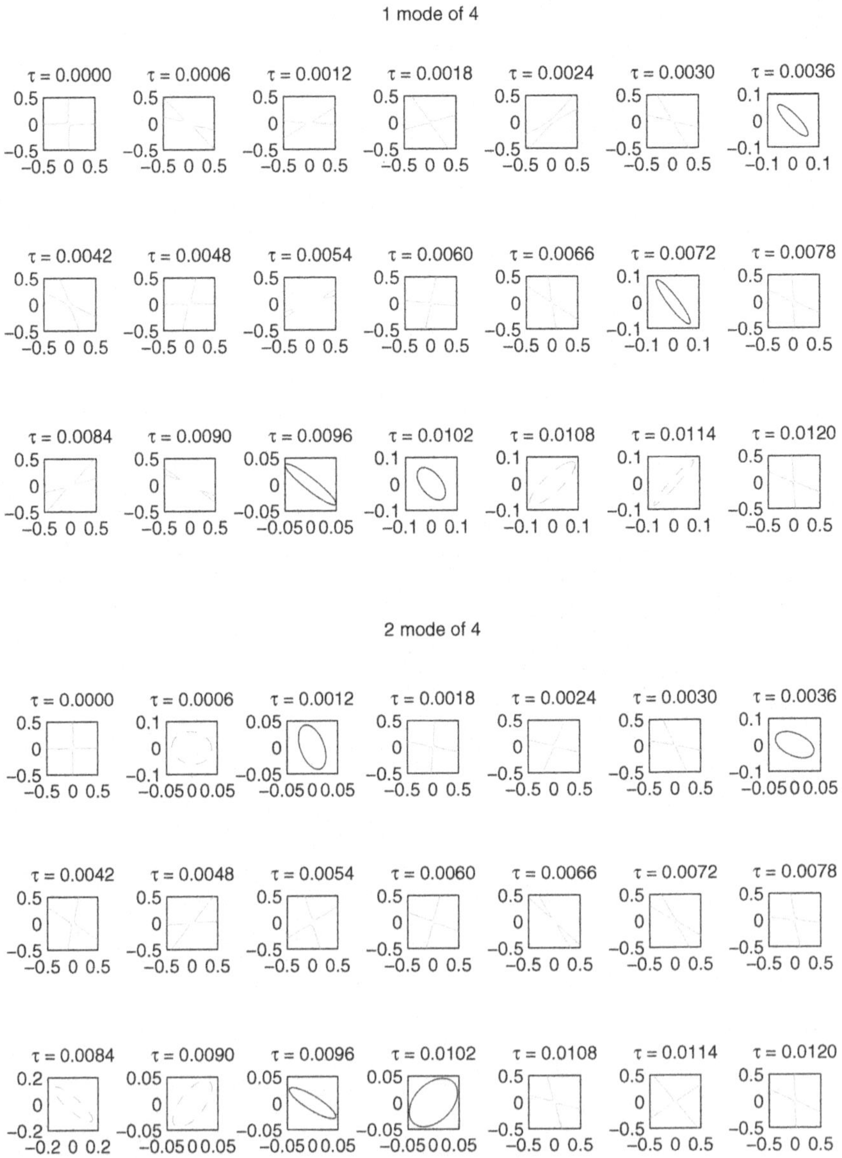

Figure 101 Evolution of wave front cross sections. Parameters: $\alpha_1 l_{01} = 0.1$, $\alpha_2 l_{02} = 0.2$, $\alpha_3 l_{03} = 0.07$, $\alpha_4 l_{04} = 0.007$, $w1_{\min}(0) = 10^{-5}$, $w1_{\max}(0) = 2 \times 10^{-5}$, $w2_{\min}(0) = 1.1 \times 10^{-5}$, $w2_{\max}(0) = 2.1 \times 10^{-5}$, $w3_{\min}(0) = 1.2 \times 10^{-5}$, $w3_{\max}(0) = 2.2 \times 10^{-5}$, $w4_{\min}(0) = 1.3 \times 10^{-5}$, $w4_{\max}(0) = 2.3 \times 10^{-5}$, ${}^1R_{11}(0) = 0$, ${}^1R_{12}(0) = -397$, ${}^1R_{22}(0) = 0$, ${}^1I_{11}(0) = 1592$, ${}^1I_{12}(0) = 0$, ${}^1I_{22}(0) = 397$, ${}^2R_{11}(0) = 0$, ${}^2R_{12}(0) = 795$, ${}^2R_{22}(0) = 0$, ${}^2I_{11}(0) = 1315$, ${}^2I_{12}(0) = 0$, ${}^2I_{22}(0) = 360$, ${}^3R_{11}(0) = 0$, ${}^3R_{12}(0) = -795$, ${}^3R_{22}(0) = 0$, ${}^3I_{11}(0) = 1105$, ${}^3I_{12}(0) = 0$, ${}^3I_{22}(0) = 328$, ${}^4R_{11}(0) = 0$, ${}^4R_{12}(0) = 397$, ${}^4R_{22}(0) = 0$, ${}^4I_{11}(0) = 941$, ${}^4I_{12}(0) = 0$, ${}^4I_{22}(0) = 300$.

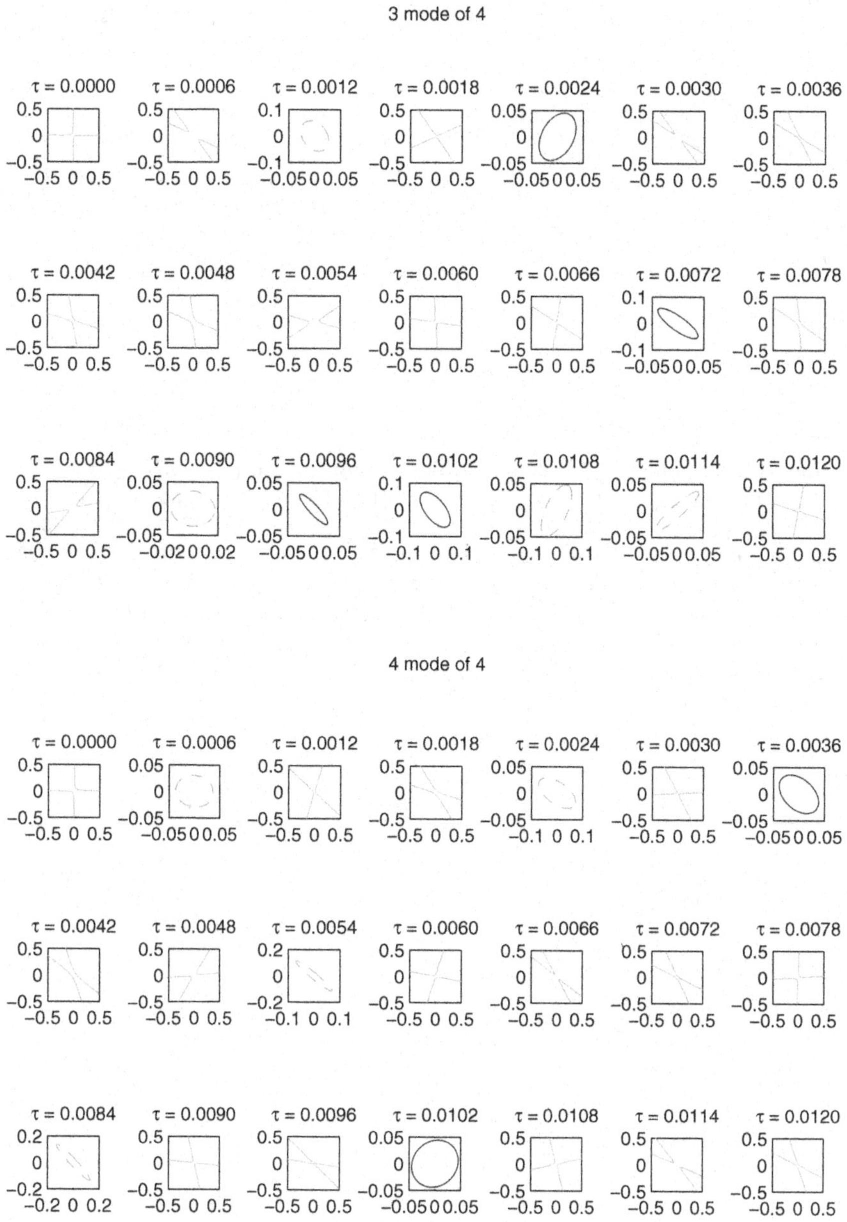

Figure 101 (*continued*)

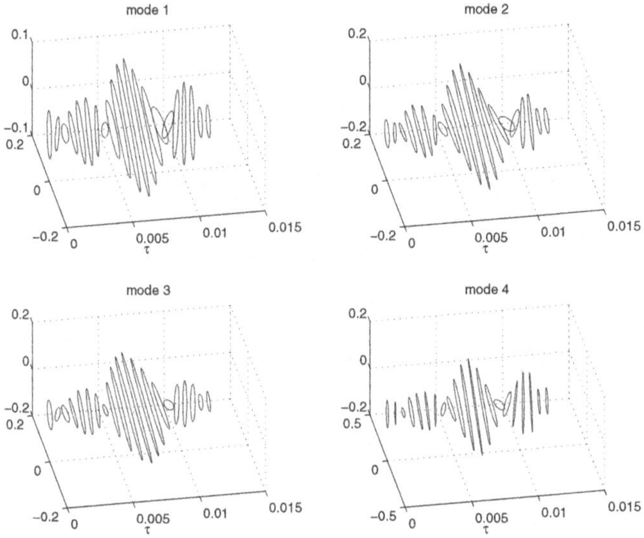

Figure 102 Evolution of GB cross sections. Parameters: $\alpha_1 l_{01} = 0.1$, $\alpha_2 l_{02} = 0.2$, $\alpha_3 l_{03} = 0.07$, $\alpha_4 l_{04} = 0.007$, $w1_{\min}(0) = 10^{-5}$, $w1_{\max}(0) = 2 \times 10^{-5}$, $w2_{\min}(0) = 1.1 \times 10^{-5}$, $w2_{\max}(0) = 2.1 \times 10^{-5}$, $w3_{\min}(0) = 1.2 \times 10^{-5}$, $w3_{\max}(0) = 2.2 \times 10^{-5}$, $w4_{\min}(0) = 1.3 \times 10^{-5}$, $w4_{\max}(0) = 2.3 \times 10^{-5}$, $^1R_{11}(0) = -707$, $^1R_{12}(0) = -397$, $^1R_{22}(0) = 0$, $^1I_{11}(0) = 1592$, $^1I_{12}(0) = 0$, $^1I_{22}(0) = 397$, $^2R_{11}(0) = 707$, $^2R_{12}(0) = 198$, $^2R_{22}(0) = 0$, $^2I_{11}(0) = 1315$, $^2I_{12}(0) = 0$, $^2I_{22}(0) = 360$, $^3R_{11}(0) = -707$, $^3R_{12}(0) = -795$, $^3R_{22}(0) = 0$, $^3I_{11}(0) = 1105$, $^3I_{12}(0) = 0$, $^3I_{22}(0) = 328$, $^4R_{11}(0) = -707$, $^4R_{12}(0) = 397$, $^4R_{22}(0) = 0$, $^4I_{11}(0) = 941$, $^4I_{12}(0) = 0$, $^4I_{22}(0) = 300$.

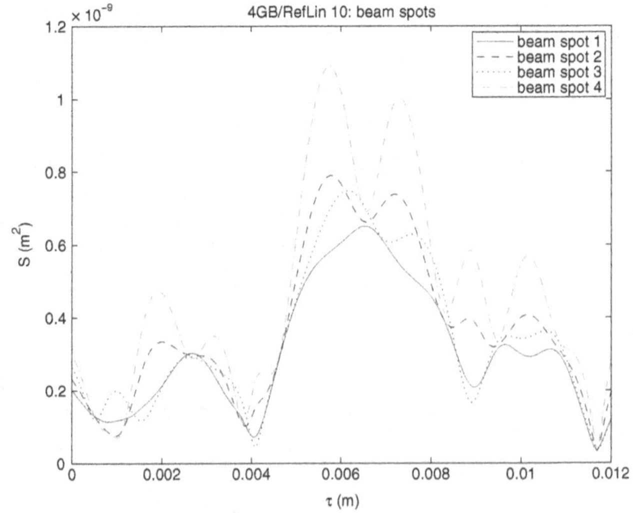

Figure 103 Evolution of GB spots. Parameters: $\alpha_1 l_{01} = 0.1$, $\alpha_2 l_{02} = 0.2$, $\alpha_3 l_{03} = 0.07$, $\alpha_4 l_{04} = 0.007$, $w1_{\min}(0) = 10^{-5}$, $w1_{\max}(0) = 2 \times 10^{-5}$, $w2_{\min}(0) = 1.1 \times 10^{-5}$, $w2_{\max}(0) = 2.1 \times 10^{-5}$, $w3_{\min}(0) = 1.2 \times 10^{-5}$, $w3_{\max}(0) = 2.2 \times 10^{-5}$, $w4_{\min}(0) = 1.3 \times 10^{-5}$, $w4_{\max}(0) = 2.3 \times 10^{-5}$, $^1R_{11}(0) = -707$, $^1R_{12}(0) = -397$, $^1R_{22}(0) = 0$, $^1I_{11}(0) = 1592$, $^1I_{12}(0) = 0$, $^1I_{22}(0) = 397$, $^2R_{11}(0) = 707$, $^2R_{12}(0) = 198$, $^2R_{22}(0) = 0$, $^2I_{11}(0) = 1315$, $^2I_{12}(0) = 0$, $^2I_{22}(0) = 360$, $^3R_{11}(0) = -707$, $^3R_{12}(0) = -795$, $^3R_{22}(0) = 0$, $^3I_{11}(0) = 1105$, $^3I_{12}(0) = 0$, $^3I_{22}(0) = 328$, $^4R_{11}(0) = -707$, $^4R_{12}(0) = 397$, $^4R_{22}(0) = 0$, $^4I_{11}(0) = 941$, $^4I_{12}(0) = 0$, $^4I_{22}(0) = 300$. (See the color plate.)

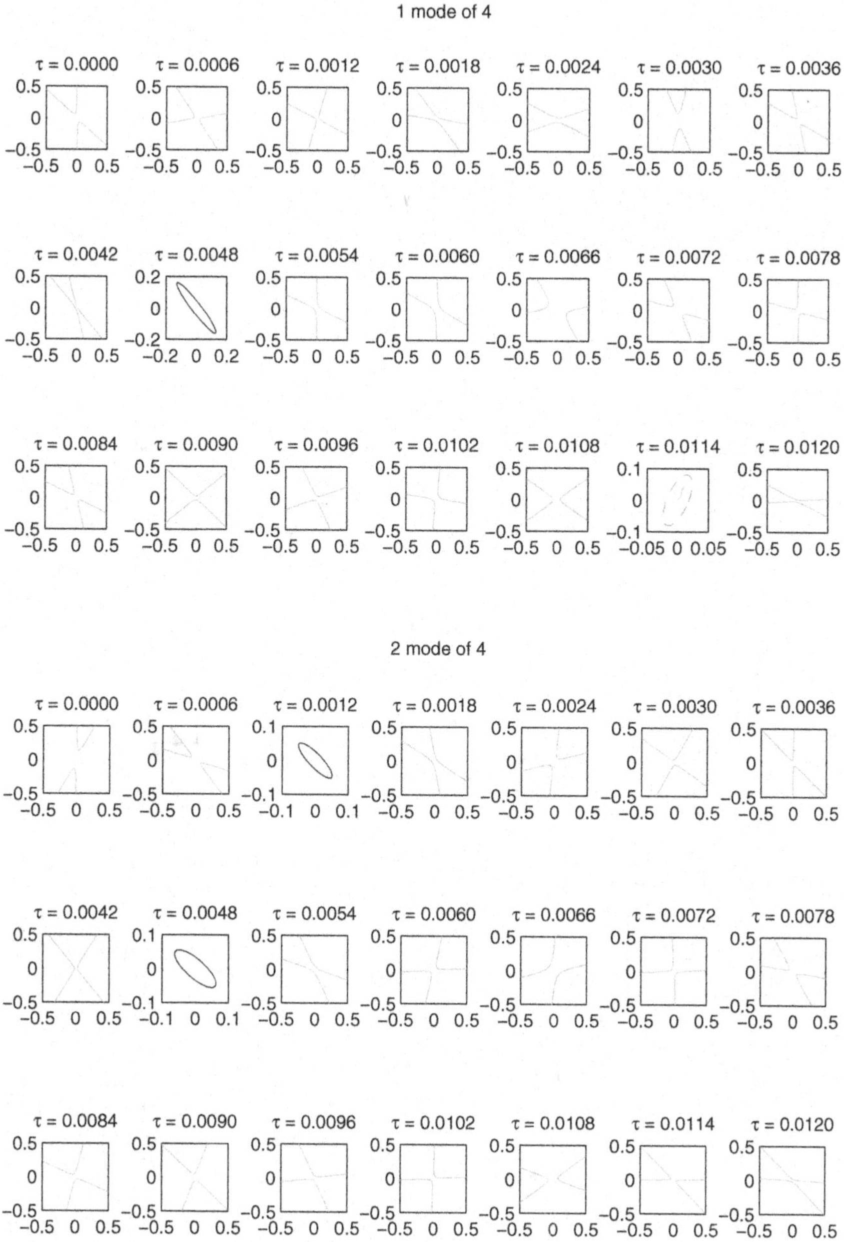

Figure 104 Evolution of wave front cross sections. Parameters: $\alpha_1 l_{01}=0.1$, $\alpha_2 l_{02}=0.2$, $\alpha_3 l_{03}=0.07$, $\alpha_4 l_{04}=0.007$, $w1_{\min}(0)=10^{-5}$, $w1_{\max}(0)=2\times10^{-5}$, $w2_{\min}(0)=1.1\times10^{-5}$, $w2_{\max}(0)=2.1\times10^{-5}$, $w3_{\min}(0)=1.2\times10^{-5}$, $w3_{\max}(0)=2.2\times10^{-5}$, $w4_{\min}(0)=1.3\times10^{-5}$, $w4_{\max}(0)=2.3\times10^{-5}$, ${}^1R_{11}(0)=-707$, ${}^1R_{12}(0)=-397$, ${}^1R_{22}(0)=0$, ${}^1I_{11}(0)=1592$, ${}^1I_{12}(0)=0$, ${}^1I_{22}(0)=397$, ${}^2R_{11}(0)=707$, ${}^2R_{12}(0)=198$, ${}^2R_{22}(0)=0$, ${}^2I_{11}(0)=1315$, ${}^2I_{12}(0)=0$, ${}^2I_{22}(0)=360$, ${}^3R_{11}(0)=-707$, ${}^3R_{12}(0)=-795$, ${}^3R_{22}(0)=0$, ${}^3I_{11}(0)=1105$, ${}^3I_{12}(0)=0$, ${}^3I_{22}(0)=328$, ${}^4R_{11}(0)=-707$, ${}^4R_{12}(0)=397$, ${}^4R_{22}(0)=0$, ${}^4I_{11}(0)=941$, ${}^4I_{12}(0)=0$, ${}^4I_{22}(0)=300$.

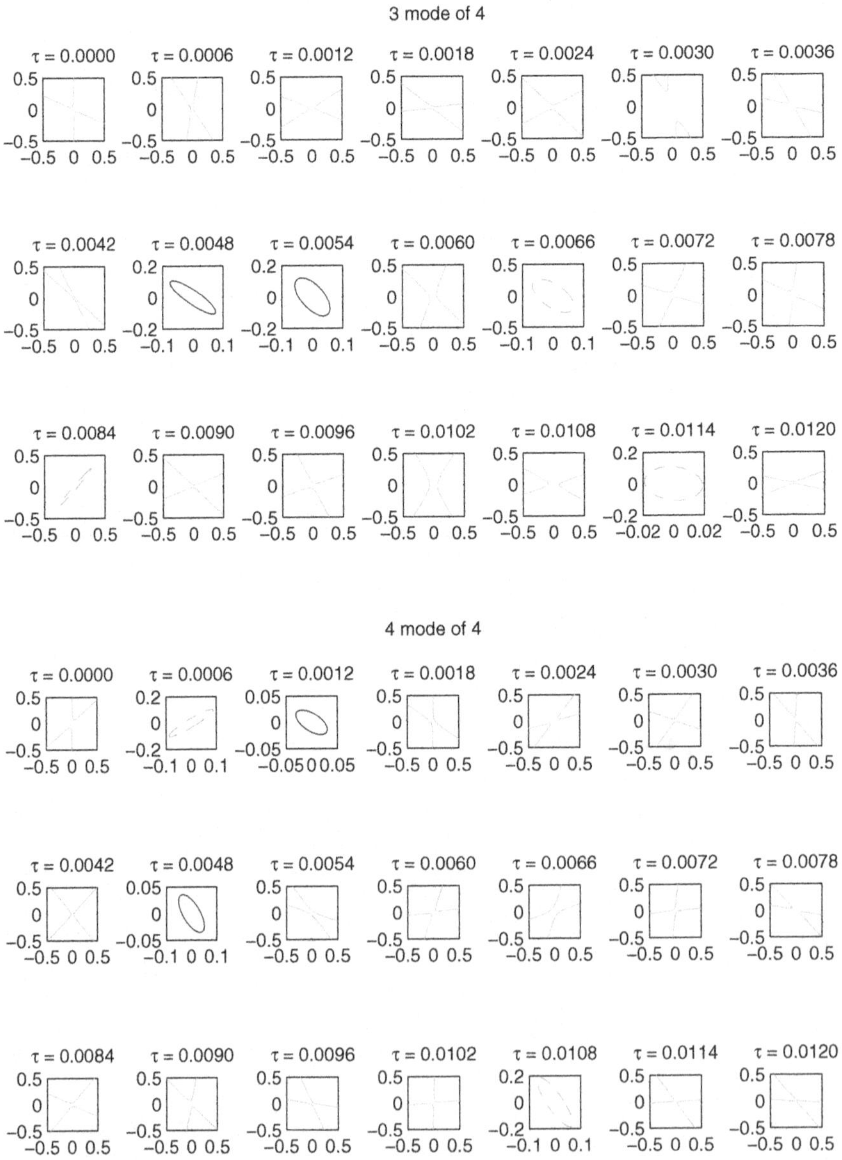

Figure 104 *(continued)*

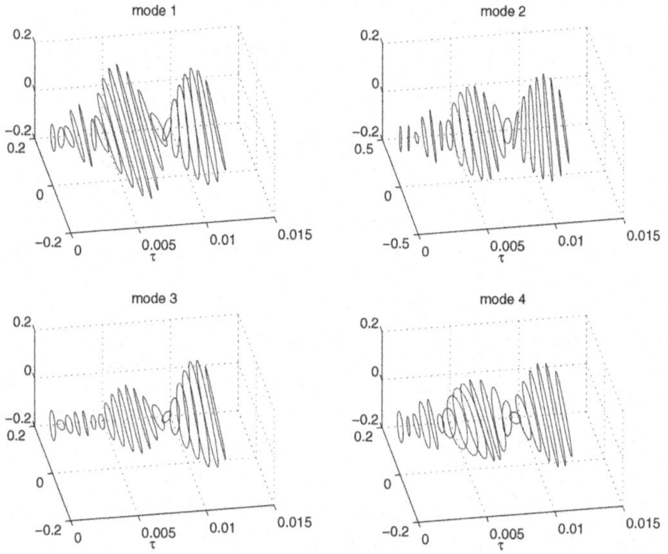

Figure 105 Evolution of GB cross sections. Parameters: $\alpha_1 l_{01} = 0.1$, $\alpha_2 l_{02} = 0.2$, $\alpha_3 l_{03} = 0.07$, $\alpha_4 l_{04} = 0.007$, $w1_{\min}(0) = 10^{-5}$, $w1_{\max}(0) = 2 \times 10^{-5}$, $w2_{\min}(0) = 1.1 \times 10^{-5}$, $w2_{\max}(0) = 2.1 \times 10^{-5}$, $w3_{\min}(0) = 1.2 \times 10^{-5}$, $w3_{\max}(0) = 2.2 \times 10^{-5}$, $w4_{\min}(0) = 1.3 \times 10^{-5}$, $w4_{\max}(0) = 2.3 \times 10^{-5}$, ${}^1R_{11}(0) = 707$, ${}^1R_{12}(0) = -397$, ${}^1R_{22}(0) = 400$, ${}^1I_{11}(0) = 1592$, ${}^1I_{12}(0) = 0$, ${}^1I_{22}(0) = 397$, ${}^2R_{11}(0) = -707$, ${}^2R_{12}(0) = 198$, ${}^2R_{22}(0) = 400$, ${}^2I_{11}(0) = 1315$, ${}^2I_{12}(0) = 0$, ${}^2I_{22}(0) = 360$, ${}^3R_{11}(0) = 707$, ${}^3R_{12}(0) = -795$, ${}^3R_{22}(0) = -400$, ${}^3I_{11}(0) = 1105$, ${}^3I_{12}(0) = 0$, ${}^3I_{22}(0) = 328$, ${}^4R_{11}(0) = -707$, ${}^4R_{12}(0) = 397$, ${}^4R_{22}(0) = -400$, ${}^4I_{11}(0) = 941$, ${}^4I_{12}(0) = 0$, ${}^4I_{22}(0) = 300$.

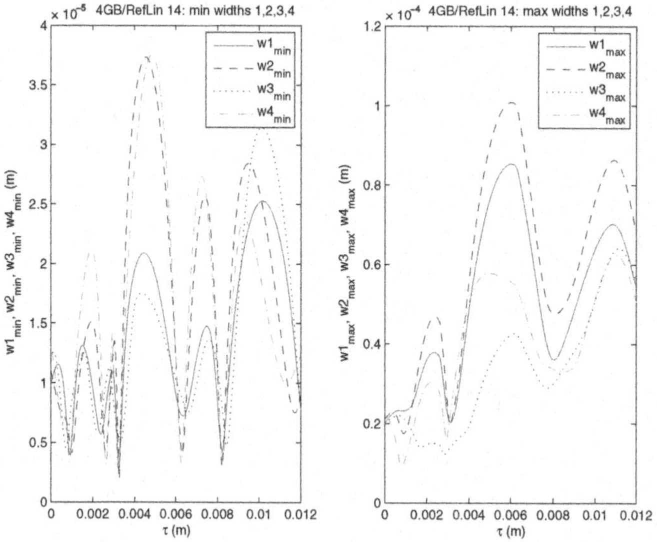

Figure 106 Evolution of GB widths. Parameters: $\alpha_1 l_{01} = 0.1$, $\alpha_2 l_{02} = 0.2$, $\alpha_3 l_{03} = 0.07$, $\alpha_4 l_{04} = 0.007$, $w1_{\min}(0) = 10^{-5}$, $w1_{\max}(0) = 2 \times 10^{-5}$, $w2_{\min}(0) = 1.1 \times 10^{-5}$, $w2_{\max}(0) = 2.1 \times 10^{-5}$, $w3_{\min}(0) = 1.2 \times 10^{-5}$, $w3_{\max}(0) = 2.2 \times 10^{-5}$, $w4_{\min}(0) = 1.3 \times 10^{-5}$, $w4_{\max}(0) = 2.3 \times 10^{-5}$, ${}^1R_{11}(0) = 707$, ${}^1R_{12}(0) = -397$, ${}^1R_{22}(0) = 400$, ${}^1I_{11}(0) = 1592$, ${}^1I_{12}(0) = 0$, ${}^1I_{22}(0) = 397$, ${}^2R_{11}(0) = -707$, ${}^2R_{12}(0) = 198$, ${}^2R_{22}(0) = 400$, ${}^2I_{11}(0) = 1315$, ${}^2I_{12}(0) = 0$, ${}^2I_{22}(0) = 360$, ${}^3R_{11}(0) = 707$, ${}^3R_{12}(0) = -795$, ${}^3R_{22}(0) = -400$, ${}^3I_{11}(0) = 1105$, ${}^3I_{12}(0) = 0$, ${}^3I_{22}(0) = 328$, ${}^4R_{11}(0) = -707$, ${}^4R_{12}(0) = 397$, ${}^4R_{22}(0) = -400$, ${}^4I_{11}(0) = 941$, ${}^4I_{12}(0) = 0$, ${}^4I_{22}(0) = 300$. (See the color plate.)

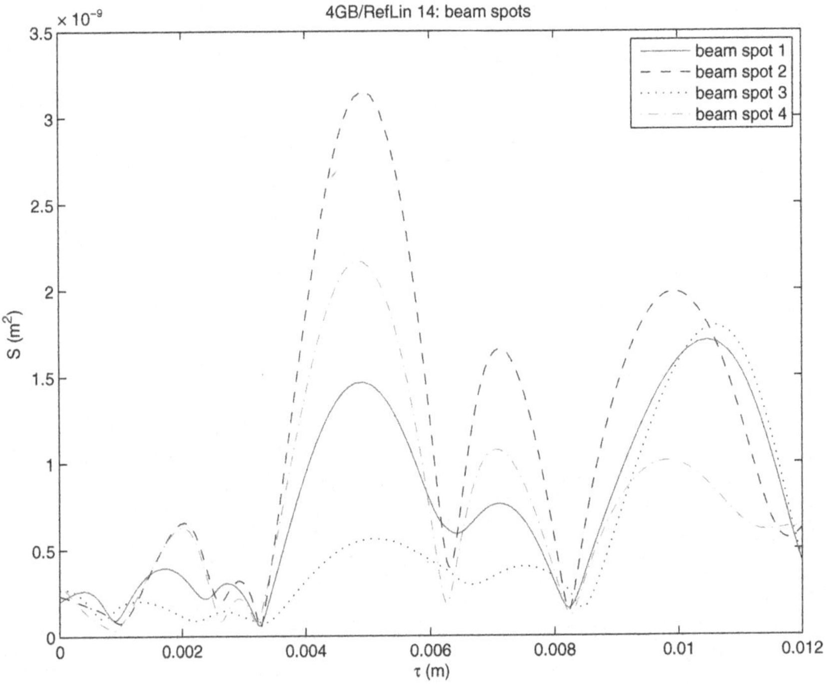

Figure 107 Evolution of GB spots. Parameters: $\alpha_1 I_{01} = 0.1$, $\alpha_2 I_{02} = 0.2$, $\alpha_3 I_{03} = 0.07$, $\alpha_4 I_{04} = 0.007$, $w1_{min}(0) = 10^{-5}$, $w1_{max}(0) = 2 \times 10^{-5}$, $w2_{min}(0) = 1.1 \times 10^{-5}$, $w2_{max}(0) = 2.1 \times 10^{-5}$, $w3_{min}(0) = 1.2 \times 10^{-5}$, $w3_{max}(0) = 2.2 \times 10^{-5}$, $w4_{min}(0) = 1.3 \times 10^{-5}$, $w4_{max}(0) = 2.3 \times 10^{-5}$, $^1R_{11}(0) = 707$, $^1R_{12}(0) = -397$, $^1R_{22}(0) = 400$, $^1I_{11}(0) = 1592$, $^1I_{12}(0) = 0$, $^1I_{22}(0) = 397$, $^2R_{11}(0) = -707$, $^2R_{12}(0) = 198$, $^2R_{22}(0) = 400$, $^2I_{11}(0) = 1315$, $^2I_{12}(0) = 0$, $^2I_{22}(0) = 360$, $^3R_{11}(0) = 707$, $^3R_{12}(0) = -795$, $^3R_{22}(0) = -400$, $^3I_{11}(0) = 1105$, $^3I_{12}(0) = 0$, $^3I_{22}(0) = 328$, $^4R_{11}(0) = -707$, $^4R_{12}(0) = 397$, $^4R_{22}(0) = -400$, $^4I_{11}(0) = 941$, $^4I_{12}(0) = 0$, $^4I_{22}(0) = 300$. (See the color plate.)

16. CONCLUSION

This chapter applies the method of complex geometrical optics (CGO) to the analysis of GB rotation in smoothly inhomogeneous and optical nonlinear saturable media. The CGO method reduces diffraction and self-focusing problems for the Gaussian beam via a solution of ordinary differential equations, describing the behavior of the amplitude, the beam width, and the curvature of the wave front. The CGO method readily provides solutions for inhomogeneous nonlinear saturable fibers in a

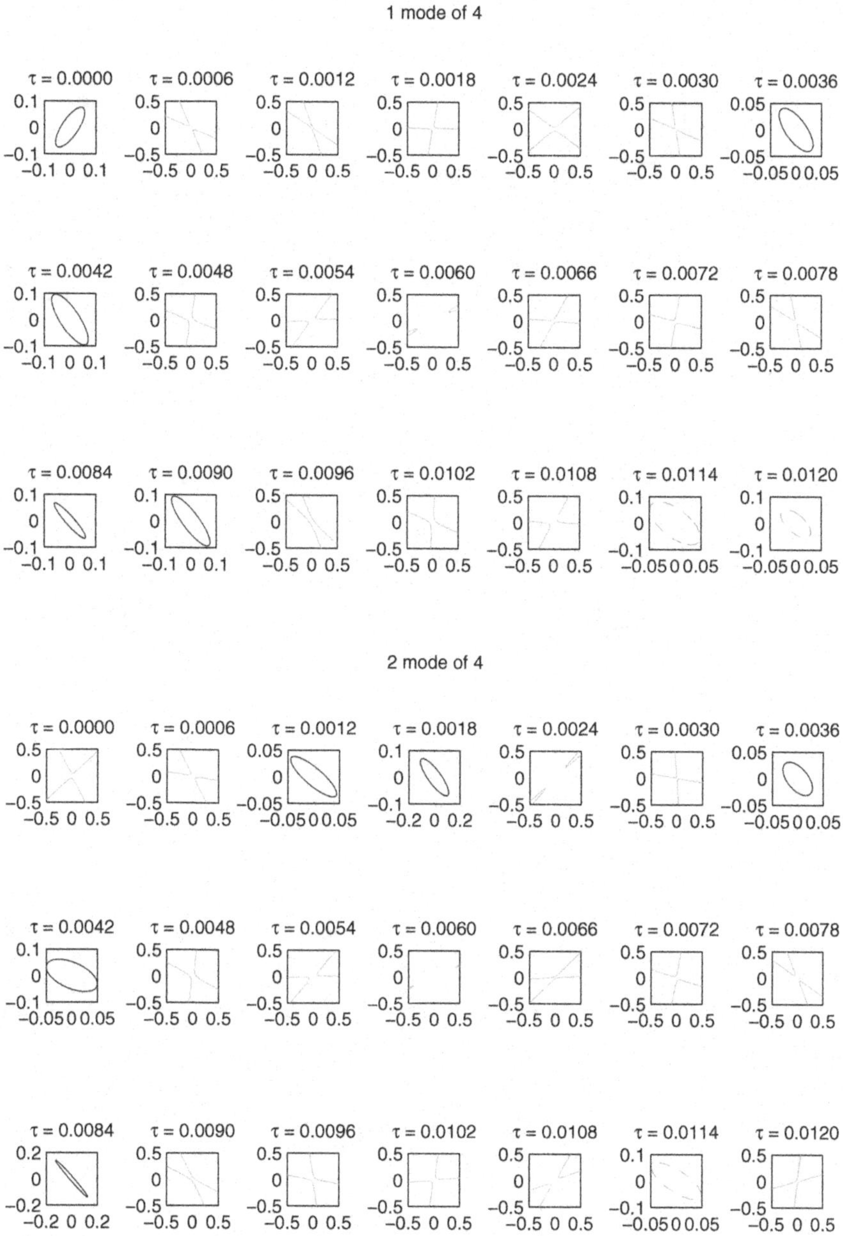

Figure 108 Evolution of wave front cross sections. Parameters: $\alpha_1 I_{01} = 0.1$, $\alpha_2 I_{02} = 0.2$, $\alpha_3 I_{03} = 0.07$, $\alpha_4 I_{04} = 0.007$, $w1_{\min}(0) = 10^{-5}$, $w1_{\max}(0) = 2 \times 10^{-5}$, $w2_{\min}(0) = 1.1 \times 10^{-5}$, $w2_{\max}(0) = 2.1 \times 10^{-5}$, $w3_{\min}(0) = 1.2 \times 10^{-5}$, $w3_{\max}(0) = 2.2 \times 10^{-5}$, $w4_{\min}(0) = 1.3 \times 10^{-5}$, $w4_{\max}(0) = 2.3 \times 10^{-5}$, ${}^1R_{11}(0) = 707$, ${}^1R_{12}(0) = -397$, ${}^1R_{22}(0) = 400$, ${}^1I_{11}(0) = 1592$, ${}^1I_{12}(0) = 0$, ${}^1I_{22}(0) = 397$, ${}^2R_{11}(0) = -707$, ${}^2R_{12}(0) = 198$, ${}^2R_{22}(0) = 400$, ${}^2I_{11}(0) = 1315$, ${}^2I_{12}(0) = 0$, ${}^2I_{22}(0) = 360$, ${}^3R_{11}(0) = 707$, ${}^3R_{12}(0) = -795$, ${}^3R_{22}(0) = -400$, ${}^3I_{11}(0) = 1105$, ${}^3I_{12}(0) = 0$, ${}^3I_{22}(0) = 328$, ${}^4R_{11}(0) = -707$, ${}^4R_{12}(0) = 397$, ${}^4R_{22}(0) = -400$, ${}^4I_{11}(0) = 941$, ${}^4I_{12}(0) = 0$, ${}^4I_{22}(0) = 300$.

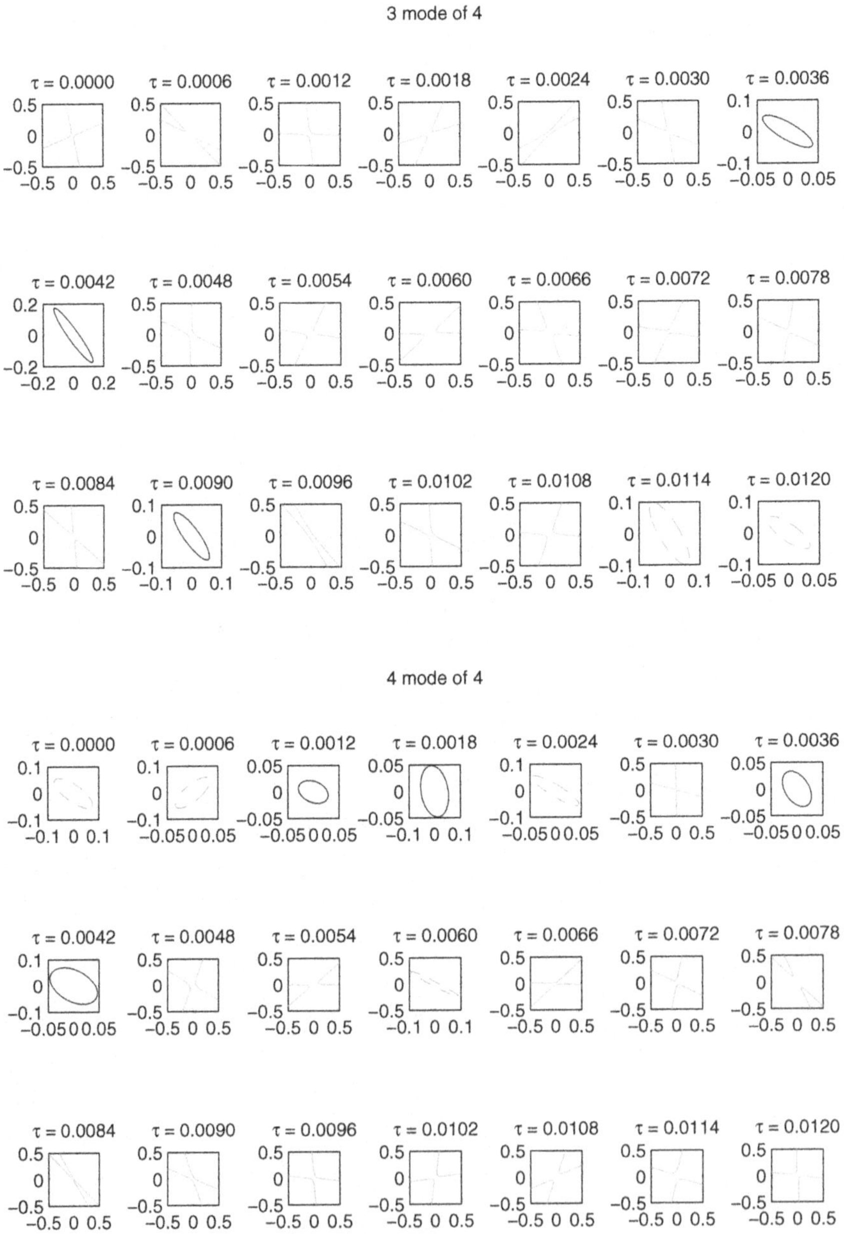

Figure 108 *(continued)*

simpler way than the standard methods of nonlinear optics, such as the variation method approach, the method of moments, and the beam propagation method. The CGO method effectively supplies the results for GBs propagating in free space, which happened to be identical to classical solutions using the diffraction theory. Besides simplicity and efficiency, this method supplies a number of new results. First, we demonstrate the ability of the CGO method to describe the evolution of arbitrary number of coupled GBs, which is a new and interesting problem for engineers in the field of electron physics. Second, we present matrix forms of ordinary differential CGO equations for N-rotating GBs, which are very convenient for numerical implementations and uses basic software such as Matlab and Mathcad. We also would like to emphasize that the numerical approach based on derived ordinary matrix differential equations is useful because the number of numerical operations increases (with number of modes) to an acceptable level for practical proposes. Moreover, we demonstrate the interesting behaviors of coupled modes, where all the waves rotate in the same manner, the modes form pairs or evolve separately. Depended on physical parameters modes may (or may not) conserve their initial direction of rotation. We also demonstrate that the CGO method enables the modeling of the evolution of wave-front cross sections of the N-rotating beams, which are very complicated dynamics when we increase the number of interacting beams. In this way, we believe that the CGO method can be an interesting and effective tool to use to solve sophisticated problems of electron physics.

REFERENCES

Agrawal, G. (1989). *Nonlinear Fiber Optics*. New York: Academic Press.

Akhmanov, S. A., & Nikitin, S. Yu. (1997). *Physical Optics*. Oxford, U.K: Clarendon Press.

Akhmanov, S. A., Khokhlov, R. V., & Sukhorukov, A. P. (1972). Self-focusing, self-defocusing, and self-modulation of laser beams. In F. T. Arecchi, & E. O. Shulz-Dubois (Eds.), *Vol. 2. Laser Handbook* (pp. 1151–1228). New York: Elsevier.

Akhmanov, S. A., Sukhorukov, A. P., & Khokhlov, R. V. (1968). Self-focusing and light diffraction in nonlinear medium. *Soviet Physics Uspekhi, 10*, 609–636.

Akhmediev, N. N. (1998). Spatial solitons in Kerr and Kerr-like media. *Optical and Quantum Electronics, 30*, 535–569.

Anderson, D. (1983). Variational approach to nonlinear pulse propagation in optical fibers. *Physical Review A, 27*, 3135–3145.

Arnaud, J. A. (1976). *Beam and Fiber Optics*. San Diego: Academic Press.

Babič, V. M., & Buldyrev, V. S. (1991). *Asymptotic Methods in Short-Wavelength Diffraction Theory*. Alpha Science International Limited. ISBN 184265232X, 9781842652329.

Babič, V. M., & Kirpichnikova, N. J. (1980). *Boundary Layer Method in Diffraction Problems, Springer Verlag*. Berlin.

Berczynski, P. (2011). Complex geometrical optics of nonlinear inhomogeneous fibres. *Journal of Optics, 13*, 035707.

Berczynski, P. (2012). Gaussian beam diffraction in inhomogeneous and logarithmically saturable nonlinear media. *Central European Journal of Physics, 10*(4), 898–905.

Berczynski, P. (2013a). Complex geometrical optics of inhomogeneous and nonlinear saturable media. *Optics Communications, 295*, 208–218.

Berczynski, P. (2013b). Gaussian beam evolution in longitudinally and transversely inhomogeneous nonlinear fibres with absorption. *Opto-Electronics Review, 21*(3), 303–319.

Berczynski, P. (2013c). Spatiotemporal complex geometrical optics for wavepacket evolution in inhomogeneous and nonlinear media of Kerr type. *Journal of Optics, 15*, 125201.

Berczynski, P. (2014). Elliptical Gaussian beam propagation in inhomogeneous and nonlinear fibres of Kerr type. *Optical and Quantum Electronics, 46*(7), 945–974. http://dx.doi.org/10.1007/s11082-013-9812-z.

Berczynski, P., & Kravtsov, Y. A. (2004). Theory for Gaussian beam diffraction in 2D inhomogeneous medium, based on the eikonal form of complex geometrical optics. *Physics Letters A, 331*(3–4), 265–268.

Berczynski, P., Bliokh, K. Y., Kravtsov, Y. A., & Stateczny, A. (2006). Diffraction of Gaussian beam in 3D smoothly inhomogeneous media: Eikonal-based complex geometrical optics approach. *Journal of the Optical Society of America A, 23*(6), 1442–1451.

Berczynski, P., Kravtsov, Y. A., & Sukhorukov, A. P. (2010). Complex geometrical optics of Kerr type nonlinear media. *Physica D: Nonlinear Phenomena, 239*(5), 241–247.

Berczynski, P., Kravtsov, Y. A., & Zeglinski, G. (2010). Gaussian beam diffraction in inhomogeneous media of cylindrical symmetry. *Optica Applicata, 40*(3), 705–718.

Červený, V. (2001). *Seismic Ray Theory*. Cambridge University Press.

Chapman, S. J., Lawry, J. M., Ockendon, J. R., & Tew, R. H. (1999). On the theory of complex rays. *SIAM Review, 41*, 417–509. http://dx.doi.org/10.1137/S0036144599352058.

Chen, Z., Segev, M., & Christodoulides, D. N. (2012). Optical spatial solitons: Historical overview and recent advances. *Reports on Progress in Physics, 75*(086401), 1–21. http://dx.doi.org/10.1088/0034-4885/75/8/086401.

Deschamps, G. A. (1971). Gaussian beam as a bundle of complex rays. *Electronics Letters, 7*(23), 684–685.

Egorchenkov, R. A., & Kravtsov, Y. A. (2001). Complex ray-tracing algorithms with application to optical problems. *Journal of the Optical Society of America A, 18*, 650–656.

Egorchenkov, R. A., & Kravtsov, Y. A. (2000). Numerical realization of complex geometrical optics method. *Radiophysics and Quantum Electronics, 43*, 512–517.

Fox, J. (1964). *Quasi-Optics*. Brooklyn, NY: Polytechnic Press.

Goncharenko, A. M., Logvin, Y. A., Sampson, A. M., & Shapovalov, P. S. (1991). Rotating elliptical Gaussian beams in nonlinear media. *Optics Communications, 81*(3–4), 225.

Hasegawa, A. (1990). *Optical Solitons in Fibres* (2nd ed.). Berlin: Springer-Verlag.

Jiang, X, Guo, Q., Li, H., & Hu, W. (2004). Induced focusing from counter-propagation of two optical beams in self-defocusing media. *Optics Communications, 233*, 1–6.

Karlsson, M., Anderson, D., & Desaix, M. (1992). Dynamics of self-focusing and self-phase modulation in a parabolic index optical fiber. *Optics Letters, 17*(1), 22–24.

Keller, J. B. (1958). A geometrical theory of diffraction. In *Vol. 8. Calculus of Variations and Its Applications, Proceedings of Symposia in Applied Mathematics* (pp. 27–52). New York: McGraw-Hill.

Keller, J. B., & Streifer, W. (1971). Complex rays with application to Gaussian beams. *Journal of the Optical Society of America, 61*, 40–43.

Kogelnik, H. (1965). On the propagation of Gaussian beams of light through lenslike media, including those with a loss or gain variation. *Applied Optics, 4*(12), 1562–1569.

Kogelnik, H., & Li, T. (1966). Laser beams and resonators. *Applied Optics, 5*(10), 1550.

Kravtsov, Y. A. (1967). Complex ray and complex caustics. *Radiophysics and Quantum Electronics, 10*, 719–730.

Kravtsov, J. A., Kravtsov, I. U. A., & Zhu, N. Y. (2010). Theory of Diffraction: Heuristic Approaches. *Alpha Science International*. ISBN 9781842653722.

Kravtsov, Y. A., & Orlov, Y. I. (1990). *Geometrical Optics of Inhomogeneous Medium*. Berlin: Springer-Verlag.

Kravtsov, Y. A., Forbes, G. W., & Asatryan, A. A. (1999). Theory and applications of complex rays. In E. Wolf (Ed.), *Progress in Optics*. Elsevier, Amsterdam, 39, 3–62.

Longhi, S., & Janner, D. (2004). Self-focusing and nonlinear periodic beams in parabolic index optical fibres. *Journal of Optics B: Quantum and Semiclassical Optics, 6*, S303.

Malomed, B. A. (2002). Variational methods in nonlinear fiber optics and related fields. *Progress in Optics, 43*, 69–191.

Manash, J. T., Baldeck, P. L., & Alfano, R. R. (1988). Self-focusing and self-phase modulation in a parabolic graded-index optical fiber. *Optics Letters, 13*(7), 589–591.

Medhekar, S., Sarkar, K. R., & Paltani, P. P (2006). Coupled spatial-soliton pairs in saturable nonlinear media. *Optics Letters, 31*(1), 77–79. http://dx.doi.org/10.1364/OL.31.000077.

Paré, C., & Bélanger, P. A. (1992). Beam propagation in a linear and nonlinear lens-like medium using ABCD ray matrices: The method of moments. *Optical and Quantum Electronics, 24*, 1051–1070.

Pereverzev, G. V. (1993). *Paraxial WKB Solution of a Scalar Wave Equation, Max-Planck-Institut für Plasmaphysik, 26, pages 31.*

Pereverzev, G. V. (1998). Beam tracing in inhomogeneous anisotropic plasmas. *Physics of Plasmas, 5*, 3529–3541.

Perez-Garcia, V. M., Torres, P., Garcia-Ripoll, J. J., & Michinel, H. (2000). Moment analysis of paraxial propagation in a nonlinear graded index fibre. *Journal of Optics B: Quantum and Semiclassical Optics, 2*, 353.

Permitin, G. V., & Smirnov, A. I. (1996). Quasioptics of smoothly inhomogeneous isotropic media. *Journal of Experimental and Theoretical Physics, 82*(3), 395–402.

Pietrzyk, M. E. (1999). On the properties of two pulses propagating simultaneously in different dispersion regimes in a nonlinear planar waveguide. *Journal of Optics A: Pure Applied Optics, 1*, 685–696.

Pietrzyk, M. E. (2001). Influence of nonlinear coupling of pulses on spatio-temporal compression. *Journal of Modern Optics, 48*(2), 303–317.

Popov, M. M. (1969). Eigen-oscillations of multi-mirrors resonators. *Vestnik Leningradskogo Universiteta, 22*, 44–54 (in Russian).

Popov, M. M. (1977). On a method of computation of geometrical spreading in ihomogeneous media containing interfaces. *Proceedings of the USSR Academy of Sciences, 237*, 1059–1062 (in Russian).

Popov, M. M. (1982). New method of computation of wave fields using Gaussian beams. *Wave Motion, 4*, 85–95.

Popov, M. M., & Pšenčik, I. (1978a). *Ray amplitudes in inhomogeneous media with curved interfaces, 24. Geofys. Sb.* (pp. 118–129). Praha: Akademia.

Popov, M. M., & Pšenčik, I. (1978b). Computation of ray amplitudes in inhomogeneous media with curved interfaces. *Studia Geophysica Geodaetica, 22*(3), 248–258. http://link.springer.com/article/10.1007%2FBF01627902.

Sarkar, R. K., & Medhekar, S. (2009). Spatial soliton pairing of two cylindrical beams in saturable nonlinear media. *Progress in Electromagnetics Research M, 9*, 53–64.

Stegeman, G. I., & Segev, M. (1999). Optical spatial solitons and their interactions: Universality and diversity. *Science, 286*, 1518–1523.

Vlasov, S. N., & Talanov, V. I. (1995). The parabolic equation in the theory of wave propagation (on the 50th anniversary of its publication). *Radiophysics and Quantumn Electronics, 38*(B, 1–2), 1–12.

CHAPTER TWO

Single-Particle Cryo-Electron Microscopy (Cryo-EM): Progress, Challenges, and Perspectives for Further Improvement

David Agard[1], Yifan Cheng[2], Robert M. Glaeser[3,*], Sriram Subramaniam[4]

[1]HHMI and the Department of Biochemistry and Biophysics, University of California, San Francisco, CA 94158, USA
[2]Department of Biochemistry and Biophysics, University of California, San Francisco, CA 94158, USA
[3]Lawrence Berkeley National Laboratory, University of California, Berkeley, CA 94720, USA
[4]Laboratory for Cell Biology, Center for Cancer Research, National Cancer Institute, National Institutes of Health (NIH), Bethesda, MD 20892, USA
*Corresponding author: E-mail: rmglaeser@lbl.gov

Contents

Advances in Imaging and Electron Physics, Volume 185
ISSN 1076-5670
http://dx.doi.org/10.1016/B978-0-12-800144-8.00002-1

1. INTRODUCTION

1.1 Single-Particle Cryo-EM has Experienced a Sudden Improvement in What Can Be Accomplished

The phrase *single-particle cryo-electron microscopy (cryo-EM)* refers to applications of electron microscopy (EM) in which specimens consist of randomly dispersed, unstained biological macromolecules. Such specimens are embedded in a thin film of vitrified buffer, thus preserving their structure in a near-native state. The use of cryo-EM to determine the structures of large macromolecular complexes has previously been described in a number of books (e.g., Frank, 2006; Glaeser *et al.*, 2007; Jensen, 2010; Schmidt-Krey & Cheng, 2013). Briefly, the images of some tens of thousands (until recently, even millions) of individual particles are recorded at very low doses to minimize radiation damage. These must be aligned translationally and assigned three Euler angles (corresponding to their orientations with respect to one another). Systematic effects, such as spherical aberration, defocusing, and astigmatism, must be quantified and corrected. Because low values of image contrast and the resulting difficulty to align particles are limiting, the best results traditionally have come from large, symmetric assemblies such as icosahedral viruses and helical assemblies.

With the introduction of "direct" electron-detection camera technology [e.g., Glaeser (2013b)], single-particle cryo-EM recently has experienced a sudden improvement in what can be accomplished. As a consequence of the better performance of this new type of camera, the already-low electron exposure used to record high-resolution images (Baker *et al.*, 2010; Glaeser, 2008) can be fractionated into 10 or more sequential subexposures, called *movie frames* (Campbell *et al.*, 2012). The improved signal-to-noise ratio (SNR) provided by this type of camera is now sufficient to allow whole frames, or even small parts of whole frames, to be aligned with one another. Not only does this technique limit the extent of motion that occurs within each frame, but subsequent alignment of frames also makes it possible to compensate, computationally, for the cumulative motion that occurs as the exposure continues (Li *et al.*, 2013). As a result, this approach substantially mitigates the beam-induced movement effect, as described in Brilot *et al.* (2012), which significantly degrades the resolution obtained with conventional cameras or photographic film.

Several examples have been reported recently that take advantage of these new technological capabilities. In the first paper to exploit the benefit of dose-fractionated exposures, Bai *et al.* (2013) reported a structure of the 80S ribosome from *S. cerevisiae* at an average resolution of 0.45 nm. The resolution achieved was sufficient, in many parts of the structure, to associate amino-acid side chains with corresponding features of the density map. In addition, these researchers obtained this improved density map by merging data from only approximately 35,000 ribosome particles, which is about 2% of the number of particles that were used in previous attempts to obtain a high-resolution structure of ribosomal particles.

At about the same time, Li *et al.* (2013) reported a structure of the approximately 700 kDa 20S proteasome from *T. acidophilum* at a resolution of 0.33 nm. In this case, the resolution and quality of the cryo-EM density map was comparable to that of the X-ray map used previously to solve the structure (Lowe *et al.*, 1995). Compensation for beam-induced motion, made possible by dose fractionation in conjunction with single-electron counting capability of the Gatan K2 camera, resulted in substantial improvement over the best cryo-EM map that had been obtained previously for this particle, which was about 0.56 nm (Rabl *et al.*, 2008).

1.2 Single-Particle Cryo-EM is Now an Attractive Complement to X-Ray Crystallography for Determining High-Resolution Structures of Large Complexes

A very attractive feature of single-particle cryo-EM is that one does not have to crystallize the macromolecule of interest to obtain a high-resolution structure. Thus cryo-EM provides an enabling strategy for studying classes of proteins, such as integral membrane proteins having large extra-membrane domains or multiprotein complexes for which crystallization has proven quite challenging. In addition, single-particle cryo-EM is more compatible with the buffer conditions that are known to optimize the biochemical function of the specimen.

Another feature of single-particle cryo-EM is that specimens often exist in different conformational states, with a relative occupancy dictated by the Gibbs free energy of each state (Frank, 2013). Such variability generally interferes with the ability to obtain high-quality crystals. By contrast, all these states may be captured in cryo-EM specimens, which may be seen as either an advantage or a disadvantage. If particle images can be assigned computationally to distinct conformational states, then it is an advantage that one can obtain much-desired structures for the most abundant intermediates in a biochemical cycle. If the conformational differences are too subtle to be computationally distinguishable, however, then merging (i.e., averaging) the data must result in the blurring of high-resolution features. In this case, it may be necessary to use substrate analogs, inhibitors, conformation-specific antibodies, or even functionally "dead" mutants to lock the macromolecule in a unique conformation, as is sometimes done in X-ray crystallography.

The unique capabilities of single-particle cryo-EM can often complement those of X-ray crystallography. One example where value is added by both methods, referred to as *hybrid structure determination,* uses the structure of a large complex, determined by cryo-EM at moderate resolution (0.4–1.5 nm), and higher-resolution structures of individual components, determined by X-ray crystallography. Then a model of the complete structure is obtained by docking the X-ray structures into the EM density map. The advantage of having higher-resolution maps of the complete macromolecular complex, in which even amino-acid side chains are visible, is that one can begin to visualize changes in the conformation of component pieces as a result of assembly.

While hybrid approaches can provide significant structural insight, they are limited to cases where high-resolution structures of component subunits are available. Fortunately, recent advances provide hope that high-resolution cryo-EM density maps can be obtained for many protein complexes for which no previous high-resolution structures are available. Of particular note is the recent determination of the structure of the tetrameric TRPV1 integral membrane protein (approximately 300 kDa) at a resolution of 0.34 nm (Liao *et al.*, 2013). Not only was it possible to build the first atomic model for this new class of membrane channels, but also to map conformational changes resulting from addition of various ligands (Cao *et al.*, 2013). This is an unusually important advance because, until now, it has been thought that further improvements in methodology (which are indeed still possible, as discussed in section 3) would be required before high-resolution structures could be obtained for particles with low symmetry and a size smaller than 1 MDa (Henderson, 1995).

1.3 Lessons Learned from the EM of Ordered Arrays and Large Particles with High Symmetry

Previous studies have established that a number of conditions must be met to obtain high-resolution maps that show clear densities for amino-acid side chain residues. These conditions include the requirement that the particles must be structurally homogeneous. In addition, it is required that particles must adopt a uniform distribution of orientations. The requirement of having a uniform coverage of different views is the same as that in X-ray crystallography, where it may be stated as the requirement to measure all unique data in three-dimensional (3-D) Fourier space. Furthermore, it goes without saying that images also must be recorded with only a limited ("safe") amount of electron exposure. The total exposure that can be recommended is approximately 20 electrons/Å^2 for 300 keV electrons, a value that is independent of the biological macromolecule being examined.

2. GOING BEYOND LARGE PARTICLES WITH HIGH SYMMETRY: DEFINING THE PROBLEM

In favorable cases, the resolution—as estimated by the "gold standard" version of the Fourier shell correlation (FSC) curve described by Scheres and Chen (2012)—can initially be expected to improve as one increases the number of particle images used to compute the average structure. With the use of dose-fractionated movies, the resolution may even continue to improve up to about 0.3 or 0.4 nm. Some projects nevertheless may "hit a wall" at a much earlier point—at a resolution of about 0.6 nm or less—and not improve further as more and more particles are added to the data set.

The purpose of this review is to discuss some of the factors that might be responsible for limiting the resolution in such cases and to suggest what sometimes can be done to improve the resolution to a better value.

2.1 Specimen Heterogeneity Can Limit Resolution

2.1.1 The Available Sample may be Conformationally Heterogeneous

It has already been noted in this chapter that macromolecular complexes used for high-resolution cryo-EM must be structurally homogeneous. When that is not the case, images of particles with different conformations or compositions must at least be distinguishable from one another.

To the extent that a degree of structural heterogeneity is present, the resolution achieved during refinement is certain to hit a wall, giving a result that stops improving at a point short of what is desired. In some cases, the largest part of a structure may be structurally homogeneous, while one or more smaller parts adopt multiple structural states. This may produce a structure that actually does become high-resolution for the homogeneous portion, but not for the other parts of the structure.

Fortunately, a number of possible improvements can be tried if it is suspected that conformational heterogeneity limits the resolution. One approach is biochemical, of course, in which buffer conditions, substrate analogs, allosteric ligands, and other options can be explored in an attempt to lock the whole structure into a single conformation, as was mentioned in section 1.2. This approach could easily be a large subject by itself, of course, and the details are beyond the scope of this discussion. Suffice it to say that a deep knowledge of the biochemistry of a system may be needed to prepare structurally homogeneous specimens.

Another approach is computational. One method of sorting images into groups that correspond to structurally more homogeneous particles is based on use of the 3-D variance map (Penczek, Frank, & Spahn, 2006). Such maps show the regions of the structure that exhibit the most significant variability in the particles making up the data set. Then this information is used to focus on the most variable regions to sort the images into more homogeneous subsets. An alternative approach to 3-D classification

(Scheres *et al.*, 2007) is used in RELION (Scheres, 2012) and FREALIGN (Lyumkis *et al.*, 2013). This approach is closely related to K-means classification, but it is implemented in 3-D using a maximum-likelihood formalism. Use of the 3-D classification strategy provided within RELION was critical for solving the structure of the TRPV1 channel, for which more than 50% of the particle images formed a coherent class (Liao *et al.*, 2013).

2.1.2 Structural Damage also Can Occur During Preparation of EM Grids

It is sometimes possible that structural heterogeneity is introduced by physical forces that are generated during the preparation of cryo-EM specimens, a point that was emphasized by Taylor and Glaeser (2008). For example, many proteins are adsorbed to the air-water interface, much as they do to a solid-water interface during hydrophobic interaction chromatography. In some cases, however, binding to the air-water interface results in partial (or even complete) denaturation. For example, rapid denaturation upon exposure to the air-water interface is the reason why protein solutions tend to foam if stirred or agitated too vigorously.

To avoid binding to the air-water interface, and also to concentrate the specimen directly on the EM grid, some researchers prefer to adsorb specimens to a thin carbon film. While this is effective for many types of specimen, there also are dangers inherent in the adsorption of macromolecules to carbon. Some specimens adopt strongly preferred orientations, for example. In addition, one might imagine that adsorption to a support film can introduce conformational heterogeneity that does not exist for particles suspended in a buffer. Carbon film also adds substantial noise to particle images, a problem that is especially severe for small particles. An alternative strategy to prevent adsorption of proteins to the air-water interface involves the addition of detergents such as NP-40 or dodecylmaltoside at concentrations below their critical micelle concentration.

In addition, one should be aware that substantial evaporation of buffers may occur in the interval between blotting excess sample from the EM grid and subsequent vitrification of the remaining aqueous sample. If the resulting changes in buffer conditions (such as pH, ionic strength, or concentrations of specific ions or other solutes) are too great, these factors themselves may cause unwanted changes in the structure of the system under study. One way to tell whether substantial evaporation, preferential adsorption at the air-water interface, or both may have occurred is to estimate whether the number of particles per unit area in the image is much greater than expected for the initial (bulk) concentration and for the thickness of the vitrified film.

2.2 The Ice Thickness of the Sample Might be too Great

2.2.1 Excessive Variation in Z-Height Position Is a Problem When the Ice Thickness Is Too Great

Variation in the Z-height of individual particles is difficult to account for during correction for the contrast transfer function (CTF). This is because there generally is not enough signal available to determine the CTF accurately on a particle-by-particle basis. Therefore, it is a serious problem if the ice thickness is greater than the depth of field (for a given resolution), and if different particles within an image are distributed at random Z-heights, limited only by the thickness of the ice. When this happens, the Thon rings in the FFT (Fast Fourier Transform) of the entire field of view disappear at high resolution (DeRosier, 2000). In addition, the phases of high-resolution structure factors will be merged with random error, a topic discussed in detail by Jensen (2001).

To illustrate with a specific example, the depth of field for 300-keV electrons, at a resolution of 0.6 nm, is approximately 90 nm. As will be discussed in section 2.5.2 (and shown in Figure 4), the error made during CTF correction becomes severe if the Z-height of a particle is different from the assumed value of defocus by an amount exceeding the depth of field. It thus is unlikely that a reconstruction will reach a resolution better than approximately 0.6 nm if the thickness of the vitrified ice is 100 nm or more. At a resolution of 0.4 nm, the depth of field decreases to only 40 nm, imposing even more stringent limitations on how thick the specimen can be.

For some specimens, it may prove problematic to prepare samples for which the ice thickness is equal to (or less than) the depth of field for the desired resolution. If that is the case, it still may be possible to confine all the particles to be at nearly the same Z-height by tethering them to some type of substrate. The commonly used technique of adsorbing particles to continuous carbon film often can be a good solution. Newer, possibly more protein-friendly solutions also can be considered, such as using the affinity grids introduced by Kelly *et al.* (2008) or using streptavidin monolayer crystals as support films, as proposed by Wang *et al.* (2008). In all cases, it is necessary to avoid that particles might be bound with one or a small number of preferred orientations. A further benefit of immobilizing the particles onto a thin support film is that they are held away from the air-water interface produced after blotting the grid, thus protecting them from possible damage as discussed previously in section 2.1.2.

2.2.2 A Shallow Depth of Field Need not be a Limitation for Highly Symmetric Structures

In apparent contradiction with the argument made in section 2.2.1, a resolution of better than 0.45 nm has been obtained with a large number of icosahedral virus particles, ranging in diameter from approximately 50 to almost 100 nm, as was reviewed by Grigorieff and Harrison (2011). It is nevertheless clear that the CTF applied to structural features at the top of such large virus particles is significantly different at high resolution from that which is applied to features at the bottom. A possible resolution of this "paradox" begins with the fact that a "best estimate" of the defocus corrects the CTF for only one slab of the virus, the thickness of which is equal to the depth of field. Multiple copies of the asymmetric unit, which lie inside that slab, thus contribute well to the signal during the merging of data. Other copies of the asymmetric unit, for which the CTF is not properly corrected, must contribute noise to the high-resolution features of the map. With enough corrected copies of the asymmetric unit in the data set, the signal eventually wins out over the noise. As a result, reconstructions of highly symmetrical structures, such as icosahedral virus particles, can still go to 0.4 nm resolution or better, even though the particles are thicker than the depth of field for that resolution.

It should be noted that an iterative algorithm (referred to as an *Ewald sphere correction*) has been developed to correct for the depth-of-field effect when processing data for virus particles. Unfortunately, it is reported that the algorithm has led to little or no improvement in the density maps, which, after all, were already very good without such correction (Leong *et al.*, 2010; Wolf *et al.*, 2006). It remains to be seen whether the use of the same algorithm would play a more important role when merging data for large objects that have little or no internal symmetry.

2.2.3 The SNR Is Diminished as the Fraction of Inelastically Scattered Electrons Increases

The SNR in images of ice-embedded specimens is adversely affected in two ways when the ice thickness is greater than it needs to be. To begin with, fewer electrons make their way through the specimen without suffering an inelastic scattering event. To explain why this decreases the SNR, first imagine that one were to use a zero-loss energy filter to remove all the inelastically scattered electrons. The effect is just as if one had intentionally used a smaller incident-electron exposure to record images for a thinner specimen. The loss of electrons that occurs with a too-thick specimen has no solution, of course, because one cannot just increase the exposure to compensate for the electrons that are lost due to inelastic scattering.

Going beyond the loss of signal, next imagine that a zero-loss energy filter is *not* used (as, indeed, often is the case). The already inferior image (as discussed in the previous paragraph) now will be degraded even further as a result of including the inelastically scattered electrons. While an accurate description of the role of the inelastically scattered electrons is too complicated to go into here, a simplified model can give at least a sense of why the SNR is further degraded. This model assumes that the inelastically scattered electrons are uniformly present throughout the image and thus contribute nothing to the signal. While the average value of this added constant intensity has no effect on the SNR, statistical fluctuation in this added term (i.e., shot noise) is nevertheless present, and this shot noise further degrades the SNR. As a result, it is always beneficial to use an energy filter to remove the background of inelastically scattered electrons. It is also worth noting that a further benefit of an energy filter is to "unmask" the amplitude contrast produced by inelastic scattering, and doing so can substantially improve the SNR at low resolution (Yonekura *et al.*, 2006).

The combined, adverse effects of inelastic scattering, mentioned previously, nevertheless will remain fairly small if the specimen is thinner than one-third of the mean-free-path for inelastic scattering, in which case the fraction of electrons transmitted with zero energy loss is greater than 70%. Since the mean-free-path for inelastic scattering of 300-keV electrons in ice is estimated to be more than 300 nm [as extrapolated from the measurements of Grimm *et al.* (1996) at 120 keV], inelastically scattered electrons have a relatively small effect on image quality if the ice thickness is less than 100 nm. Because of the depth-of-field issue discussed previously, one should actually aim for a sample thickness of about less than 50 nm—i.e., approximately one-sixth of the mean-free-path for 300-keV electrons. As a result, the loss of signal can be minimal for such thin specimens, and the shot noise associated with the inelastically scattered electrons also can be fairly minimal.

Accurate measurement of the sample thickness can be too time-consuming to perform while one is collecting image data. A reasonable alternative, illustrated in Figure 1, is simply to select areas of suitable thickness based on the contrast presented by the specimen in Search mode. One imaging condition employed in Search mode uses a highly defocused electron

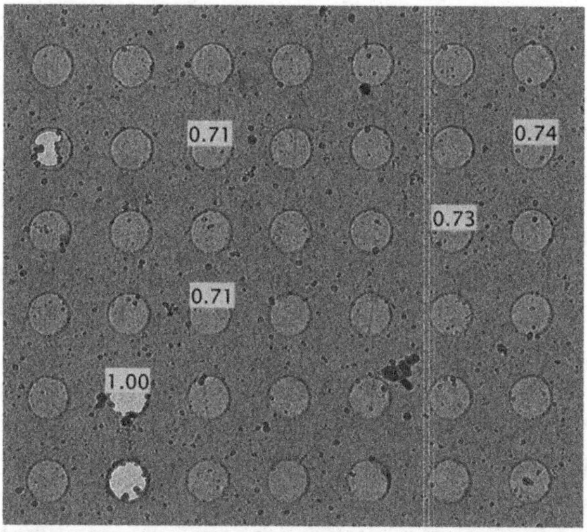

Figure 1 Example of a low-magnification image of a cryo-EM specimen recorded with the low-dose software of the electron microscope set to the SEARCH condition. Most of the holes in the Quantifoil carbon film shown here are filled with a thin film of vitrified sample, but a few of them are empty. The electron intensity, as registered by the camera, is used to compute the apparent transmission of electrons through a hole "filled" with sample, relative to that of an open hole. The numbers in the white boxes show the ratios of the local intensity within the adjacent holes, relative to that in an open hole. We thank Dr. Roseann Csencsits for writing the DigitalMicrograph script used to compute the apparent electron transmission, and Dr. Alison Killilea for providing this example of estimating the thickness of the ice.

diffraction pattern, so the central spot becomes enlarged sufficiently to produce an image of a suitable size. Alternatively, a low-magnification image can be used if it is underfocused enough to give nearly the maximum contrast possible. Either way, it can be recommended that the intensity in an area chosen for data collection should not be less than approximately 70% (and preferably between 80% and 90%) of that in an open hole in the specimen. If, on the other hand, the intensity in a chosen area of the specimen is much above 90% of that in an open hole, one may have to worry whether the sample has been essentially air-dried before being frozen.

2.3 The SNR in Images of Smaller Particles Can Be Inherently too Low to Support Refinement to Higher Resolution

It is now clear that almost any structurally homogeneous macromolecular complex greater than a certain size (e.g., >1 MDa) is a good candidate for high-resolution cryo-EM. As the size of the complex of interest is reduced, however, the SNR that is available to distinguish different views of the particle begins to decrease. Although high-resolution structures have been obtained for the 700-kDa (20S) proteasome (Li *et al.*, 2013) and for the < 300 kDa TRPV1 ion channel (Liao *et al.*, 2013), initial work with both the 464-kDa β-galactosidase homotetramer and the octahedral, 450-kDa apoferritin molecule proved to be more challenging (Henderson and McMullan, 2013). In such cases, it may be that additional efforts nevertheless can result in getting a successful, high-resolution structure.

As the size of a given protein complex decreases, its shape takes on increasing importance in determining the relative ease with which different orientations (views) can be distinguished. Assigning correct Euler angles will clearly be more challenging for a specimen whose shape is nearly spherical (or nearly cylindrical) than it is for one whose low-resolution features vary quite strongly with orientation. One thus expects that, for a given particle size, it may be possible to solve some structures at high resolution while others may "hit a wall" at a lower resolution.

If a given particle proves to be too small and symmetrical for high-resolution cryo-EM to succeed, the situation might be improved by adding a structural tag to make it easier to assign Euler angles. As is shown in Figure 2, monoclonal antibodies can be one good choice to use as an orientation-specifying tag (Wu *et al.*, 2012), as they can interact with a significant surface area on the protein, and thus be rigidly oriented. Tagging a given subunit of a multiprotein complex with a fusion protein, a technique that has been useful in crystallizing some membrane proteins (Zou *et al.*, 2012), also may prove to be effective (Park *et al.*, 2013). It still remains to be shown for cryo-EM samples, however, how often fusion proteins can adopt a sufficiently unique orientation relative to the native complex without interfering with proper folding of the subunit to which they have been fused.

2.4 Nonideal Imaging Conditions Can Limit Resolution

2.4.1 High-Resolution Features Can Be Partially or Completely Lost due to "Delocalization"

Cryo-EM images must be recorded with a substantial amount of defocus to produce enough contrast to see the particles and extract them from the original micrographs. In this regard, it is especially noteworthy that the improved Detective Quantum Efficiency (DQE) at low frequencies

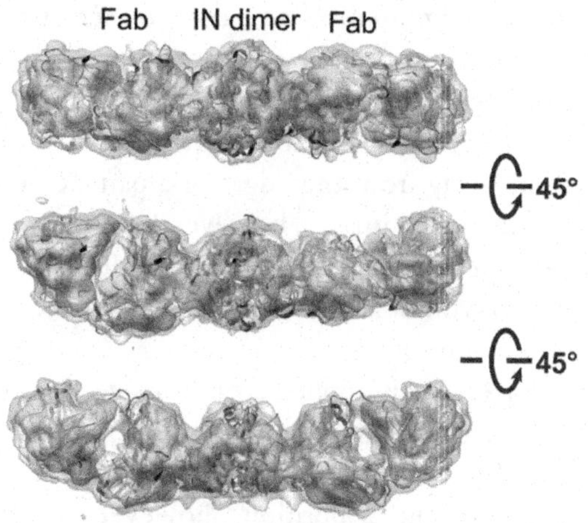

Figure 2 Three views of a 3-D cryo-EM density map of an HIV-1 integrase-Fab complex, modified from Figure 6 from Wu *et al.* (2012). The integrase dimer (labeled "IN dimer") is in the center of the complex, flanked on either side by the monoclonal Fab fragments. This image illustrates how monoclonal antibodies can make it possible to align, as well as assign Euler angles to, an otherwise very small, nearly spherical particle. In addition, to the extent that the cryo-EM structure of the approximately 50 kDa Fab portion of the complex agrees with the previously known X-ray structure, the Fab fragment provides an internal standard to validate the entire cryo-EM reconstruction. (See the color plate.)

provided by the new (direct-detection) cameras makes it possible to record images with less defocus than before. As a result of defocusing the image, however, high-resolution Fourier components become delocalized (Downing and Glaeser, 2008)—i.e., shifted to either side of the particle—by a distance equal to

$$a = C_S\lambda^3 g^3 - (\Delta Z)\lambda g, \quad (1)$$

where a is the amount of delocalization, λ is the electron wavelength, g is the spatial frequency of the Fourier component, C_S is the coefficient of spherical aberration, and ΔZ is the amount of defocus. (At a resolution of 0.3 nm, the delocalization due to spherical aberration is rather small, and thus it will be ignored here.)

For the specific case that the defocus is 2 μm, the Fourier component associated with a scattered ray at 0.4-nm resolution (and the one associated with its Friedel mate) are delocalized to either side of a particle by a distance of approximately 10 nm, which is a significant fraction of the size of macromolecular complexes of interest. Even the Fourier components at 0.8-nm resolution are delocalized by approximately 5 nm, an amount that remains too large to be ignored. When the Fourier components—each delocalized by a different amount—are superimposed on the image, the main features in the object appear blurred, as illustrated in Figure 3(e). In noise-free simulations, such as those shown in this figure, Fresnel fringes also are seen to propagate away from the edges of a particle. These fringes are usually not visible in low-dose images, however, because of the high level of shot noise. (The blurring and emergence of Fresnel fringes can be explained equally well by the convolution of the image wave function by an appropriate point-spread function.)

When a CTF correction is applied to the image data, the delocalized Fourier components are partially returned to their initial position, thus *deblurring* the image. Up to half of the delocalizing information is shifted even further from the initial position, however, as is illustrated in Figures 3(c–d). The net result is that only approximately half of the nonoverlapping delocalized Fourier components are returned to their correct location, while the other half is delocalized by the same distance once again, as is explained by Downing and Glaeser (2008).

In order to recover the delocalized information as fully as possible during CTF correction, one must extract the particles with a box that is large enough, or, alternatively, one might apply the CTF correction to the entire image before extracting particles. A smaller mask (which will cut off the delocalized information) can be applied temporarily during particle alignment and assignment of Euler angles, in order to reduce the influence of noise, as well as of parts of adjacent particles. Similarly, a smaller mask can be permanently applied, if desired, after the alignment and orientation parameters have been frozen and the CTF correction has been made.

2.4.2 Off-Axis Image Coma Can Introduce Significant Phase Errors at High Resolution

In the past, coma did not receive nearly as much attention in cryo-EM as have defocus, astigmatism, and spherical aberration. This is probably because the effect of coma is usually negligible at a resolution below 1 nm or so. Nevertheless, because it causes a systematic phase error that increases as the cube of the resolution, coma can be of major importance at a resolution higher than about 0.6 nm. Attention was not initially given to using coma-free alignment in high-resolution imaging of 2-D crystals because the systematic phase error due to beam tilt could be corrected during image processing (Henderson *et al.*, 1986). Unfortunately, the same correction is

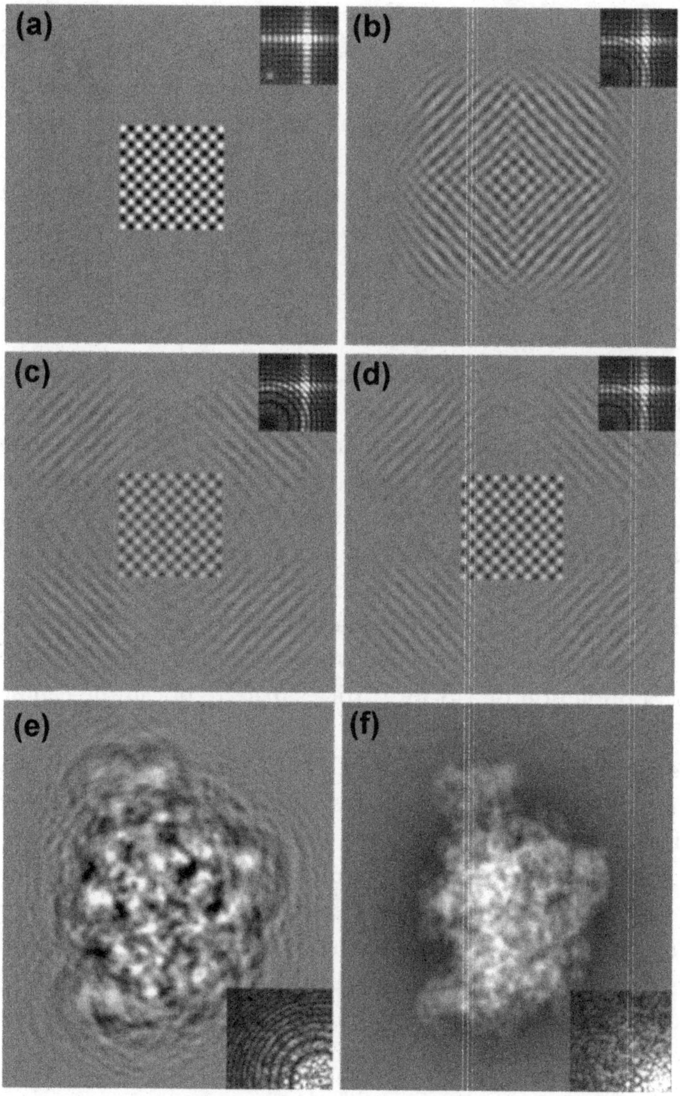

Figure 3 Illustration of the effects of delocalization of signal caused by intentionally defocusing the image. (a) The square at the center represents a "model" particle, and the cross-grating within the square represents two Fourier components of interest. The square is assumed to be 17 nm on the edge, and the spacing of the grating corresponds to 1.3 nm. (b) Image obtained with a simulated defocus of 2 μm. (c) Partial restoration of approximately half the signal by phase flipping is shown, but this is accompanied by the other half of the signal being delocalized by twice the original distance. (d) Similar partial-restoration results obtained by multiplication by the CTF. (e) Noise-free simulation of an image of the ribosomal large subunit, again obtained with a defocus of 2 μm. (f) Restoration obtained with a Weiner filter, assuming that the SNR parameter is 900. Note the dark halo that is still present at low resolution due to the poor recovery of low-frequency information, where the CTF goes asymptotically to zero. This Figure is modified from Figure 2 and part of Figure 3 from Downing and Glaeser (2008).

much more difficult to implement for single-particle images. Adjusting the beam alignment to the coma-free condition is actually a relatively easy thing to do on modern electron microscopes, and some high-end electron microscopes also provide illumination that remains parallel over a wide range of beam diameters and intensities.

Although it should be standard practice to align the illumination to eliminate axial coma, this ensures only that phase errors caused by beam tilting are minimized for particles that are close to the coma-free axis. Unless the electron beam is also sufficiently parallel to this axis over the entire field of view, the image still will be corrupted by off-axis coma (Glaeser *et al.*, 2011; Zhang & Zhou, 2011). As a result, particles that are some distance from the center still suffers significant phase errors that affect the high-resolution structure factors. These phase errors increase in proportion to the beam-tilt angle, which increases linearly with the distance from the center of the beam (due to slight convergence or divergence of the illumination), and in proportion to the third power of the resolution. As a result, an amount of convergence or divergence of the illumination that can be tolerated in order to get accurate phases at 8 Å, for all particles in the field of view, may cause essentially random phases at a resolution of 4 Å for particles that are not in the center of the field of view.

As is explained by Glaeser *et al.* (2011), it is not easy to know when the illumination is sufficiently parallel across the field of view. Perhaps the easiest method is to know which value of the beam diameter (relative to the condenser-aperture diameter) corresponds to parallel illumination. This information can be obtained from electron-optical calculations performed by the manufacturer. Even better is to use a microscope equipped with three condenser lenses rather than the usual two. In this case, the area of illumination can be zoomed over a wide range of sizes, while the beam remains parallel everywhere within the illuminated area.

2.5 Data Processing and Computation are Always Important Issues

2.5.1 Beam-Induced Movement Should be Corrected as Fully as Possible

When an image is recorded without dose fractionation (or when frames of a movie are summed without alignment), it is quite likely that the Fourier transform of the summed image will show the effect of specimen movement; i.e., a partial loss of one or more Thon rings in one direction

(Li *et al.*, 2013). It also can happen that some specimen movement remains after an attempt is made to align successive frames of the movie. If that is the case, it may be that the algorithm used to align the frames was not accurate enough to compensate for beam-induced movement.

Perhaps the simplest alignment algorithm treats the successive frames as rigid bodies. The second frame is shifted in X and Y relative to the first in order to maximize the cross-correlation between the two images. Similarly, the third frame is aligned to the second, the fourth frame is aligned to the third, etc., and all frames are summed. A more sophisticated algorithm computes the pairwise shifts needed to align every frame to every other frame in the movie (i.e., 1 with 2, 1 with 3, 2 with 3, etc.). An improved model of pairwise shifts then can be calculated that minimizes the mean-square residual between the model and the data (Li *et al.*, 2013). Even the least-squares model has limitations because it treats each frame as a rigid body.

In practice, beam-induced motion seems to occur to a different extent in locally coherent patches (Glaeser & Hall, 2011). As a result, it may be an improvement to align independently smaller sections of a frame, but currently, there is a limit on how small such sections can be. For example, Li *et al.* (2013) found that image areas of 20S proteasomes smaller than 2k × 2k pixels contained insufficient information for accurate alignment. The best possible protocol would be to align independently each particle from one frame to the next, of course. Because images of single particles are very noisy, Bai *et al.* (2013) found that they had to sum between four and six adjacent frames to get enough SNR for single-particle tracking of the large ribosome particle to work. Since summing four or more frames partially undoes the goal of minimizing the movement that occurs within a single frame, it is clear that some further workaround is wanted. One approach is to use a running average of four to six adjacent frames to estimate the amount of particle motion that is assigned to each frame. Another approach is to align entire frames (or subsections of the frames) as rigid bodies (as discussed previously) and then, once partially aligned, proceed to use running averages of four to six frames to align each particle. However, such an approach may be problematic for particles less than 1 MDa. Indeed, as mentioned above, Li *et al.* (2013) found that image areas of 20S proteasomes smaller than 2k × 2k contained insufficient information for accurate alignment.

2.5.2 CTF Correction Must be Made with Sufficient Accuracy

An unwanted but unavoidable consequence of recording highly defocused images is that the contrast transfer function oscillates quite rapidly at high

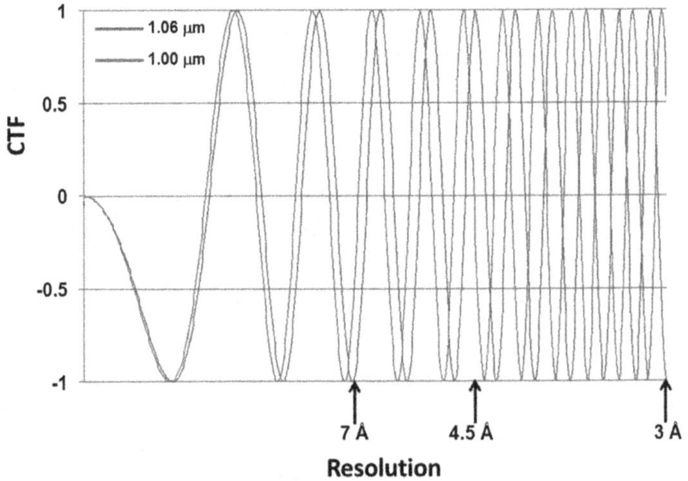

Figure 4 Comparison of the CTF curves simulated for defocus values of 1.0 and 1.06 μm, respectively. This simulation shows (1) that significant errors would occur in the CTF correction made at a resolution of 0.7 nm if the assumed value of defocus is in error by only 60 nm, and (2) that a completely wrong CTF "correction" would be made at a resolution of 0.4 nm or better. (See the color plate.)

resolution—for example, see Figure 4. Since the amount of defocus required to see smaller macromolecules is greater than that required to see larger ones, the problem of CTF oscillations is generally worse for smaller particles. It thus is important that the amount of defocus and the amount of astigmatism must be determined accurately for each image. Errors in estimating these parameters can cause the structure factors from individual particles to be merged with random signs.

As a rule of thumb, the error of estimating the amount of defocus should be less than $\Delta Z = \frac{d^2}{2\lambda}$, the depth of field for d, the highest resolution of interest. For 300-keV electrons and a resolution of 4 Å, for example, the depth of field is approximately 40 nm, as was mentioned previously. Of course, the poor SNR of low-dose cryo-EM images makes it difficult to estimate the defocus and astigmatism with the desired accuracy. Fortunately, direct-detection cameras greatly improve the SNR, and this fact significantly improves the accuracy with which the CTF can be estimated. Even so, it still can be challenging to achieve the desired accuracy in estimating defocus and astigmatism.

3. PERSPECTIVES FOR FURTHER IMPROVEMENT OF SINGLE-PARTICLE CRYO-EM

3.1 The Current State-of-the-Art Is Still Well Short of the Physical Limit

As mentioned in section 1.2, the resolution achieved with the yeast 80S ribosome was approximately 4.5 Å from 35,000 particles (Bai *et al.*, 2013), but that is still significantly worse than the target value of approximately 3 Å that physics would allow for the same number of particles. Similarly, in studies involving the 20S proteasome (Li *et al.*, 2013) and TRPV1 (Liao *et al.*, 2013), both of which are much smaller particles than the 80S ribosome, a much better resolution (approximately 3.3 Å) was achieved, but only after merging data from about five times more particles than were estimated to be necessary if everything were ideal. While all these recent results are a major advance over the previous state of the art, the point here is that there remains considerable room for improvement. In particular, a resolution approaching 3 Å continues to be challenging (but not impossible) to achieve for particles smaller than 0.5 MDa with low symmetry, although it has been suggested that it should be possible for particles that are 10 times smaller than that (Henderson, 1995).

In what way, then, does current cryo-EM technology still fall short of being ideal? What improvements might be made that would come closer to achieving the physically allowed level of performance?

3.1.1 Cameras Can Be Improved

Perhaps the first issue that should be mentioned is that camera technology can be improved considerably. While electron counting can bring the camera output noise down to the same, irreducible level of the input shot noise, the value of the modulation-transfer function (for signal) could still be improved by a factor of 2 or more at Nyquist frequency. One way to improve the situation would be to combine a larger pixel size (as in the FEI Falcon camera) with electron counting (as in the Gatan K2 Summit camera). A second improvement in camera technology that would be of great value would be to increase the field of view (i.e., the number of pixels), but to do so while maintaining the higher SNR noted previously. This would mean that more particles could be imaged at once, thus reducing the data-collection time as well as indirectly improving the accuracy of CTF determination.

3.1.2 Phase Plates Can Improve Contrast at Low Resolution

The currently imperfect contrast transfer achieved in cryo-EM represents a second area where physics can allow significant improvement. At present, images must be recorded with a defocus value of approximately 0.7 μm or more (at 300 keV) just to be able to see where particles are located in the image. Even then, the contrast at low resolution (which is important in aligning particles, assigning Euler angles, and recognizing conformational heterogeneity) remains far below what physics allows. The visibility of particles thus would be greatly improved if one or more of the methods for obtaining in-focus images of weak-phase objects, reviewed in Glaeser (2013a), could be developed as a reliable and robust tool.

Improving the contrast transfer at low resolution would be especially welcome when imaging particles that are smaller than approximately 1 MDa, of course. In addition, cryo-EM images would require little or no CTF correction if a phase-plate device were used. Such images thus would not suffer a loss of up to half of the signal that is delocalized outside the envelope of the particle (Downing & Glaeser, 2008), as do the highly defocused images described previously.

The SNR at the band limit of the specimen itself thus could be improved over what is currently achieved (1) by avoiding the loss of up to half of the high-resolution signal (caused by imperfect recovery of delocalized information) and (2) by a further improvement in the DQE of the camera at high resolution. It thus is possible that most of the gap can be closed between what is theoretically possible versus what is accomplished at present.

3.2 Two Further Issues Need to be Addressed in Order to Achieve Resolutions Higher than 0.3 nm

It is now reasonable to think that obtaining cryo-EM structures at a resolution of approximately 0.3–0.35 nm soon will become as routine as obtaining X-ray structures at this resolution. It thus is natural to look ahead and ask what it might take to make structure determination by cryo-EM routine at even higher resolutions—say, approaching 0.2 nm and beyond. The electron-optical performance of modern electron microscopes is already very capable of near-perfect performance at this level of resolution.

3.2.1 Beam-Induced Tilt May Limit the Achievable Resolution

In addition to the beam-induced translational movement that is corrected well by the use of dose-fractionated imaging, there is known to be some additional degree of beam-induced change in the tilt angle of local areas of the specimen (Brilot *et al.*, 2012). This is still a poorly characterized effect, and thus it is unknown how it might be best mitigated. If, for example, the local tilt angle changes during data collection, and if that is not accounted for, it will limit the achieved resolution for many particles, especially at the outer radius of the particle. To illustrate, for a particle diameter of 20 nm, data at a resolution of 0.3 nm are mathematically independent from one another when particle orientations are separated by an angle of less than 0.9 degrees, as was explained by Crowther *et al.* (1970). Averaging images of a particle that rotates by this much or more during image acquisition therefore will smear out the resulting density map at 0.3 nm or better resolution.

One solution, if it were possible, would be to index the information from each frame (or a number of adjacent frames) of a dose-fractionated image at Euler angles that continue to change slightly as the exposure continues. The ability to do this will depend upon the SNR of the frames, of course. Another solution might be to gain sensitivity in tracking changes at a local tilt angle that is as small as 1 degree by including information from an appropriate support film, such as streptavidin monolayer crystals, or from other fiducial particles added to the specimen.

3.2.2 Data Collection and Merging Data May Have to be Modified to Deal with Curvature of the Ewald Sphere

Curvature of the Ewald sphere (discussed previously in section 2.2) ultimately may be the most important issue that must be dealt with when going to higher resolutions. In this case, it is the size of the particle itself, rather than the Z-height positions of different particles within the field of view, that poses an irreducible limitation. In other words, once the particle thickness is greater than the depth of field, one can no longer use the projection approximation (i.e., flat Ewald sphere approximation) to analyze images. At a resolution of 0.2 nm, this will be the case for macromolecular complexes whose diameter is 10 nm or more.

There are three approaches that can be considered for any particle, which do not exploit the special symmetry of the particle that may be present. The first is the iterative algorithm already described in section 2.2.2. The second approach is to collect data as focal pairs. Although DeRosier (2000) correctly emphasized that there is no loss of SNR involved in collecting data as a focal pair, it nevertheless will be a problem that the high-resolution signal in the second image is reduced by the radiation dose used to record the first image. As a result, there may be little information available to separate the cosine-modulated and sine-modulated structure factors at high resolution, which is where the Ewald sphere compensation is needed. The third approach might be to use an objective-aperture shape that blocks scattered electrons at high resolution on one side of the electron diffraction pattern, but not on the other side. This approach would provide double-sideband contrast (perhaps improved by using a phase plate) at low and intermediate contrast, but single-sideband contrast at higher resolutions, where curvature of the Ewald sphere begins to be important. The structure-factor amplitudes and phases in the single-sideband region thus could be uniquely indexed in the high-resolution region of the 3-D Fourier space, as well as in the low- and intermediate-resolution regions. This approach is reminiscent of using a screened precession camera (which for many years was the preferred way to collect macromolecular X-ray diffraction patterns) in that, to avoid confusion, some of the diffraction pattern is not allowed to contribute to the data that are collected.

4. SUMMARY: HIGH-RESOLUTION STRUCTURE ANALYSIS BY CRYO-EM SEEMS TO BE RAPIDLY APPROACHING ITS FULL POTENTIAL

As is described by Marton (1968), imaging biological specimens was among the earliest ambitions of the physicists and engineers who developed the first electron microscopes. Although the idea that electrons severely damage biological specimens was discussed even at that time, it also has been appreciated for a long time that the potential exists to determine macromolecular structures of such radiation-sensitive specimens by merging data from a sufficiently large number of identical copies of the molecule (Glaeser, 1999; Henderson, 1995).

The latest progress toward this goal has involved major new developments in camera technology and in data analysis, as has been reviewed in this chapter. This progress was achieved with commercial instruments that are designed for the highest level of performance. Issues such as beam-induced (translational) movement, proper CTF correction, and even coma no longer pose as much of a problem as they did only a few years ago. In addition, the latest advances highlight how useful it is for the total ice thickness to be kept smaller than the depth of field for the desired resolution. As a result, it suddenly has become possible to obtain 3-D density maps of structures as small as 300 kDa at a high enough resolution to be interpreted in terms of an atomic model of the structure. As always, however, it remains crucial that the specimen must be structurally homogeneous. Alternatively, images of structurally heterogeneous particles may be sorted into distinct, more homogeneous subsets to some degree.

It now is perhaps more realistic than ever to think that cryo-EM ultimately will produce structures at even higher resolutions (approaching 0.2 nm, for example), as well as for even smaller protein complexes (possibly down to 100 kDa, or even less). Progress in these directions can be accelerated by further improvement in camera performance. It also will help if in-focus phase contrast technology can be made robust and useful for routine work, as this would improve the SNR at low spatial frequency without introducing both oscillations in the CTF and delocalization of the high-frequency information. Other developments may include experimental techniques to deal with beam-induced changes in specimen tilt, and computational (or experimental) techniques that address the issue of curvature of the Ewald sphere. While cryo-EM is thus in the midst of rapid improvement, it also appears that it is not likely to stop making such improvements for many years to come.

ACKNOWLEDGMENTS

The preparation of this review has been supported in part by funds from the Lawrence Berkeley National Laboratory (LBNL) under Contract No. DE-AC02-O5CH11231 (RMG); the Howard Hughes Medical Institute (DAA); NIH grants GM098672 and GM082250 (YC); National Institutes of Health (NIH) grant GM083039 (RMG); and funds from the intramural research program of the National Cancer Institute, NIH, Bethesda (SS).

REFERENCES

Bai, X.-C., Fernandez, I. S., McMullan, G., & Scheres, S. H. (2013). Ribosome structures to near-atomic resolution from 30,000 cryo-EM particles. *eLife Sciences, 2*.

Baker, L. A., Smith, E. A., Bueler, S. A., & Rubinstein, J. L. (2010). The resolution dependence of optimal exposures in liquid nitrogen temperature electron cryomicroscopy of catalase crystals. *Journal of Structural Biology, 169*, 431–437.

Brilot, A. F., et al. (2012). Beam-induced motion of vitrified specimen on holey carbon film. *Journal of Structural Biology, 177*, 630–637.

Campbell, M. G., et al. (2012). Movies of ice-embedded particles enhance resolution in electron cryo-microscopy. *Structure, 20*, 1823–1828.

Cao, E., Liao, M., Cheng, Y., & Julius, D. (2013). TRPV1 structures in distinct conformations reveal activation mechanisms. *Nature, 504*, 113–118.

Crowther, R. A., Amos, L. A., Finch, J. T., Derosier, D. J., & Klug, A. (1970). 3-dimensional reconstructions of spherical viruses by Fourier synthesis from electron micrographs. *Nature, 226*, 421–425.

DeRosier, D. J. (2000). Correction of high-resolution data for curvature of the Ewald sphere. *Ultramicroscopy, 81*, 83–98.

Downing, K. H., & Glaeser, R. M. (2008). Restoration of weak phase-contrast images recorded with a high degree of defocus: The "twin image" problem associated with CTF correction. *Ultramicroscopy, 108*, 921–928.

Frank, J. (2006). *Three-Dimensional Electron Microscopy Of Macromolecular Assemblies— Visualization of Biological Molecules in Their Native State* (2nd ed.). New York: Oxford University Press.

Frank, J. (2013). Story in a sample—the potential (and limitations) of cryo-electron microscopy applied to molecular machines. *Biopolymers, 99*, 832–836.

Glaeser, R. M. (1999). Review: Electron crystallography: Present excitement, a nod to the past, anticipating the future. *Journal of Structural Biology, 128*, 3–14.

Glaeser, R. M. (2008). Retrospective: Radiation damage and its associated "information limitations." *Journal of Structural Biology, 163*, 271–276.

Glaeser, R. M. (2013a). Invited review article: Methods for imaging weak-phase objects in electron microscopy. *Review of Scientific Instruments, 84*, 111101.

Glaeser, R. M. (2013b). Stroboscopic imaging of macromolecular complexes. *Nature Methods, 10*, 475–476.

Glaeser, R. M., & Hall, R. J. (2011). Reaching the information limit in cryo-EM of biological macromolecules: Experimental aspects. *Biophysical Journal, 100*, 2331–2337.

Glaeser, R. M., Downing, K., DeRosier, D., Chiu, W., & Frank, J. (2007). *Electron Crystallography of Biological Macromolecules*. New York: Oxford University Press.

Glaeser, R. M., Typke, D., Tiemeijer, P. C., Pulokas, J., & Cheng, A. C. (2011). Precise beam-tilt alignment and collimation are required to minimize the phase error associated with coma in high-resolution cryo-EM. *Journal of Structural Biology, 174*, 1–10.

Grigorieff, N., & Harrison, S. C. (2011). Near-atomic resolution reconstructions of icosahedral viruses from electron cryo-microscopy. *Current Opinion in Structural Biology, 21*, 265–273.

Grimm, R., Typke, D., Barmann, M., & Baumeister, W. (1996). Determination of the inelastic mean free path in ice by examination of tilted vesicles and automated most probable loss imaging. *Ultramicroscopy, 63*, 169–179.

Henderson, R. (1995). The potential and limitations of neutrons, electrons, and X-rays for atomic-resolution microscopy of unstained biological molecules. *Quarterly Reviews of Biophysics, 28*, 171–193.

Henderson, R., & McMullan, G. (2013). Problems in obtaining perfect images by single-particle electron cryomicroscopy of biological structures in amorphous ice. *Microscopy, 62*, 43–50.

Henderson, R., Baldwin, J. M., Downing, K. H., Lepault, J., & Zemlin, F. (1986). Structure of purple membrane from halobacterium-halobium—recording, measurement, and evaluation of electron-micrographs at a 3.5 resolution. *Ultramicroscopy, 19*, 147–178.

Jensen, G. J. (2001). Alignment error envelopes for single-particle analysis. *Journal of Structural Biology, 133*, 143–155.

Jensen, G. J. (2010). *Cryo-EM*. San Diego: Academic Press.

Kelly, D. F., Abeyrathne, P. D., Dukovski, D., & Walz, T. (2008). The affinity grid: A pre-fabricated EM grid for monolayer purification. *Journal of Molecular Biology, 382*, 423–433.

Leong, P. A., Yu, X. K., Zhou, Z. H., & Jensen, G. J. (2010). Correcting for the Ewald sphere in high-resolution single-particle reconstructions. In G. J. Jensen (Ed.), *Vol. 482. Methods in Enzymology*. Cryo-EM, Part B: 3-D Reconstruction, pp. 369–380.

Li, X., et al. (2013). Electron counting and beam-induced motion correction enable near-atomic-resolution single-particle cryo-EM. *Nature Methods, 10*, 584–590.

Liao, M., Cao, E., Julius, D., & Cheng, Y. (2013). Structure of the TRPV1 ion channel determined by electron cryo-microscopy. *Nature, 504*, 107–112.

Lowe, J., Stock, D., Jap, R., Zwickl, P., Baumeister, W., & Huber, R. (1995). Crystal structure of the 20S proteasome from the archaeon T. acidophilum at 3.4-angstrom resolution. *Science, 268*, 533–539.

Lyumkis, D., Brilot, A. F., Theobald, D. L., & Grigorieff, N. (2013). Likelihood-based classification of cryo-EM images using FREALIGN. *Journal of Structural Biology, 183*, 377–388.

Marton, L. (1968). *Early History of the Electron Microscope.* San Francisco: San Francisco Press.

Park, S., et al. (2013). Reconfiguration of the proteasome during chaperone-mediated assembly. *Nature, 497*, 512–516.

Penczek, P. A., Frank, J., & Spahn, C. M. T. (2006). A method of focused classification, based on the bootstrap 3D variance analysis, and its application to EF-G-dependent translocation. *Journal of Structural Biology, 154*, 184–194.

Rabl, J., Smith, D. M., Yu, Y., Chang, S. C., Goldberg, A. L., & Cheng, Y. (2008). Mechanism of gate opening in the 20S proteasome by the proteasomal ATPases. *Molecular Cell, 30*, 360–368.

Scheres, S. H. W. (2012). RELION: Implementation of a Bayesian approach to cryo-EM structure determination. *Journal of Structural Biology, 180*, 519–530.

Scheres, S. H. W., & Chen, S. X. (2012). Prevention of overfitting in cryo-EM structure determination. *Nature Methods, 9*, 853–854.

Scheres, S. H. W., et al. (2007). Disentangling conformational states of macromolecules in 3D-EM through likelihood optimization. *Nature Methods, 4*, 27–29.

Schmidt-Krey, I., & Cheng, Y. (2013). *Electron crystallography of soluble and membrane proteins: Methods and protocols.* New York: Humana Press.

Taylor, K. H., & Glaeser, R. M. (2008). Retrospective on the development of cryoelectron microscopy of macromolecules. *Journal of Structural Biology, 163*, 214–223.

Wang, L. G., Ounjai, P., & Sigworth, F. J. (2008). Streptavidin crystals as nanostructured supports and image-calibration references for cryo-EM data collection. *Journal of Structural Biology, 164*, 190–198.

Wolf, M., DeRosier, D. J., & Grigorieff, N. (2006). Ewald sphere correction for single-particle electron microscopy. *Ultramicroscopy, 106*, 376–382.

Wu, S., et al. (2012). Fabs enable single-particle cryo-EM studies of small proteins. *Structure (Cambridge), 20*, 582–592.

Yonekura, K., Braunfeld, M. B., Maki-Yonekura, S., & Agard, D. A. (2006). Electron energy filtering significantly improves amplitude contrast of frozen-hydrated protein at 300 kV. *Journal of Structural Biology, 156*, 524–536.

Zhang, X., & Zhou, Z. H. (2011). Limiting factors in atomic resolution cryo-electron microscopy: No simple tricks. *Journal of Structural Biology, 175*, 253–263.

Zou, Y. Z., Weis, W. I., & Kobilka, B. K. (2012). N-terminal T4 lysozyme fusion facilitates crystallization of a G protein–coupled receptor. *Plos One, 7*, e46039.

CHAPTER THREE

Morphological Amoebas and Partial Differential Equations

Martin Welk[1], **Michael Breuß**[2]
[1]UMIT, Biomedical Image Analysis Division, Eduard-Wallnoefer-Zentrum 1, 6060 HALL (Tyrol), Austria
[2]Mathematical Image Analysis Group, Saarland University, Campus E1.1, 66041 Saarbrücken (Germany)

Contents

Advances in Imaging and Electron Physics, Volume 185
ISSN 1076-5670
http://dx.doi.org/10.1016/B978-0-12-800144-8.00003-3

1. INTRODUCTION

In this chapter, we study the relations between two classes of adaptive image filters: on one hand, iterative filters arising from a space-discrete morphological framework, and on the other hand, partial differential equations (PDEs).

Starting with Matheron (1967), the field of mathematical morphology has developed into a powerful theory that provides useful tools for many fundamental image-processing tasks such as image denoising, structure enhancement, and shape simplification, see Serra (1982, 1988) and Heijmans (1994). At the heart of mathematical morphology, operations are applied in a pixelwise fashion; examples are dilation, erosion, or median filtering. These operations use *tonal* (i.e., intensity) information within a geometric shape that contains the origin pixel. This usually predefined geometric shape is the *structuring element.*

In contrast to the pixelwise use of such a fixed window with the same filter operation, adaptive approaches allow for varying the filtering rule in order to preserve important features such as homogeneous regions, edges, or texture. Since adaptive methods have been recognized for achieving superior results, many works in the literature have explored the inherent freedom in designing such filters (e.g., Verly & Delanoy, 1993; Braga-Neto, 1996; Shih & Cheng, 2004; Grazzini & Soille, 2009).

Morphological amoebas are a class of structuring elements for morphological image filters that adapt to image structures with maximal flexibility. Lerallut, Decencière, and Meyer (2005, 2007) introduced them for the purpose of structure-adaptive denoising. In amoeba construction, the structuring elements adapt locally to the variation of gray (or color) values, and they also take into account the spatial distance to the origin pixel. While the maximal size of structuring elements is constrained in this way, one may emphasize that large deviations in the tonal values are penalized. Therefore, amoebas have the property that they may grow around corners or along anisotropic image structures.

Using in each pixel the amoeba shape computed for this particular pixel as a structuring element, many filtering procedures can be applied on it. In this paper, we are particularly interested in the use of the median filter and some related filters, such as M-smoothers.

Median filtering in its nonadaptive form goes back to Tukey (1971) and became common as a structure-preserving image filter in the 1990s (see Klette and Zamperoni, 1996; Dougherty and Astola, 1999). When applying the median filter, it is of practical interest to do this iteratively. This is also true for amoeba median filtering (AMF).

Building the bridge: Discrete and continuous scales. For iterated median filtering with a fixed structuring element, Guichard and Morel (1997) have shown that in the continuous-scale limit, the iteration process approximates the PDE:

$$u_t = |\nabla u| \operatorname{div}\left(\frac{\nabla u}{|\nabla u|}\right), \tag{1}$$

known as *mean curvature motion (MCM)* – in 2D also: curvature motion (see Alvarez, Lions, & Morel, 1992). Herein, $u := u(x,y,t)$ describes an image evolving in time t, which is initialised with the input image at time $t = 0$, and ∇u denotes the spatial gradient. On the other hand, from the PDE-based point of view, iterated discrete median filtering with a fixed structuring element can be understood as a specific discretization of Eq. (1) (see Jalba and Roerdink, 2009).

A major theme of this paper is the relationship between discrete AMF and related processes and their continuous-scale, PDE-based counterparts. Since the seminal paper by Guichard and Morel (1997), similar cross-relationships have been studied. For example, van den Boomgaard (2002) proved a PDE approximation result for the Kuwahara-Nagao operator (Kuwahara *et al.*, 1976; Nagao & Matsuyama, 1979). Didas and Weickert (2007) studied correspondences between adaptive averaging and a class of generalized curvature motion filters. Barash (2001) and Chui and Wang (2009) considered PDE limits of bilateral filters introduced by Tomasi and Manduchi (1998).

It is useful to consider important properties of AMF- and PDE-based filters that are known to produce results of comparable quality. Iterating the AMF process simplifies an image towards a cartoonlike appearance; i.e., qualitatively, one receives homogeneous regions separated by sharp contours. Using AMF, even corners are preserved fairly well. This is in contrast to standard median filtering with a fixed structuring element, where corners usually appear rounded after filtering. Standard iterated median filtering also tends to displace curved edges in the inward direction of the curvature.

From a PDE viewpoint, a class of segmentation methods with edge-enhancing properties that can give similar cartoonlike results are the so-called *self-snakes* (see Sapiro, 1996; Whitaker and Xue, 2001). This class of PDE-based filters is closely related to the concept of curvature motion. Comparing the self-snakes equation,

$$u_t = |\nabla u| \operatorname{div}\left(g(|\nabla u|) \frac{\nabla u}{|\nabla u|} \right), \tag{2}$$

and the mean curvature motion PDE [Eq. (1)], the difference is given by the *edge-stopping function g* that modulates the evolution in dependence on the spatial image gradient.

Focussing on possible candidates for edge-stopping functions, such a function g should be nonnegative and monotonically decreasing. A popular example is the function

$$g(s) = \frac{1}{1 + s^2/\lambda^2} \tag{3}$$

with a contrast parameter λ. This function is also used as a diffusivity function by Perona and Malik (1990). By introducing such an edge-stopping function, (curved) edges are no longer displaced, a quality also found with AMF filtering. For suitable g, including Eq. (3), edges are even sharpened.

Geodesic active contours (GAC). Another PDE that is of interest in the discussed context which is very similar to Eq. (1), and especially Eq. (2), is the evolution equation for geodesic active contours (GAC) used for image segmentation. GAC were formulated by Caselles, Kimmel, and Sapiro (1995) and Kichenassamy *et al.* (1995), based on earlier work on active contours (e.g., Kass, Witkin, & Terzopoulos 1988; Caselles *et al.*, 1993; Malladi, Sethian, & Vemuri, 1993).

The geodesic active contour PDE is given by

$$u_t = |\nabla u| \operatorname{div}\left(\mathrm{g}(|\nabla f|) \frac{\nabla u}{|\nabla u|} \right) \tag{4}$$

and describes the evolution of a level-set function u of some *contour* under the influence of the gradients of the fixed given image f that is to be segmented. This is in contrast to Eq. (2), where the evolution relies on the gradients of the evolved image u and where f just defines the initial state of the evolution.

The level-set evolution described by Eq. (4) corresponds to each of the level lines of u to the contour evolution:

$$c_t = \left(g(|\nabla f|)k - \langle \nabla g(|\nabla f|), \boldsymbol{n}\rangle\right) \boldsymbol{n}, \tag{5}$$

in which $\boldsymbol{n}$ denotes a unit inward normal vector of the contour curve $\boldsymbol{c}$, whereas k is its curvature. Let us emphasize that g denotes again an edge-stopping function, but now the argument is the local gradient magnitude of the input image f, as is typical for active contour methods. Among a multitude of active contour schemes, the name "*geodesic* active contours" singles out the evolution [Eqs. (4) and (5)] as one for which the resulting contour minimizes (locally) the arc length in some image-dependent metric; i.e., the contour becomes a geodesic in that metric.

When using an active contour method like this for segmenting an image, one is actually interested in the evolution of one single contour $\boldsymbol{c}$. This contour is often initialized by user input and evolves towards regions with high contrast, which are assumed to yield plausible segment boundaries. Particularly if the initial contour is far from the sought segment boundary, and if the topology of the segment boundary is very complex, the GAC evolution might stop at an undesired location away from the desired contour. For such cases, it is recommended by several sources (e.g., Cohen, 1991; Kichenassamy *et al.*, 1995) that Eq. (5) be modified by an additional force term, such as $\pm \nu \boldsymbol{n}$. The mathematical model for this so-called balloon force resembles morphological dilation or erosion. Its purpose is to push the evolution in a chosen direction, thereby preventing it from stopping prematurely in regions with little contrast.

Rewriting the self-snakes equation in the form

$$u_t = g \cdot |\nabla u| \operatorname{div}\left(\frac{\nabla u}{|\nabla u|}\right) + \langle \nabla g, \nabla u\rangle \tag{6}$$

draws attention to a significant difference between the MCM equation [Eq. (1)] and self-snakes: only the latter involve the contribution $\langle \nabla g, \nabla u\rangle$. Whereas this summand is essential for the favorable edge-enhancing property of self-snakes, it is related to a shock filter (Osher & Rudin, 1990; Sapiro, 1996), or backward diffusion (see Sapiro, 1996; Breuß and Welk, 2007). From an analytical viewpoint, this has the consequence that the self-snakes PDE is ill posed. In particular, it displays staircasing behavior (You *et al.*, 1994). On the numerical side, when approaching the discretization of the PDE, the shock filter component needs specific consideration. In a finite difference setting, it is usually treated by an upwind discretization, as in Osher and Sethian (1988). However, even when employing such approximations, the discrete process is not perfect since severe numerical dissipation artifacts appear. As demonstrated in Welk, Breuß, and Vogel (2011), computational results depend heavily on the grid mesh size. As a consequence, standard discretizations based directly on the given PDE [Eq. (6)] are unreliable.

One remedy against these difficulties is to modify the PDE itself: By the introduction of presmoothing, one obtains a well-posed PDE that can be properly numerically approximated (Feddern *et al.*, 2006). To achieve presmoothing, the gradient ∇u in the argument of g is replaced with $\nabla u_\sigma := K_\sigma * \nabla u$; i.e., the gradient smoothed by convolution with a Gaussian K_σ of standard deviation σ.

On the other hand, as pointed out in Welk, Breuß, and Vogel (2011), the approximation results that are also in the focus of this chapter allow to consider iterated AMF as an unconventional, nonstandard discretization of self-snakes. Experiments presented in Welk, Breuß, and Vogel (2011) show that staircasing artifacts do not occur in an iterated AMF, which indicates that the AMF procedure itself acts in a regularizing way. Therefore, it is of interest to study this regularization effect and its connection to presmoothing.

Contributions of this work. Motivated by the above mentioned result by Guichard and Morel (1997), the question arises whether a similar correspondence can be established between a continuous-scale limit case of amoeba filters and a PDE of the self-snakes type.

The answer to this question is in the focus of this chapter; we summarize and extend thereby our previous work (e.g., Welk, Breuß, and Vogel, 2011; Welk, 2013a). It turns out that iterated amoeba filters can be interpreted as discrete approximations of suitable curvature-based PDE image filters. In particular, iterated AMF corresponds to self-snakes. Different amoeba metrics translate to different edge-stopping functions in the self-snakes equation. In particular, the amoeba metric induced from the Euclidean metric in the embedding space corresponds to the well-known Perona-Malik function [Eq. (3)] as an edge-stopping function.

As a novelty, we also study PDE limit cases for quantile and mode filters with amoeba-structuring elements. Based on Welk (2007), we also discuss the case of multichannel images (such as color images). In this case, however, already nonadaptive median filtering leads to a very complicated PDE limit, as we will demonstrate. Therefore, we do not deal with the case of multi-channel amoeba filters with this PDE limit instrumentary.

In a further step, the relationship between AMF and self-snakes gives rise to an active contour algorithm based on AMF of level-set functions. This algorithm has been formulated initially in Welk (2012) and studied further in Welk (2013a). In Welk (2013a), an approximation property in the same spirit as in Guichard and Morel (1997) and Welk, Breuß, and Vogel (2011) was proved that links amoeba active contours (AAC) to a PDE evolution similar to GAC; see also Welk (2013b). In a special case, it was already shown (Welk, 2012) that the exact geodesic active contour PDE is approximated. In this chapter, we summarize the results from both sources and place them in the context of the other amoeba image processing methods.

Finally, we address the role of presmoothing in the self-snakes equation and the abovementioned regularizing effect of AMF. Following Welk (2013a), we demonstrate by a simple model case that indeed presmoothing in the self-snakes PDE can be linked to the nonzero structuring element radius used in practical computations with amoeba models.

We complement the theoretical discussions with a number of experiments that support the validity of the model and conclusions. Our results extend the framework of known correspondences between discrete and PDE formulations of morphological filters. The study of these relationships helps to gain a unified view on image filtering methods and to combine the advantages of both approaches. In particular, they also allow the use of amoeba procedures as discretizations of structure-adaptive PDE filters; a similar strategy was used by Jalba and Roerdink (2009).

Further notes on previous and related work. While curvature motion smooths in the direction of level-lines only, Caselles, Morel, and Sbert (1998) defined for image interpolation purposes a process that smooths in gradient flow line direction, called *adaptive monotone Lipschitz extension (AMLE)*. The general principle to write curvature- and diffusion-based image filter PDEs as mixtures of smoothing along the directions of level lines and gradients has been established by Carmona and Zhong (1998). This viewpoint also will be important in this analysis.

The space-continuous description of amoebas relies heavily on the representation of an image by an image manifold. This setup was used in the context of the Beltrami framework by Kimmel, Sochen, and Malladi (1997) and Yezzi (1998) and also underlay the bilateral filter in Tomasi and Manduchi (1998) and Barash (2001). Image patches analogous to the amoebas considered here also were used by Spira, Kimmel, and Sochen (2007) for short-time Beltrami kernels.

In the literature, there are numerous approaches to creating structuring elements for morphological filters that adapt to image structures. These approaches can be divided roughly into two groups: One group relies on parametric models that allow for varying the scale and directional parameters of structuring elements across the image (e.g., Yang *et al.*, 1995; Breuß, Burgeth, & Weickert, 2007; Burgeth *et al.*, 2011; Verdú-Monedero, Angulo, & Serra, 2011) The other group combines spatial and intensity information in various ways to generate structuring elements that can attain more or less arbitrary (connected) shapes. While the amoebas by Lerallut, Decencière, & Meyer (2005) are the focus of this specific study, the idea of such a combination can be traced back as far as Nagao and Matsuyama (1979). Since then, several variants of this concept have been formulated and given names such as *adaptive neighborhoods* or *distance-based approach* [e.g., Braga-Neto (1996); Tomasi & Manduchi (1998); Debayle & Pinoli (2006); Grazzini & Soille (2009); Angulo (2011) and Ćurić, Hendriks, & Borgefors (2012)], and these studies rely on the salience distance transform by Rosin and West (1995). Further approaches in this direction are mentioned in the overview by Maragos and Vachier (2009). Aspects of morphological axiomatics related to such adaptive approaches have been discussed by Roerdink (2009).

Using a discrete filter for the discretization of a PDE by virtue of an equivalence result like that of Guichard and Morel (1997) also can be seen in the context of other unconventional discretizations of continuous filters that are tailored to preserve certain important qualitative properties of PDEs. This includes, for example, mimetic discretizations (Hyman *et al.*, 2002; Hyman and Shashkov, 1997), as well as so-called nonstandard schemes (Mickens, 1994; Weickert, Welk, & Wickert 2013).

Some preliminary results on the relation between iterated AMF and PDEs have been published (Welk, Breuß, & Vogel, 2009), where some coefficients in the derived PDEs were flawed due to a mistake in the derivation. This prevented a correct interpretation of the results.[1] This mistake was corrected in Welk, Breuß, and Vogel (2011).

Structure. The remainder of this chapter is structured as follows. In section 2, we describe the discrete amoeba construction and amoeba image

[1] In Eq. (9) of Welk, Breuß, and Vogel (2009), the integrands on both sides lacked a factor called $\partial x/\partial z$; see the corrected Eq. (24) in this chapter and the erratum for Welk, Breuß, and Vogel (2009) at http://www.mia.uni-saarland.de/publications.

filters based upon it. Section 3 covers continuous formulations of these filters, providing the basis for subsequent analysis. The latter is presented in section 4, which concentrates on deriving PDEs that are approximated by iterated amoeba filters when the radius of structuring elements approaches zero. The interplay between the nonzero structuring element radius in actual computations of amoeba filters and presmoothing on the side of PDEs is the subject of section 5. Results presented here represent the current stage of work that is still in progress. Section 6 demonstrates the relationships between discrete and continuous filters derived in the preceding sections. Conclusions are found in section 7. Finally, an appendix collects several lengthy calculations that were deferred in the discussions in the main sections.

2. DISCRETE AMOEBA ALGORITHMS

In this section, we describe the construction of the image-adaptive structuring elements called *morphological amoebas*, which were introduced by Lerallut, Decencière, & Meyer (2005, 2007). Also, we describe median filtering algorithms that employ these structuring elements, which underlie the analysis in later sections.

2.1 Discrete Amoeba Construction

We start by describing the amoeba filter construction in the discrete form in which it is algorithmically realized, following Lerallut, Decencière, & Meyer (2005, 2007) with slight modifications.

Assume that we are given a digital image f whose pixels are indexed by some index set I. The gray-value of the pixel with index $i \in I$ is denoted by f_i, whereas its spatial coordinates are given by (x_i, y_i).

To construct a structuring element around a given pixel i_0 of that image, consider pixels i^* within a Euclidean neighborhood of radius ϱ. For each such pixel, consider paths $P = (i_0, i_1, \ldots, i_k \equiv i^*)$, which start at i_0 and end at i^*, and in which each two subsequent pixels i_j, i_{j+1} are adjacent in the discrete image grid, where adjacency can be defined either via a 4-neighborhood or via an 8-neighborhood. For each such path, we define its length $L(P)$ with respect to a *discrete amoeba metric* d, still to be specified, as

$$L(P) = \sum_{j=0}^{k-1} d\big((x_{i_j}, y_{i_j}, f_{i_j}), (x_{i_{j+1}}, y_{i_{j+1}}, f_{i_{j+1}})\big). \tag{7}$$

Pixel i^* is included in the structuring element if and only if there exists a path P from i_0 to i^* such that $L(P) \leq \varrho$.

The discrete amoeba metric d between adjacent pixels (x_i, y_i, f_i) and (x_j, y_j, f_j) is defined using both the *spatial distance* between the grid locations (x_i, y_i) and (x_j, y_j), as well as the contrast between the gray-values f_i, f_j, which also is called *tonal distance.*

In the original setting of Lerallut, Decencière, and Meyer (2005), only 4-neighborhoods in the grid are used such that the spatial distance is always 1. The discrete amoeba distance is then computed as an l_1 sum of spatial distance and the contrast rescaled by a parameter σ; i.e.,

$$d_L\big((x_i, y_i, f_i), (x_j, y_j, f_j)\big) = 1 + \sigma |f_i - f_j|. \tag{8}$$

In this setting, an 8-neighborhood is used, and distances in space are measured by the Euclidean distance $||(x_i, y_i) - (x_j, y_j)||_2$. More sophisticated choices would be possible in order to achieve better approximations of continuous arc-lengths; for example, see works on digital distance transforms by Borgefors (1986, 1996); Ikonen and Toivanen (2005), and Ikonen (2007), but also the exact Euclidean distance calculation by Fabbri *et al.* (2008). However, since the focus of this discussion is on theoretical analysis in a continuous setting rather than on discrete approximation, we do not detail these more elaborate approaches any further.

Again, we use a scaling factor σ to weight between the spatial and tonal distances, and combine both via a first-order homogeneous C^2 function $\varphi : {\mathbb{R}^+}^2 \to \mathbb{R}^+$ of two arguments that is increasing in both arguments and satisfies the triangle inequality $\varphi(s+v, t+w) \leq \varphi(s,t) + \varphi(v, w)$. In this way, we obtain as our discrete amoeba metric

$$d_\varphi\big((x_i, y_i, f_i), (x_j, y_j, f_j)\big) = \varphi\big(||(x_i, y_i) - (x_j, y_j)||_2, \ \sigma|f_i - f_j|\big). \tag{9}$$

One possible choice for φ, staying close to Lerallut, Decencière, and Meyer (2005), is the l_1 sum $\varphi_1(s,t) = ||(s,t)||_1 = s + t$. Another attractive possibility is given by the Pythagorean sum $\varphi_2(s,t) = ||(s,t)||_2 = \sqrt{s^2 + t^2}$. We will denote the resulting amoeba metrics as d_1 (for φ_1) and d_2 (for φ_2). These will be the two cases implemented in our experiments.

A further generalization of these two choices would lead to l_p sums $\varphi_p(s,t) = (s^p + t^p)^{1/p}$ for $p > 0$, whose limit case for $p \to \infty$ is $\varphi_\infty(s, t) = \max\{s, t\}$; however, this is no longer C^2.

2.2 Iterated Amoeba Median Filtering Algorithm

Using the previously constructed morphological amoebas as structuring elements, various filter operations can be applied to digital images. As our first and central candidate, we introduce iterated AMF.

Let us mention first that iterated filter application can be carried out in several ways. In the original study by Lerallut, Decencière, and Meyer (2005), a *pilot image* is used to steer a specific iterated process, as follows: First, a smoothed version of the input image f is used for computing the amoebas for all pixels. Then, median filtering is applied, with these amoebas as structuring elements. In subsequent iterations, new amoebas are computed from the previous *filtered* image. These amoebas then are used as structuring elements to refilter the *original* image f. For the purpose of our investigations, we employ instead a classic procedure for iterated AMF in the spirit of the works by Klette and Zamperoni (1996); Tukey (1971). In each iteration, the following two steps are carried out pixelwise on the previous *filtered* image: (1) amoeba construction, and (2) median filtering, using the amoeba as structuring element.

Conventional median filtering introduced by Tukey (1971) uses a fixed structuring element that is moved across the image as a sliding window. One filtering step consists in computing for each pixel i a filtered gray-value as the median of the gray-values of all pixels within the structuring element centered at pixel i. Median filtering is known as a robust denoising filter (Klette and Zamperoni, 1996; Dougherty and Astola, 1999) because it is less sensitive to outliers than smoothing filters based on averaging gray-values. A broader discussion of robustness has been given by Huber (1981). Additional reasons for the popularity of the median filter are its simplicity and its ability to preserve sharp edges in images.

AMF differs from conventional median filtering just by using morphological amoebas as introduced in the preceding section for the role of structuring elements. Like standard median filtering, this procedure can be iterated. The computation of amoebas has to be repeated for each iteration.

To state the algorithm, we introduce a sequence of images $u^{(n)}$, where $n = 0, 1, 2,...$ denotes the iteration number. The resulting algorithm reads as follows:

Algorithm for Iterated AMF

1. Let $u^{(0)} := f$ and $n = 0$.
2. Compute the amoeba structuring elements $\mathcal{A}_i$ for all pixels i from the image $u^{(n)}$.

3. For each pixel i, compute $u_i^{(n+1)} := \underset{j \in \mathcal{A}_i}{\text{med}}\, u_j^{(n)}$.
4. Increase n by 1. Stop if desired number of iterations is reached; otherwise, repeat from step 2.

For the median computation itself, efficient algorithms like the median-of-medians algorithm introduced by Blum *et al.* (1973) can be used. Its per-pixel complexity in this case is linear in the number of pixels in the structuring element. Note that in the classical case of fixed (sliding-window) structuring elements, the more efficient algorithms discussed by Weiss (2006) reduce the computational expense even more, such that the per-pixel complexity becomes logarithmic in the structuring element size. Future investigation may discover whether similar ideas also can be applied to achieve even more efficient AMF.

Iterated AMF, like classical iterated median filtering, behaves as a robust denoising filter with edge-preserving properties. However, the amoeba construction that governs the selection of pixels included in the median computation has the effect that in the vicinity of edges, pixels on the same side of the edge as the pixel being processed will dominate. This has two consequences. First, edges can even be enhanced in this way. Second, the displacement of edges is largely suppressed, keeping image structures in place even over multiple iterations of the filter.

2.3 Dilation, Erosion, and Quantiles

Dilation is a morphological operation that extends bright structures in the image. To this end, the new gray-value of each pixel is defined as the maximum of the old gray-values of all pixels in the structuring element centered at this pixel. Also, dilation filtering can be iterated. Multiple steps of dilation with a given structuring element are equivalent to a single step of dilation with a larger structuring element (for convex structuring elements, this is just an upscaled version of the small structuring element).

Similarly, erosion extends dark structures in the image by replacing the gray-value of each pixel with the minimum of old gray-values of all pixels within the structuring element.

It is straightforward to use amoebas as structuring elements for both dilation and erosion. The resulting structure-adaptive filters, amoeba dilation and amoeba erosion, extend bright or dark regions more slowly across high-contrast boundaries than do classic dilation and erosion. They are similar in spirit to anisotropic dilation and erosion approaches in the literature (e.g., Breuß, Burgeth, & Weickert 2007; Burgeth *et al.*, 2011).

The maximum, minimum, and median of a finite set of values can be considered as the 100%-, 0%-, and 50%-quantile of that set, respectively. Thus, filters that choose as the new gray-value of each pixel the q-quantile of the gray-values from the structuring element (where q is any prescribed value between 0 and 1) form a natural generalization of median filter, dilation, and erosion. We will call these filters *quantile filters;* see also Welk *et al.* (2007). As in the preceding cases, combination with amoeba-structuring elements into amoeba q-quantile filters is straightforward.

Algorithmically, for each pixel, another selection problem has to be solved, which can be treated using similar algorithms as for the median.

2.4 M-Smoothers

The median of a set $X = \{x_i \in \mathbb{R} \mid i = 1, 2, \ldots, n\}$ is also the minimizer of the sum of absolute differences to all given values; i.e.,

$$\mathrm{med}(X) = \underset{x \in \mathbb{R}}{\operatorname{argmin}} \sum_{i=1}^{n} |x - x_i| \,. \tag{10}$$

(In the case of an even number n of values, this minimizer is nonunique since the two middle values x_i and all values in between satisfy the minimization condition.)

It is also known that the average of the given data minimizes a similar objective function in which the differences $|x - x_i|$ are squared. This gives rise to the following generalization, which can be traced to Barral Souto (1938).

Definition 1 *For a set* $X = \{x_i \in \mathbb{R} | i = 1, 2, \ldots, n\}$ *and a real* $p > 0$, *define the quantity*

$$m_p(X) := \underset{x \in \mathbb{R}}{\operatorname{argmin}} \sum_{i=1}^{n} |x - x_i|^p. \tag{11}$$

Then, $m_p(X)$ *is called the* M-estimator *for X.*

The concept of M-estimators can be generalized by allowing other functions of $|x - x_i|$ than power functions in the objective function. We do not consider such generalizations here.

It is straightforward to construct image filters by applying M-estimators to gray-values within structuring elements. The filters obtained in this way are called *M-smoothers* (Torroba *et al.*, 1994; Winkler *et al.*, 1999). In this work, we will use morphological amoebas as structuring elements, thus obtaining amoeba M-smoothers.

For $p \geq 1$, the objective functions for M-estimators are convex and thereby easy to minimize. In the limit case $p \rightarrow \infty$, the M-estimator $m_p(X)$ converges to the mid-range value $\frac{1}{2}(\max(X) + \min(X))$, as noted by Barral Souto (1938).

M-estimators for $p < 1$ are algorithmically more challenging because they are not only nonconvex, but they even feature a local minimum at each single $x_i \in X$. This is caused by the kink of the root function $|x - x_i|^p$ at zero, where its derivative becomes infinite. In the limit case $p \rightarrow 0$, one is led to minimize the discontinuous function $\psi(x) = \sum_{i=1}^{n} d(x - x_i)$, where $d(y) = 0$ for $y = 0$, $d(y) = 1$; otherwise, $\psi(x) = \#\{i \in \{1, ..., n\} \mid x_i \neq x\}$. Its minimizer is obviously the most frequent value of X (i.e., its *mode*). In the case of a discrete set of distinct values, of course, each $x_i \in X$ is an equally good minimizer in the limit. One way to distinguish a unique minimizer for $p = 0$ is to consider the limit process $p \rightarrow 0$.

2.5 Multivariate Median Filters

Apart from gray-value images, multivariate images play an important role in image processing. This includes color images, but also vector or tensor fields that emerge from diffusion tensor imaging (to cite one example); see Pierpaoli *et al.* (1996).

The crucial point in transferring median filters to these sorts of images consists of defining medians for sets of multivariate data points. To this end, it is useful to take up the characterization of the median already mentioned in subsection 2.4 as the minimizer of the absolute distance sum to all given data points. Generalizing this property, the median of a set of points

$$X = \{x_i \in \mathbb{R}^d \mid i = 1, ..., n\} \tag{12}$$

in d-dimensional real vector space is defined as

$$\operatorname{med}(X) := \underset{x \in \mathbb{R}^d}{\operatorname{argmin}} \sum_{i=1}^{n} \| x - x_i \| \tag{13}$$

with a suitable vector norm $\| \bullet \|$. In our considerations, we will assume that $\| \bullet \|$ is the Euclidean norm in $\mathbb{R}^d$.

This notion of multivariate median has been introduced by Austin (1959); see also the follow-up by Seymour (1970) and discussion by Barnett (1976). In image processing, it has been advocated for color images by

Spence and Fancourt (2007), as well as for tensor field filtering (Welk *et al.*, 2003, 2007). In Welk *et al.* (2007), multivariate M-smoothers and quantile filters also were discussed.

For the computation of multivariate medians, an algorithm has already been proposed by Weiszfeld (1937); however, this can be trapped in nonminima in exceptional cases, as pointed out by Kuhn (1973). The algorithm proposed by Austin (1959) is different, but it also runs into problems in some configurations, as noted by Seymour (1970). These problems can be overcome by a convex programming approach (Welk *et al.*, 2007, Section 3.2).

We want to mention at this point a different notion of vector median that is occasionally proposed in the literature, which restricts the values to the set of input data (see Astola, Haavisto, & Neuvo 1990; Barni *et al.*, 2000; Caselles, Sapiro, and Chung 2000; Koschan and Abidi, 2001). While such a definition is advantageous from the algorithmic point of view, it fails to choose a good representative of the data set in cases in which no data point is located sufficiently well "in the middle" of the others, and it is even unstable in these situations. These shortcomings have been discussed (e.g., Spence & Fancourt, 2007; Welk *et al.*, 2007). Spence and Fancourt (2007) also demonstrates the superiority of the unrestricted minimization approach [Eq. (13)] in the case of color image denoising.

2.6 Amoeba Active Contour Algorithm

There are different ways to represent an evolving contour curve as needed in active contour methods (Caselles *et al.*, 1993; Malladi, Sethian, & Vemuri, 1993). Of course, a straightforward approach is a parametric curve represented by a sequence of sample points. While this proceeding is computationally efficient since the curve is represented as a one-dimensional (1-D) object, it bears some disadvantages: Not only are the sampling points in generally not aligned to the grid, thus necessitating interpolation. More problematically, the sampling density along the curve undergoes changes when the length of curve segments changes during the subsequent updates, leading to undersampling or oversampling of parts of the curve. Resampling steps, therefore, must be inserted to avoid imprecisions and artifacts. Last but not least, if segments with multiple connected components are to be handled, appropriate topology changes must be carried out explicitly.

In contrast, a *level-set approach* (Osher & Sethian, 1988) represents the curve implicitly as the zero-level set of some function over the two-dimensional (2-D) image domain. The latter function, the *level-set function* is the object whose updates are actually computed. Given some initial contour $c^{(0)}$, a suitable level-set function $u^{(0)}$ can be constructed as a *signed distance function*.

Then subsequent level-set functions $u^{(1)}$, $u^{(2)}$,... are computed from $u^{(0)}$; at a desired iteration number n, a contour $c^{(n)}$ can be extracted again as a zero-level set of $u^{(n)}$. In this proceeding, the sampling resolution remains fixed at the image resolution such that no resampling steps are necessary. Moreover, topology changes for multiple connected components are processed implicitly. On the other hand, a naive implementation involves a computation in the entire 2-D domain, whereas only the 1-D contour is of interest. This can often be mitigated by heuristics that restrict the actual computation to a narrow region around the actual contour, dynamically extending or clipping the level-set function as appropriate; see the narrow-band method by Adalsteinsson and Sethian (1995) for curve evolutions governed by PDEs.

The level-set approach is similar to the image filters considered previously because the level-set functions $u^{(n)}$ that are updated assume the role of the image being filtered. The difference to image filtering is that the image f itself remains unchanged, while the filtering of $u^{(n)}$ takes place under the influence of f.

An active contour method using AMF for the update of the level-set functions is the following *AAC algorithm*:

AAC Algorithm

1. For all pixels i, compute amoeba-structuring elements $\mathcal{A}_i$ from the input image f.
2. For the given initial contour, construct the level-set function $u^{(0)}$ as signed distance function.
3. For $n = 1, 2,...$, compute $u^{(n)}$ by median filtering $u^{(n-1)}$ using the structuring elements $\mathcal{A}_i$ from step 1.

Note that iterations of the AAC algorithm are computationally less expensive than those of iterated AMF since amoebas are computed only once for all iterations. For simplicity, we have formulated the algorithm here without any narrow-band strategy. Of course, such a strategy will speed up the computation significantly.

To introduce a force term (as mentioned in section 1 for GAC) into the AAC method, one can bias the filtering step. One possibility for doing this is to use a quantile filter instead of the median. Another possibility is to apply a fixed offset b, such that one selects the element with index $m/2 + b$ from the ordered sequence $g_0, \ldots, g_m$ of gray-values inside the amoeba instead of that with index $m/2$. Following Welk (2012), we will use the latter modification in one of our experiments. For an additional discussion of biasing, see also Welk (2013b).

3. CONTINUOUS AMOEBA FILTERING

In our analysis of amoeba filters, we are interested in the approximation of PDEs. To this end, we need a space-continuous formulation of amoeba filtering.

3.1 Continuous Amoeba Construction

Here, we consider a space-continuous image; i.e., a sufficiently smooth function $u : \Omega \to \mathbb{R}$. The amoeba construction then will rely on the representation of this image by its (rescaled) graph:

$$\Gamma(u) := \{(x, y, \sigma u(x, y)) | (x, y) \in \Omega\} \subseteq \mathbb{R}^3. \tag{14}$$

As mentioned previously, this embedding has been used in the image-processing context in the Beltrami framework by Kimmel *et al.* (1997) and Yezzi (1998).

On the 2-D manifold Γ embedded in $\mathbb{R}^3$, a continuous amoeba metric is constructed as an infinitesimal (Riemannian or Finslerian) metric ds^2. It is again combined with the standard (Euclidean) metric ds_Ω^2 on Ω and the standard metric $ds_\mathbb{R}^2$ on the gray-value range via

$$ds = \varphi(ds_\Omega, ds_\mathbb{R}), \tag{15}$$

where φ is as in subsection 2.1, such that the length of a curve $C : [0, 1] \to \Gamma$ is measured as

$$\begin{aligned} L(C) &= \int_0^1 \varphi\left(\sqrt{\left(\frac{dx(p)}{dp}\right)^2 + \left(\frac{dy(p)}{dp}\right)^2}, \sigma\left|\frac{du(x(p), y(p))}{dp}\right|\right) dp \\ &= \int_0^1 \varphi\left(\sqrt{x'(p)^2 + y'(p)^2}, \sigma|u_x x'(p) + u_y y'(p)|\right) dp, \end{aligned} \tag{16}$$

and the distance between two points $\boldsymbol{p}_1, \boldsymbol{p}_2 \in \Gamma$ is the length of the shortest curve between them:

$$d(\boldsymbol{p}_1, \boldsymbol{p}_2) = \min_{\substack{C:[0,1]\to\Gamma \\ C(0)=\boldsymbol{p}_1 \\ C(1)=\boldsymbol{p}_2}} L(C). \tag{17}$$

As in the discrete case, two choices of particular interest for φ are $\varphi_2(s,t) = \sqrt{s^2+t^2}$, for which the amoeba metric $d \equiv d_2$ is induced by the Euclidean metric of the embedding space $\mathbb{R}^3$, and $\varphi_1(s,t) = s + t$ with the resulting amoeba metric $d \equiv d_1$. Typical amoeba shapes for both cases are shown in Figure 1. Note that in the case of d_2, the contour of the amoeba is smooth. In contrast, for d_1, the contour has a digon-like overall shape, with kinks at the very points where it is hit by the level line through the point of reference (x_0, y_0).

Then an amoeba-structuring element $\mathcal{A}(x_0, y_0)$ is obtained by taking a closed ϱ-neighborhood of $(x_0, y_0, u(x_0, y_0))$ on Γ with regard to the metric ds, and projecting this set back onto Ω. Such neighborhoods have also been used in the construction of short-time Beltrami kernels by Spira, Kimmel, and Sochen (2007). By construction, the structuring elements are compact sets in the image domain Ω.

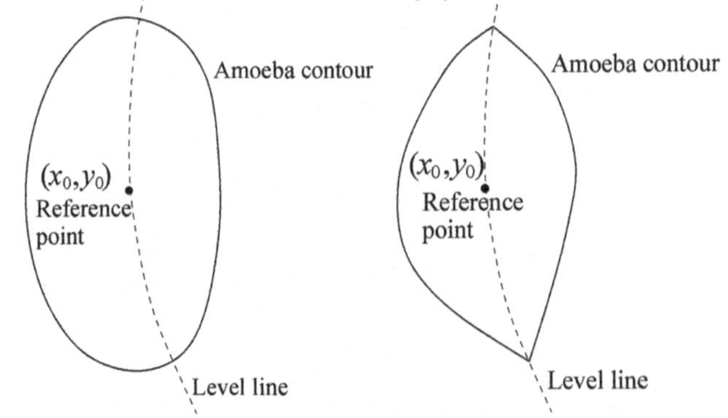

Figure 1 Amoeba structuring elements. (a) Typical amoeba with metric $d \equiv d_2$. (b) Typical amoeba with metric $d \equiv d_1$. *From Welk, Breuß, and Vogel (2011), ©Springer 2010. With kind permission from Springer Science and Business Media.*

3.2 Continuous Amoeba Median Filtering

Given the notion of space-continuous amoeba-structuring elements, the complete translation of AMF to a space-continuous setting requires as its second component the notion of the median for the distribution of gray-values within the structuring element. With the area measure of the image domain Ω underlying the distribution, one step of median filtering can be stated as follows: Given a smooth function $u : \Omega \rightarrow \mathbb{R}$, and compact structuring elements $\mathcal{A}(x_0, y_0)$ centered at all points $(x, y) \in \Omega$, assign to each location $(x_0, y_0) \in \Omega$ as its filtered function value the value $\mu := \widetilde{u}(x_0, y_0)$, for which the level line $\{(x, y) | u(x, y) = \mu\}$ cuts the amoeba $\mathcal{A} = \mathcal{A}(x_0, y_0)$ in half; i.e., for which the areas of $\mathcal{A}_{\mu,+} := \{(x, y) \in \mathcal{A} | u(x, y) \geq \mu\}$ and $\mathcal{A}_{\mu,-} := \{(x, y) \in \mathcal{A} | u(x, y) \leq \mu\}$ are equal.

This procedure can equally be applied in the settings of image filtering (see subsection 2.2) and of AAC (see subsection 2.6), just by assuming that the amoebas depend either on the function u itself or on a fixed function f, respectively.

Even if the obtained filter is space-continuous, it acts in discrete steps.

3.3 Continuous Dilation, Erosion, and Quantile Filters

In the same way as the distribution of function values within the amoeba gives rise to a median value, it possesses a minimum, maximum, and generally q-quantiles for $q \in [0,1]$. The q-quantile of function values within the structuring element $\mathcal{A}$ can be characterized by the splitting ratio in which the associated level line cuts $\mathcal{A}$; i.e., it will be the function value z for which the ratio of the areas of $\mathcal{A}_{z,+}$ and $\mathcal{A}_{z,-}$ is $(1 - q) : q$.

With these notions, the definitions of amoeba dilation, erosion, and quantile filtering transfer directly from the space-discrete to the space-continuous setting.

3.4 Continuous M-Smoothers

To establish space-continuous M-smoothers, one has to determine M-estimators of value distributions on the structuring elements. A continuous version of the M-estimators from subsection 2.4 reads as

$$m_p(u, \mathcal{A}) := \underset{z \in \mathbb{R}}{\operatorname{argmin}} \iint_{\mathcal{A}} |z - u(x, y)|^p \mathrm{d}x\mathrm{d}y. \tag{18}$$

As in the discrete case, the objective function is convex for $p \geq 1$, but not for $p < 1$. Nevertheless, for $p > 0$, the functions $|z - u|^p$ (which are not differentiable at $z = u$) will integrate into a differentiable objective function in generic cases. Thus, the objective function may have multiple minima, but in general, these are discrete points. In particular, unlike in the space-discrete setting, not every single function value occurring within the amoeba is a minimum. For a smooth function u and sufficiently small radius ϱ of the structuring elements, the minimizer can be expected to be unique.

3.5 Multivariate Median Filtering on a Continuous Domain

Like M-smoothers, multivariate median filters were introduced in subsection 2.5 by a minimization approach. Thus, translation to the space-continuous setting again consists of replacing discrete sums by integrals over structuring elements: Given a multivariate smooth function $\boldsymbol{u} : \Omega \rightarrow \mathbb{R}^d$, a compact structuring element $\mathcal{A} \subset \Omega$, and a vector norm $\|.\|$, the median of $\boldsymbol{u}$ in $\mathcal{A}$ is

$$\operatorname*{med}_{(x,y)\in\mathcal{A}} \boldsymbol{u}(x,y) := \operatorname*{argmin}_{\boldsymbol{z}\in\mathbb{R}^d} \iint\limits_{\mathcal{A}} \|\boldsymbol{z} - \boldsymbol{u}(x,y)\| \, dxdy. \tag{19}$$

A median filtering step for $\boldsymbol{u}$ assigns to each location $(x_0, y_0) \in \Omega$ as its new function value $[\widetilde{\boldsymbol{u}}(x_0, y_0)]$ the median of values within the respective amoeba; i.e.,

$$\widetilde{\boldsymbol{u}}(x_0, y_0) := \operatorname*{med}_{(x,y)\in\mathcal{A}(x_0,y_0)} u(x,y). \tag{20}$$

3.6 Amoeba Active Contours Filtering

It is straightforward to translate the AAC algorithm from subsection 2.6 to a space-continuous setting: Again, it comes down to just performing iterated amoeba median filtering, but with the amoeba structuring elements being derived from the fixed given image f, whereas a level-set function u is filtered.

Referring to the force-term modifications to AAC mentioned in subsection 2.6, the quantile filter translates straightforwardly into them, whereas the fixed index offset b corresponds to a fixed area difference between $\mathcal{A}_{\mu,+}$ and $\mathcal{A}_{\mu,-}$ from subsection 3.2.

4. SPACE-CONTINUOUS ANALYSIS OF AMOEBA FILTERS

In this section, we analyze the continuous versions of amoeba filters introduced in section 3. In particular, we derive PDEs that these filters approximate asymptotically for amoeba radius $\varrho \to 0$.

4.1 PDE for Amoeba Median Filtering

We start by analyzing iterated AMF. The following theorem states an approximation property similar to the seminal result by Guichard and Morel (1997), which links nonadaptive iterated median filtering to curvature motion.

Theorem 1 *For a smooth function u, one step of AMF with amoeba radius of ϱ asymptotically approximates for $\varrho \to 0$ a time step of size $\tau = \varrho^2/6$ of an explicit time discretization of the self-snakes PDE (Sapiro, 1996; Whitaker and Xue, 2001),*

$$u_t = |\nabla u| \operatorname{div}\left(g(|\nabla u|) \frac{\nabla u}{|\nabla u|} \right) = g(|\nabla u|) u_{\xi\xi} + \langle \nabla g(|\nabla u|), \nabla u \rangle \qquad (21)$$

with a decreasing edge-stopping function $g: \mathbb{R}_0^+ \to \mathbb{R}_0^+$ that depends on the particular choice of the amoeba metric.

For this theorem, we will provide two different proofs (one of them restricted to the L^2 amoeba metric given by $\varphi \equiv \varphi_2$), which use different approximation strategies and can be generalized in different ways.

In these proofs and later in this chapter, we will repeatedly use local coordinates adapted to the gradient and level-line directions of the evolving image u. At a given location $\boldsymbol{x}_0$ in the image domain, we will denote the normalized gradient vector $\nabla u(\boldsymbol{x}_0)/|\nabla u(\boldsymbol{x}_0)|$ *by* $\boldsymbol{\eta}$. The local level line direction of u then is indicated by the perpendicular unit vector $\boldsymbol{\xi} \perp \boldsymbol{\eta}$.

4.1.1 First Proof of Theorem 1

We start by presenting the proof given in Welk, Breuß, and Vogel (2011), with amended details in several steps.

We assume that the amoeba metric is of the type described by Eq. (15), and analyze the action of the amoeba median filter step at the location $\boldsymbol{x}_0 = (x_0, y_0) = (0, 0)$.

Without loss of generality, we may assume that σu possesses the Taylor expansion

$$\sigma u(x, y) = \alpha x + \beta x^2 + \gamma xy + \delta y^2 + \mathcal{O}(\varrho^3), \tag{22}$$

within the ϱ-neighborhood of $\boldsymbol{x}_0$; i.e., we have $u(0,0) = 0$, $\boldsymbol{\eta} = (1,0)^{\mathrm{T}}$, $\sigma u_x(0,0) = \alpha$, $\sigma u_{xx}(0,0) = \beta/2$, $\sigma u_{xy}(0,0) = \gamma$, and $\sigma u_{yy}(0,0) = \delta/2$. In particular, Eq. (22) holds throughout the sought amoeba $\mathcal{A} = \mathcal{A}(x_0, y_0)$.

Proof strategy. In the following, we will analyze the length and density of level lines for different values $z = \mathcal{O}(\varrho)$ of the function $\sigma\, u$ within the amoeba $\mathcal{A}$. These parameters together capture the density of the distribution of function values within the amoeba. Besides the median that is the focus of our attention here, further measures can be derived from this density, which will be used in later sections of this chapter.

To determine the median μ, we notice that μ is characterized by the property that the corresponding level line $\{\boldsymbol{x} | u(\boldsymbol{x}) = \mu\}$ cuts the amoeba $\mathcal{A} = \mathcal{A}(\boldsymbol{x}_0)$ into two equally large parts; i.e., the areas of $\mathcal{A}_{\mu,+} := \{\boldsymbol{x} \in \mathcal{A} | u(\boldsymbol{x}) \geq \mu\}$ and $\mathcal{A}_{\mu,-} := \{\boldsymbol{x} \in \mathcal{A} | u(\boldsymbol{x}) \leq \mu\}$ are equal:

$$|\mathcal{A}_{\mu,+}| = |\mathcal{A}_{\mu,-}|. \tag{23}$$

Note that a similar bisection argument was used in a segmentation context by Kimmel (2003).

Clearly, $\mathcal{A}_{\mu,-}$ and $\mathcal{A}_{\mu,+}$ are covered by the level lines of all z with $z < \sigma\mu$ and $z > \sigma\mu$, respectively. We can use the lengths $L(z)$ of level lines with their inverse density to express the area contributions. Assuming that the inverse density of level lines is approximately constant along a particular level line, it can be stated as a function $D(z)$. Then, Eq. (23) is rewritten as

$$\int_{Z_-}^{\sigma\mu} L(\zeta) D(\zeta) \mathrm{d}\zeta = \int_{\sigma\mu}^{Z_+} L(\zeta) D(\zeta) \mathrm{d}\zeta \tag{24}$$

where Z_- and Z_+ denote the smallest and largest function values of $\sigma\, u$ within the amoeba, respectively.

Length of a level line. Given a value $z = \mathcal{O}(\varrho)$, we are interested first in describing the corresponding level line of $\sigma\, u$ within $\mathcal{A}$; i.e., the curve on which $\sigma\, u(x, y) = z$ holds.

By our choice of local coordinates, the level lines of u within $\mathcal{A}$ are approximately oriented in the y-direction. The desired level line therefore can be described by using y as a parameter and x as a function of y. To obtain this function, we solve the equation $\sigma\, u(x, y) = z$ for x. The latter equation is in higher-order terms a quadratic equation with two solutions as follows:

$$x_{1,2} = \frac{1}{2\beta}\left(- \alpha - \gamma y \pm \sqrt{(\alpha + \gamma y)^2 - 4\beta(\delta y^2 - z)}\right) + \mathcal{O}(\varrho^3), \quad (25)$$

from which only the + solution needs further consideration because the − solution is bounded away from zero for small ϱ and therefore does not refer to a point within the amoeba. Taking into account the Taylor expansion of the square root function around 1,

$$\sqrt{1+v} = 1 + \frac{1}{2}v - \frac{1}{8}v^2 + \mathcal{O}(v^3), \quad (26)$$

we have

$$\begin{aligned} x &= \frac{1}{2\beta}\left(-\alpha - \gamma y + \alpha\sqrt{1 + \frac{2\gamma}{\alpha}y + \frac{4\beta}{\alpha^2}z + \frac{\gamma^2}{\alpha^2}y^2 - \frac{4\beta\delta}{\alpha^2}y^2}\right) + \mathcal{O}(\varrho^3) \\ &= \frac{1}{2\beta}\left(-\alpha - \gamma y + \alpha + \gamma y + \frac{2\beta z}{\alpha} + \frac{\gamma^2}{2\alpha}y^2 - \frac{2\beta\delta}{\alpha}y^2\right. \\ &\quad \left. - \frac{\gamma^2}{2\alpha}y^2 - \frac{2\beta\gamma}{\alpha^2}yz - \frac{2\beta^2}{\alpha^3}z^2\right) + \mathcal{O}(\varrho^3), \end{aligned} \quad (27)$$

and eventually the desired explicit representation of the level line:

$$x = x(y) = x(y, z) = \left(\frac{z}{\alpha} - \frac{z^2\beta}{\alpha^3}\right) - \frac{z\gamma}{\alpha^2}y - \frac{\delta}{\alpha}y^2 + \mathcal{O}(\varrho^3). \quad (28)$$

As pointed out previously, we are interested in the length of the level line segment within the amoeba $\mathcal{A}$. This segment is delimited by the two points that fulfill the condition $d_\varphi(p(x_0, y_0), p(x(y), y)) = \varrho$.

To evaluate this condition, we approximate the amoeba distance d_φ (the length of an arc on the image graph) by the distance in the embedding space $\mathbb{R}^3$ (the secant of this arc), which introduces a higher-order error that can be neglected. The condition then takes the form $\varphi\left(\sqrt{x(y)^2+y^2}, |z|\right) = \varrho$. Exploiting its homogeneity, φ can be rewritten as $\varphi(s, t) = t\psi(s/t)$, where $\psi(q) := \varphi(q, 1)$. Substituting this into the condition, we have $|z|\psi\left(\sqrt{x(y)^2+y^2}/|z|\right) = \varrho$, and therefore,

$$x(y)^2 + y^2 = z^2 \psi^{-2}\left(\frac{\varrho}{|z|}\right). \tag{29}$$

Note that ψ^{-2} in Eq. (29) denotes the square of the inverse function of ψ.

With Eq. (28) and $y,\ z = \mathcal{O}(\varrho)$, this becomes a quadratic equation for y,

$$\left(1 - \frac{2\delta}{\alpha^2} z\right) y^2 - \frac{2\gamma}{\alpha^3} z^2 y + \left(\frac{1}{\alpha^2} - \psi^{-2}\left(\frac{\varrho}{|z|}\right)\right) z^2 - \frac{2\beta}{\alpha^4} z^3 + \mathcal{O}(\varrho^4) = 0. \tag{30}$$

The two solutions $y_1,\ y_2$ of this equation fulfill $y_1, y_2 = \mathcal{O}(\varrho)$ and by Vieta's theorem, $(y_1 + y_2)/2 = \mathcal{O}(\varrho^2)$. The length $L(z)$ of the level line segment under consideration equals up to $\mathcal{O}(\varrho^3)$ the difference $|y_1 - y_2|$. This difference is calculated via the solution rule for quadratic equations. As a result, $L(z)$ equals up to an $\mathcal{O}(\varrho^3)$ summand the value

$$\widetilde{L}(z) = 2|z| \sqrt{\psi^{-2}\left(\frac{\varrho}{|z|}\right) - \frac{1}{\alpha^2}} \left(1 + \frac{\delta}{\alpha^2} z + \frac{\beta}{\alpha^4 \psi^{-2}\left(\frac{\varrho}{|z|}\right) - \alpha^2} z\right), \tag{31}$$

with $\widetilde{L}(0) = 2\varrho$ by continuity.

Density of level lines. The second ingredient needed to evaluate Eq. (24) is the inverse density of level lines. At the midpoint of the relevant level line segment, $(y_1 + y_2)/2 = \mathcal{O}(\varrho^2)$ (which is also used next), this inverse density is given by $\frac{\partial x}{\partial z}((y_1 + y_2)/2, z)$, and it deviates from that value along the entire level line only by higher-order terms, thus justifying the previous assumption that it is a function $D(z)$.

By Eq. (28), therefore, we have

$$D(z) = \frac{\partial x}{\partial z}\left(\frac{y_1 + y_2}{2}, z\right) = \frac{1}{\alpha} - \frac{2\beta}{\alpha^3} z + \mathcal{O}(\varrho^2). \tag{32}$$

Note that due to the averaging $(y_1 + y_2)/2$, the derivative of the summand $\delta y^2/\alpha$ from Eq. (28) contributes only to the $\mathcal{O}(\varrho^2)$ term.

The integration bounds Z_-, Z_+ in Eq. (24) are the largest negative and smallest positive zero of $L(z)$. Since $\alpha > 0$, they are easily found to fulfill

$$Z_+ = Z_* + \mathcal{O}(\varrho^3), \quad -Z_- = Z_* + \mathcal{O}(\varrho^3) \text{ where } Z_* = \frac{\varrho}{\psi(1/\alpha)}. \tag{33}$$

Abbreviating $\widehat{Z} := \min\{Z_+, Z_*\}$, we can rewrite the right side of Eq. (24) as

$$\int_{\sigma\mu}^{Z_+} L(\zeta)D(\zeta)\mathrm{d}\zeta = \int_0^{Z_*} \widetilde{L}(\zeta)D(\zeta)\mathrm{d}\zeta - \int_0^{\sigma\mu} \widetilde{L}(\zeta)D(\zeta)\mathrm{d}\zeta - \int_{\widehat{Z}}^{Z_*} \widetilde{L}(\zeta)D(\zeta)\mathrm{d}\zeta$$
$$+ \int_{\sigma\mu}^{\widehat{Z}} \left(L(\zeta) - \widetilde{L}(\zeta)\right)D(\zeta)\mathrm{d}\zeta + \int_{\widehat{Z}}^{Z_+} L(\zeta)D(\zeta)\mathrm{d}\zeta. \tag{34}$$

All but the first two integrals on the right side are $\mathcal{O}(\varrho^4)$: from the third and fifth integrals, one is zero (which one this is depends on whether $\widehat{Z}$ equals Z_* or Z_+), while the other one integrates an $\mathcal{O}(\varrho)$ integrand over a $\mathcal{O}(\varrho^3)$ interval; the fourth integral has an $\mathcal{O}(\varrho^3)$ integrand and a $\mathcal{O}(\varrho)$ interval.

Noticing further that the second integral equals $\sigma\mu\widetilde{L}(0)D(0)$ up to $\mathcal{O}(\varrho^4)$, we have

$$\int_{\sigma\mu}^{Z_+} L(\zeta)D(\zeta)\mathrm{d}\zeta = \int_0^{Z_*} \widetilde{L}(\zeta)D(\zeta)\mathrm{d}\zeta - \sigma\mu\,\widetilde{L}(0)D(0) + \mathcal{O}(\varrho^4). \tag{35}$$

Proceeding similarly with the left side of Eq. (24) and substituting $-\zeta$ for ζ in the integral, we obtain

$$\int_{Z_-}^{\sigma\mu} L(\zeta)D(\zeta)\mathrm{d}\zeta = \int_0^{Z_*} \widetilde{L}(-\zeta)D(-\zeta)\mathrm{d}\zeta + \sigma\mu\,\widetilde{L}(0)D(0) + \mathcal{O}(\varrho^4), \tag{36}$$

and after combining both sides, eventually we get

$$\int_0^{Z_*} \left(\widetilde{L}(\zeta)D(\zeta) - \widetilde{L}(-\zeta)D(-\zeta)\right)\mathrm{d}\zeta = 2\sigma\mu\, L(0)D(0) + \mathcal{O}(\varrho^4). \tag{37}$$

Using Eq. (32) and $D(0) = 1/\alpha$, this can be rewritten as

$$\int_0^{Z_*} \left(\widetilde{L}(\zeta)\left(1 - \frac{2\beta}{\alpha^2}\zeta\right) - \widetilde{L}(-\zeta)\left(1 + \frac{2\beta}{\alpha^2}\zeta\right)\right) \mathrm{d}\zeta = 2\sigma\mu\widetilde{L}(0) + \mathcal{O}(\varrho^4). \tag{38}$$

After inserting Eq. (31), the left side of Eq. (38) becomes up to terms of higher order,

$$\begin{aligned} &\frac{4(\delta - 2\beta)}{\alpha^2} \int_0^{Z_*} \zeta^2 \sqrt{\psi^{-2}\left(\frac{\varrho}{|\zeta|}\right) - \frac{1}{\alpha^2}}\,\mathrm{d}\zeta + \frac{4\beta}{\alpha^4} \int_0^{Z_*} \frac{\zeta^2}{\sqrt{\psi^{-2}\left(\frac{\varrho}{|\zeta|}\right) - \frac{1}{\alpha^2}}}\,\mathrm{d}\zeta \\ &= \frac{4\varrho^3(\delta - 2\beta)}{\alpha^2\psi^3(1/\alpha)} I_1(\alpha) + \frac{4\varrho^3\beta}{\alpha^4\psi^3(1/\alpha)} I_2(\alpha), \end{aligned} \tag{39}$$

where I_1 and I_2 are functions of α only. They are given by the integrals

$$I_1(\alpha) := \int_0^1 \xi^2 R_\psi(\xi, \alpha)\,\mathrm{d}\xi \tag{40}$$

$$I_2(\alpha) := \int_0^1 \frac{\xi^2}{R_\psi(\xi, \alpha)}\,\mathrm{d}\xi, \tag{41}$$

both of which depend on

$$R_\psi(\xi, \alpha) := \sqrt{\psi^{-2}\left(\frac{1}{\xi}\psi\left(\frac{1}{\alpha}\right)\right) - \frac{1}{\alpha^2}}. \tag{42}$$

Recalling $\widetilde{L}(0) = 2\varrho$, it follows that

$$\mu = \frac{\varrho^2}{3\sigma}\left(\frac{3\delta I_1(\alpha)}{\alpha^2\psi^3(1/\alpha)} + \frac{3\beta\big(I_2(\alpha) - 2\alpha^2 I_1(\alpha)\big)}{\alpha^4\psi^3(1/\alpha)}\right) + \mathcal{O}(\varrho^3). \tag{43}$$

Translating back the coefficients of the Taylor expansion [Eq. (22)] occurring in Eq. (43) to spatial derivatives of σ u, the median turns out to be

$$\mu = \frac{\varrho^2}{6}\left(\frac{3u_{yy}I_1(\sigma\, u_x)}{(\sigma\, u_x)^2\psi^3(\sigma^{-1}u_x^{-1})} + \frac{3u_{xx}\left(I_2(\sigma\, u_x) - 2(\sigma\, u_x)^2 I_1(\sigma\, u_x)\right)}{(\sigma\, u_x)^4\psi^3(\sigma^{-1}u_x^{-1})}\right.$$
$$\left. + \mathcal{O}(\varrho)\right). \tag{44}$$

Lifting the constraint $u(x_0, y_0) = 0$, the value $u(x_0, y_0)$ appears as an additional summand on the right side. Interpreting $\varrho^2/6$ as time-step size and the term in the bracket as u_t, we see that indeed, one step of AMF approximates a time step of an explicit scheme for a PDE. When ϱ tends to zero, iterated AMF converges to this PDE. Rewritten in the geometric $\boldsymbol{\xi}$, $\boldsymbol{\eta}$ coordinates, the PDE reads as

$$u_t = g(|\nabla u|)u_{\xi\xi} + h(|\nabla u|)u_{\eta\eta}, \tag{45}$$

with the functions g and h given by

$$g(s) = \frac{3I_1(\sigma s)}{(\sigma s)^2\psi^3(1/(\sigma s))}, \tag{46}$$

$$h(s) = \frac{3I_2(\sigma s)}{(\sigma s)^4\psi^3(1/(\sigma s))} - 2g(s). \tag{47}$$

Here, g and h act as weight functions controlling the influence of the second derivatives of u in the gradient direction $\boldsymbol{\eta}$ and level line direction $\boldsymbol{\xi}$. The first summand can already be interpreted straightforwardly as the right side of curvature motion $u_t = u_{\eta\eta}$ multiplied by the edge-stopping function $g(|\nabla u|)$. For the second summand, however, we have to take into account that h takes negative values. Thus, it corresponds rather to a 1-D backward diffusion in gradient direction; i.e., the inverse of the AMLE evolution investigated by Caselles, Morel, and Sbert (1998).

As the first summand already agrees with the claim of Theorem 1, it remains to investigate the second summand. To verify the theorem, we need to show that $h(s)$ is the same function as $s\,g'(s)$.

For general functions ψ, the integrals I_1 and I_2 may not be solvable in closed form. Moreover, the integrands have a (weak) singularity at $\xi = 1$, which must be taken care of in numerical computations as well as in further analysis.

By a calculation detailed in section A.1 in the Appendix, one can show the equality as

$$sg'(s) - h(s) = \frac{3\psi'\left(\frac{1}{\sigma s}\right)}{(\sigma s)^3 \psi^4\left(\frac{1}{\sigma s}\right)} I_3(\sigma s). \tag{48}$$

Herein, I_3 denotes the integral

$$I_3(\alpha) := \int_0^1 \left(3\xi^2 R_\psi(\xi, \alpha) - \frac{\xi \psi^{-1}\left(\frac{1}{\xi}\psi\left(\frac{1}{\alpha}\right)\right)\psi\left(\frac{1}{\alpha}\right)}{\psi'\left(\psi^{-1}\left(\frac{1}{\xi}\psi\left(\frac{1}{\alpha}\right)\right)\right) R_\psi(\xi, \alpha)} \right) \mathrm{d}\xi. \tag{49}$$

By direct calculation, one can verify that the derivative of $\xi^3 R_\psi(\xi, \alpha)$ with respect to ξ equals the integrand on the right side; thus, $I_3(\alpha) = R_\psi(1, \alpha)$. Now it can be read from the definition [Eq. (42)] of R_ψ that $R_\psi(1, \alpha) = \sqrt{\psi^{-2}(\psi(1/\alpha)) - 1/\alpha^2} = \sqrt{1/\alpha^2 - 1/\alpha^2} = 0$, which means that $I_3(\alpha) = 0$, and indeed, $h(s) = sg'(s)$ holds. This completes the proof of Theorem 1.

Remark. The same technique as in this proof was used in Welk (2012) to study AAC in the special case of a rotationally symmetric input image f and evolving level-set function u. The following result was proved in Welk (2012).

Theorem 2 *Consider AMF of a smooth function u in the plane, which is rotationally symmetric with respect to the origin $(0, 0)$. Assume that the amoeba-structuring elements are generated from a possibly different function f, which is also rotationally symmetric with respect to $(0, 0)$, with an amoeba radius of ϱ. Then one step of the amoeba median filter asymptotically approximates for $\varrho \to 0$ a time step of size $\tau = \varrho^2/6$ of an explicit time discretization for the GAC PDE*

$$u_t = \frac{u_{\xi\xi}}{1 + |\nabla f|^2} - \frac{2 f_{\eta\eta} |\nabla u|\,|\nabla f|}{\left(1 + |\nabla f|^2\right)^2} = |\nabla u| \operatorname{div}\left(\frac{1}{1 + |\nabla f|^2} \frac{\nabla u}{|\nabla u|} \right). \tag{50}$$

4.1.2 Second Proof of Theorem 1

The principle of the second proof goes back to Welk (2013a).

Proof strategy. In contrast to the first proof, we do not investigate in this second proof individual level lines, except for the one through the amoeba center. Instead, the amoeba is represented in a polar coordinate representation in order to study its deviations from symmetry (note that for sufficiently small ϱ, the amoeba is convex). This means that the amoeba is cut into *sectors* instead of *segments.*

For convenience, we formulate this second proof only for the L^2 amoeba metric; i.e., Eq. (15) with $\varphi \equiv \varphi_2$.

Description of the amoeba contour. The first step is to describe the shape of the amoeba $\mathcal{A} := \mathcal{A}(\boldsymbol{x}_0)$ centered at a point $\boldsymbol{x}_0 \in \Omega$ in a polar coordinate representation. To this end, we look first at a 1-D version of the problem. In this case, a function $u : \mathbb{R} \to \mathbb{R}$ is given, and we are interested in $r_\pm \in \mathbb{R}$, for which the image graph of u between x_0 and $x_0 + r_+$ has arc length ϱ, and the same is true for the arc length between x_0 and $x_0 - r_-$. Clearly, $r_\pm \le \varrho$.

By Taylor expansions of u around x_0 and of the square root function around 1, the arc length between x_0 and $x_0 + r$ (with $r > 0$) is given by

$$\int_{x_0}^{x_0+r} \sqrt{1+\sigma^2 u'(x)^2}\,\mathrm{d}x = r\sqrt{1+\sigma^2 u'(x_0)^2} + \frac{r^2}{2}\frac{\sigma^2 u'(x_0)u''(x_0)}{\sqrt{1+\sigma^2 u'(x_0)^2}} + \mathcal{O}(\varrho^3). \tag{51}$$

Equating this expression to ϱ yields a quadratic equation for r. Its solutions are

$$r_{1,2} = \frac{1+\sigma^2 u'(x_0)^2}{\sigma^2 u'(x_0)u''(x_0)}\left(-1 \pm \sqrt{1+2\varrho\frac{\sigma^2 u'(x_0)u''(x_0)}{(1+\sigma^2 u'(x_0))^{3/2}}}\right) + \mathcal{O}(\varrho^3), \tag{52}$$

from which only the + case is relevant because of $r > 0$. This case gives the desired value r_+. In the same way, r_- is derived. After rewriting once more the square root function by its Taylor expansion, both results can be combined in the equation

$$r_\pm = \frac{\varrho}{\sqrt{1+\sigma^2 u'(x_0)^2}} \mp \frac{\varrho^2\sigma^2 u'(x_0)u''(x_0)}{2\left(1+\sigma^2 u'(x_0)^2\right)^2} + \mathcal{O}(\varrho^3) \tag{53}$$

In the 2-D case, each shortest path in the amoeba metric from x_0 to a contour point of the amoeba differs from a Euclidean straight line in the image plane by not more than an $\mathcal{O}(\varrho^3)$ error. Consider now such a straight line through $\boldsymbol{x}_0$, whose direction is given by a unit vector $\boldsymbol{v}$. Applying Eq. (53) on this line gives its intersection points with the amoeba contour as

$$r_{\pm}(\boldsymbol{v}) = \frac{\varrho}{\sqrt{1+\sigma^2\langle\boldsymbol{v},\nabla u(\boldsymbol{x}_0)\rangle^2}} \mp \frac{\varrho^2\sigma^2\langle\boldsymbol{v},\nabla u(\boldsymbol{x}_0)\rangle\boldsymbol{v}^T\mathrm{D}^2u(\boldsymbol{x}_0)\boldsymbol{v}}{2\left(1+\sigma^2\langle\boldsymbol{v},\nabla u(\boldsymbol{x}_0)\rangle^2\right)^2} + \mathcal{O}(\varrho^3); \tag{54}$$

note that the first and second directional derivatives of u in direction $\boldsymbol{v}$ read as $\partial_{\boldsymbol{v}}u(\boldsymbol{x}_0) = \langle\boldsymbol{v},\nabla u(\boldsymbol{x}_0)\rangle$ and $\partial_{\boldsymbol{v}\boldsymbol{v}}u(\boldsymbol{x}_0) = \boldsymbol{v}^T\mathrm{D}^2u(\boldsymbol{x}_0)\boldsymbol{v}$.

Two contributions to the amoeba median. As in the first proof, we start from the bisection property [Eq. (23)] of the sought median. For convenience, assume again that $u(\boldsymbol{x}_0) = 0$.

An obvious special case is when (1) $\mathcal{A}$ is point-symmetric with regard to its reference point $\boldsymbol{x}_0$, and (2) u has straight level lines. In this situation, the level line $u(\boldsymbol{x}) = 0$ through the central point cuts the amoeba into two parts, $\mathcal{A}_{0,+}$ and $\mathcal{A}_{0,-}$, that are even congruent, entailing $\mu = 0$. Deviations from both prerequisites (1) and (2) create area differences between $\mathcal{A}_{0,+}$ and $\mathcal{A}_{0,-}$.

The level line $u(\boldsymbol{x}) = \mu$ for the median is approximately obtained by a small shift of the level line $u(\boldsymbol{x}) = 0$. Assuming that $\mu > 0$, $\mathcal{A}_{\mu,+}$ is $\mathcal{A}_{0,+}$ minus the small stripe between the two level lines, whereas $\mathcal{A}_{\mu,-}$ is $\mathcal{A}_{0,-}$ plus the same stripe. The double area of the stripe must compensate for the area difference between $\mathcal{A}_{0,+}$ and $\mathcal{A}_{0,-}$.

Turning to analyze the area differences caused by deviations from (1) and (2), we notice first that each of these effects is of order $\mathcal{O}(\varrho^3)$, while cross-effects

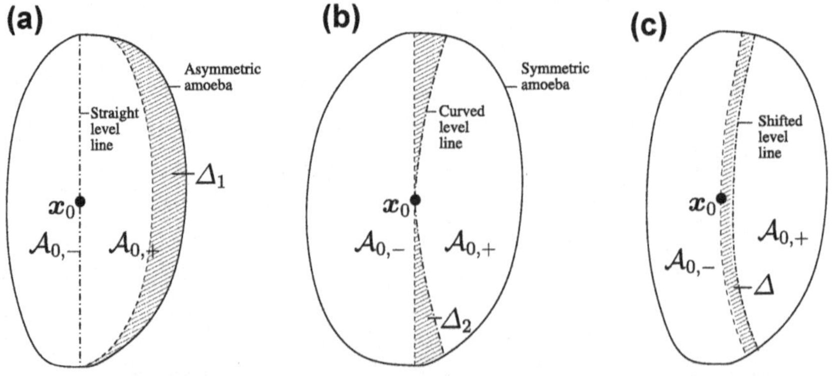

Figure 2 Left to right: (a) Area difference Δ_1 in an asymmetric amoeba with straight level lines. (b) Area difference Δ_2 in a symmetric amoeba with curved level lines. (c) Compensation of the area difference Δ by shifting the central level line (schematic). Adapted from Welk (2013a).

between the two are of order $\mathcal{O}(\varrho^4)$ or even higher, and thus negligible for this analysis. Therefore, we consider both contributions separately.

Asymmetry of amoeba shape. The first contribution to be analyzed comes from the asymmetry of the structuring element $\mathcal{A}$; see Figure 2(a). Assuming that the unit vector $\boldsymbol{\eta}$ in gradient direction of u is given by $\boldsymbol{\eta} = (\cos\varphi, \sin\varphi)^{\mathrm{T}}$, we can rewrite the first and second derivatives of u at the point $\boldsymbol{x}_0$ in the direction of an arbitrary unit vector $\boldsymbol{v} = (\cos(\varphi + \vartheta), \sin(\varphi + \vartheta))^{\mathrm{T}}$ in terms of derivatives in the $\boldsymbol{\xi}, \boldsymbol{\eta}$ local coordinates as

$$u_{\boldsymbol{v}}(\boldsymbol{x}_0) = \langle \boldsymbol{v}, \nabla u(\boldsymbol{x}_0)\rangle = |\nabla u(\boldsymbol{x}_0)| \cos\vartheta, \tag{55}$$

$$u_{\boldsymbol{vv}}(\boldsymbol{x}_0) = \boldsymbol{v}^{\mathrm{T}}\mathrm{D}^2 u(\boldsymbol{x}_0)\boldsymbol{v} = u_{\xi\xi}\sin^2\vartheta + 2u_{\xi\eta}\cos\vartheta\sin\vartheta + u_{\eta\eta}\cos^2\vartheta. \tag{56}$$

Inserting these equations into Eq. (54) yields

$$r_\pm(\boldsymbol{v}) = \frac{\varrho}{\sqrt{1 + \sigma^2|\nabla u(\boldsymbol{x}_0)|^2\cos^2\vartheta}} \mp \frac{\varrho^2\sigma^2|\nabla u(\boldsymbol{x}_0)|\cos\vartheta\left(u_{\xi\xi}\sin^2\vartheta + 2u_{\xi\eta}\cos\vartheta\sin\vartheta + u_{\eta\eta}\cos^2\vartheta\right)}{2\left(1 + \sigma^2|\nabla u(\boldsymbol{x}_0)|^2\cos^2\vartheta\right)^2} + \mathcal{O}(\varrho^3). \tag{57}$$

For convenience, we will also denote $r_\pm(\boldsymbol{v})$ by $r_\pm(\varphi + \vartheta)$ as follows. With the angle as the parameter, $r_\pm$ is a polar coordinate representation of the amoeba shape.

Assume for now that the level lines of u are straight, such that φ is the direction angle of the gradient direction throughout the amoeba. Then the area difference between the two parts of the amoeba is obtained via two standard area integrals in polar coordinates:

$$\begin{aligned}\Delta_1 := |\mathcal{A}_{0,+}| - |\mathcal{A}_{0,-}| &= \int_{\varphi-\pi/2}^{\varphi+\pi/2} \frac{1}{2}(r_+(\vartheta))^2\mathrm{d}\vartheta - \int_{\varphi-\pi/2}^{\varphi+\pi/2} \frac{1}{2}(r_-(\vartheta))^2\mathrm{d}\vartheta + \mathcal{O}(\varrho^4)\\ &= \int_{\varphi-\pi/2}^{\varphi+\pi/2} (r_+(\vartheta) - r_-(\vartheta))\frac{r_+(\vartheta) + r_-(\vartheta)}{2}\mathrm{d}\vartheta + \mathcal{O}(\varrho^4),\end{aligned} \tag{58}$$

where the last integral is equal to

$$-\varrho^3\sigma^2|\nabla u|\int\limits_{-\pi/2}^{+\pi/2}\frac{u_{\xi\xi}\cos\vartheta\sin^2\vartheta+2u_{\xi\eta}\cos^2\vartheta\sin\vartheta+u_{\eta\eta}\cos^3\vartheta}{\left(1+\sigma^2|\nabla u|^2\cos^2\vartheta\right)^{5/2}}\,\mathrm{d}\vartheta \tag{59}$$

and finally evaluates to

$$-\frac{2}{3}\varrho^3\sigma^2|\nabla u|\left(\frac{u_{\xi\xi}}{1+\sigma^2|\nabla u|^2}+\frac{2u_{\eta\eta}}{\left(1+\sigma^2|\nabla u|^2\right)^2}\right). \tag{60}$$

Curvature of level lines. The second contribution to the area difference between $\mathcal{A}_{0,+}$ and $\mathcal{A}_{0,-}$ originates from the curved shape of the level line of u through $\boldsymbol{x}_0$; see Figure 2(b). The resulting area difference is twice the area ringed by the level line and its tangent at $\boldsymbol{x}_0$. In local $\boldsymbol{\xi}$, $\boldsymbol{\eta}$ coordinates the tangent is the $\boldsymbol{\xi}$-axis, while the level line is, up to higher-order terms, a parabola with apex $\boldsymbol{x}_0$ and curvature $u_{\xi\xi}/(2|\nabla u|)$. Assuming a symmetric amoeba, the area difference thus equals

$$\Delta_2 := |\mathcal{A}_{0,+}|-|\mathcal{A}_{0,-}| = -2\int\limits_{-r_-(\varphi+\pi/2)}^{r_+(\varphi+\pi/2)} -\frac{u_{\xi\xi}}{2|\nabla u|}z^2\mathrm{d}z+\mathcal{O}(\varrho^4)$$
$$=\frac{2}{3}\varrho^3\frac{u_{\xi\xi}}{|\nabla u|}+\mathcal{O}(\varrho^4). \tag{61}$$

Median calculation. For the sought median μ of u within the amoeba $\mathcal{A}$, the corresponding level line $u(x,y)=\mu$ is approximately a copy of the level line $u(x,y)=0$ shifted by $\mu/|\nabla u(\boldsymbol{x}_0)|$ in the $\boldsymbol{\eta}$ (gradient) direction. The area of $\mathcal{A}_{\mu,+}$ is the area of $\mathcal{A}_{0,+}$ with the stripe between the two level lines removed, while the area of $\mathcal{A}_{\mu,-}$ corresponds to that of $\mathcal{A}_{0,-}$ with the same stripe added; see Figure 2(c). The double area of the stripe therefore must compensate for the area imbalance between $\mathcal{A}_{0,+}$ and $\mathcal{A}_{0,-}$, whose contributions $\boldsymbol{\Delta}_1$ and $\boldsymbol{\Delta}_2$ have been calculated before; i.e.,

$$2\frac{\mu}{|\nabla u|}\cdot(r_+(\varphi+\pi/2)+r_-(\varphi+\pi/2))=\Delta_1+\Delta_2+\mathcal{O}(\varrho^4). \tag{62}$$

Inserting Eqs. (60) and (61) and $r_+(\varphi+\pi/2)+r_-(\varphi+\pi/2) = 2\varrho+\mathcal{O}(\varrho^2)$ leads to

$$\mu=\frac{\varrho^2}{6}\left(\frac{u_{\xi\xi}}{1+\sigma^2|\nabla u(\boldsymbol{x}_0)|^2}-\frac{2\sigma^2|\nabla u(\boldsymbol{x}_0)|^2u_{\eta\eta}}{\left(1+\sigma^2|\nabla u(\boldsymbol{x}_0)|^2\right)^2}\right)+\mathcal{O}(\varrho^3). \tag{63}$$

Abandoning the condition that $u(\boldsymbol{x}_0)=0$ leads to an additional summand $u(\boldsymbol{x}_0)$ on the right side. Interpreting the term in the bracket as u_t, the claim of Theorem 1 with $g(s^2) = 1/(1 + \sigma^2 s^2)$ follows.

Remark. An advantage of the approach of the second proof is that the abundant number of Taylor coefficients to be considered in more complex situations is reduced. In fact, in the generalization to AAC, this approach turns out to be much easier to manage than the one from the first proof. In Welk (2013a), the following fact about PDE approximation by the AAC evolution has been proven.

Theorem 3 *One step of AMF of a smooth function u, where the amoeba structuring elements are generated from a possibly different function f with an amoeba radius of* ϱ*, asymptotically approximates for* $\varrho\to 0$ *a time step of size* $\tau=\varrho^2/6$ *of an explicit time discretization for the PDE*

$$\begin{aligned}u_t={}&\frac{u_{\xi\xi}}{1+|\nabla f|^2\sin^2\alpha}-\frac{|\nabla f||\nabla u|}{1+|\nabla f|^2\sin^2\alpha}\cdot\left(\frac{f_{\zeta\zeta}\cos^3\alpha}{1+|\nabla f|^2}\right.\\&\left.+2f_{\zeta\chi}\sin^3\alpha+\frac{f_{\chi\chi}\cos\alpha\left(2+\sin^2\alpha+3|\nabla f|^2\sin^2\alpha\right)}{\left(1+|\nabla f|^2\right)^2}\right),\end{aligned}\tag{64}$$

where $\boldsymbol{\chi}=\nabla f/|\nabla f|$ *and* $\boldsymbol{\zeta}\perp\boldsymbol{\chi}$ *denote unit vectors in the local gradient and level line direction of f, respectively, and* α *is the angle between the gradient directions of u and f.*

A generalized PDE approximation result for AAC with arbitrary amoeba metrics is found in Welk (2013b).

On the other hand, a disadvantage of the second proof is that it is less flexible in adapting to related filters like quantiles and M-smoothers because level lines other than the central one are no longer directly accessible.

4.2 Discussion of Amoeba Metrics

The equivalence result stated in Theorem 1 has been proven (by the first proof) for a general amoeba metric of the type in Eq. (15). The specific edge-stopping function $g(s)$ in the PDE [Eq. (21)] depends on the underlying amoeba metric according to Eq. (46), and thus, via the integral I_1 from Eq. (40).

Similarly, in the representation of Eq. (21) in the form of Eq. (45), the weight function $h(s) = sg'(s)$ that governs the edge-enhancing term is given by Eq. (47) and involves the integral I_2 from Eq. (41).

As mentioned previously, for amoeba metrics [Eq. (15)] with general functions φ, the integrals I_1 and I_2 will not be solvable in closed form. In order to solve Eq. (21) numerically, one has to compute I_1, I_2 and the resulting weights g and h numerically, taking into account the singularity of the integrands at $\xi = 1$.

We believe that the L^2 amoeba metric ($\varphi \equiv \varphi_2$) that was considered in the second proof and the L^1 amoeba metric ($\varphi \equiv \varphi_1$) constitute the most interesting cases. Next, we discuss these two amoeba metrics and corresponding edge-stopping functions in more detail.

L^2 amoeba metric. For this case, we have $\psi(v) = \sqrt{1 + v^2}$, such that the integrals I_1 and I_2 can be solved in closed form:

$$I_1(\alpha) = \sqrt{1 + \frac{1}{\alpha^2}} \int_0^1 \xi\sqrt{1 - \xi^2}\mathrm{d}\xi = \frac{1}{3}\sqrt{1 + \frac{1}{\alpha^2}}, \tag{65}$$

$$I_2(\alpha) = \frac{1}{\sqrt{1 + \frac{1}{\alpha^2}}} \int_0^1 \frac{\xi^3}{\sqrt{1 - \xi^2}}\mathrm{d}\xi = \frac{2}{3\sqrt{1 + \frac{1}{\alpha^2}}}. \tag{66}$$

This leads to the weight functions

$$g(s) = \frac{1}{1 + \sigma^2 s^2}, \qquad h(s) = \frac{-2\sigma^2 s^2}{(1 + \sigma^2 s^2)^2}, \tag{67}$$

as shown in Figure 3(a), which are also in agreement with the findings from the second proof in subsection 4.1. Here, $g(s)$ is an edge-stopping function of the Perona-Malik type [introduced by Perona and Malik (1990), with $1/\sigma$ chosen for the threshold] that is also a popular choice with self-snakes.

L^1 amoeba metric. If we set instead $\varphi \equiv \varphi_1$, and therefore $\psi(v) = 1 + v$, the integrals I_1 and I_2 do not reduce as nicely as before. Numerical evaluation leads to the weight functions g, h displayed in Figure 3(b). In contrast to the L^2 case, we notice a faster decay of the edge-stopping function g, which has a negative derivative at its starting point,

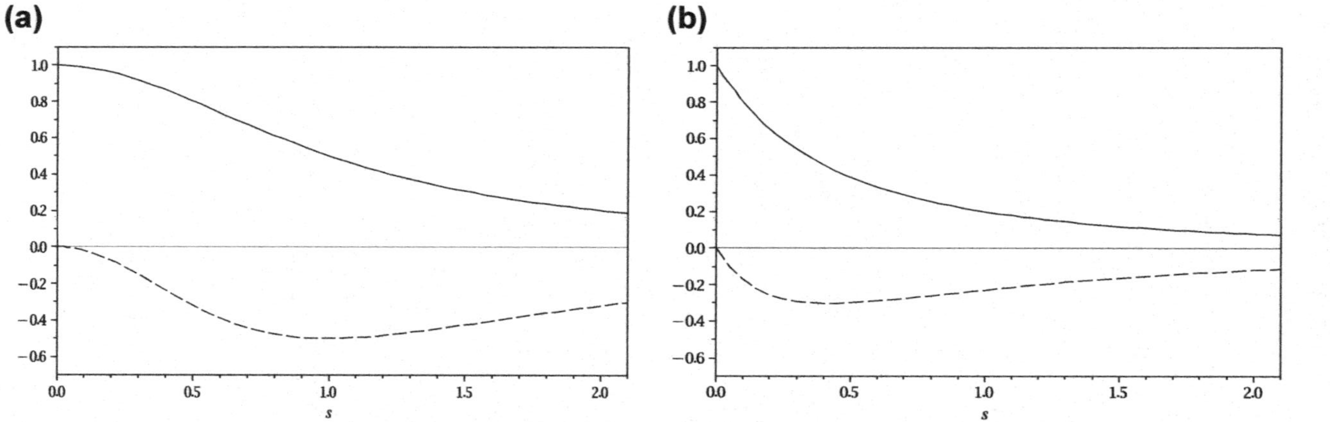

Figure 3 Edge-stopping functions in PDEs approximated by iterated AMF. For visualization, σ is fixed at a value of 1. (a) Weight functions $g(s) = 1/(1+s^2)$ for the curvature motion term (solid line), $h(s) = -2s^2/(1+s^2)^2$ for the shock term (dashed line) from the PDE [Eq. (45)] using the Euclidean amoeba metric given by Eq. (15) with $\varphi \equiv \varphi_2$. (b) Corresponding weight functions for the L^1 amoeba metric, $\varphi \equiv \varphi_1$. *From Welk, Breuß, and Vogel (2011), ©Springer 2010. With kind permission from Springer Science and Business Media.*

$g(0) = 1$. This implies that the dampening effect of even a small amount of image contrast on the evolution speed of level lines is more pronounced.

4.3 Analysis of Amoeba Dilation, Erosion, and Quantile Filters

In this subsection, we first prove an approximation theorem for amoeba quantile filtering.

Theorem 4 *For a smooth function u, one step of amoeba quantile filtering with parameter $q \neq 0.5$ with amoeba radius ϱ and L^2 amoeba metric asymptotically approximates for $\varrho \to 0$ a time step of size $\tau = \varrho\left|\theta^{-1}(2q-1)\right|$ of an explicit time discretization of the PDE*

$$u_t = \frac{\pm|\nabla u|}{\sqrt{1+\sigma^2|\nabla u|^2}} \tag{68}$$

where the + sign applies for $q > 0.5$, and the − sign for $q < 0.5$. The strictly monotonically increasing function $\theta : [-1,1] \to [-1,1]$ is given by

$$\theta(s) = \frac{2}{\pi}\left(\arcsin s + s\sqrt{1-s^2}\right). \tag{69}$$

Remark. Dilation and erosion filters are included in this statement by setting $q = 1$ for dilation and $q = 0$ for erosion.

Proof. Using Eqs. (31) and (32) from Subsection 4.1 with ψ set according to the L^2 amoeba metric, we have

$$\begin{aligned} L(z)\frac{\partial x}{\partial z}(z) &= \frac{2}{\alpha}\sqrt{\varrho^2 - \left(1+\frac{1}{\alpha^2}\right)z^2} \\ &\quad\times\left(1+\frac{\delta-2\beta}{\alpha^2}z+\frac{\beta}{\alpha^4\left(\varrho^2-\left(1+\frac{1}{\alpha^2}\right)z^2\right)}z^3\right)+\mathcal{O}(\varrho^3). \end{aligned} \tag{70}$$

By substituting $\omega := z/Z_*$ with $Z_* = \varrho/\psi(1/\alpha)$ as in Eq. (33), this becomes up to terms of higher order, as follows:

$$L(z)\frac{\partial x}{\partial z}(z) = \frac{2\varrho}{\alpha}\sqrt{1-\omega^2}\left(1+\frac{(\delta-2\beta)\varrho}{\alpha\sqrt{1+\alpha^2}}\omega+\frac{\beta\varrho}{\alpha(1+\alpha^2)^{3/2}(1-\omega^2)}\omega^3\right). \tag{71}$$

Similar to our characterization [Eq. (24)] of the median, the q-quantile z_q of the values within $\mathcal{A}$ corresponds to the level line that cuts $\mathcal{A}$ in the ratio $(1-q):q$; i.e.,

$$(1-q)\int\limits_{Z_-}^{z_q} L(\zeta)\frac{\partial x}{\partial z}(\zeta)\mathrm{d}\zeta = q\int\limits_{z_q}^{Z_+} L(\zeta)\frac{\partial x}{\partial z}(\zeta)\mathrm{d}\zeta. \tag{72}$$

As in subsection 4.1, the integration bounds $Z_\pm$ can be replaced with $\pm Z_*$, where $Z_* = \varrho/\sqrt{1+1/\alpha^2}$ (note that we are restricted to the L^2 amoeba metric here) with a negligible error.

Setting $\omega_q := z_q/Z_*$ and substituting Eq. (71) into Eq. (72) yields

$$\begin{aligned}0 &\overset{!}{=} (1-q)\int\limits_{-1}^{\omega_q}\sqrt{1-\omega^2}\mathrm{d}\omega - q\int\limits_{\omega_q}^{1}\sqrt{1-\omega^2}\mathrm{d}\omega\\ &+\frac{(\delta-2\beta)\varrho}{\alpha\sqrt{1+\alpha^2}}\left((1-q)\int\limits_{-1}^{\omega_q}\omega\sqrt{1-\omega^2}\mathrm{d}\omega - q\int\limits_{\omega_q}^{1}\omega\sqrt{1-\omega^2}\mathrm{d}\omega\right)\\ &+\frac{\beta\varrho}{\alpha(1+\alpha^2)^{3/2}}\left((1-q)\int\limits_{-1}^{\omega_q}\frac{\omega^3}{\sqrt{1-\omega^2}}\mathrm{d}\omega - q\int\limits_{\omega_q}^{1}\frac{\omega^3}{\sqrt{1-\omega^2}}\mathrm{d}\omega\right).\end{aligned} \tag{73}$$

All integrals evaluate to finite values, and the first pair of integrals does not cancel out for $q \neq 0.5$. Thus, the condition reduces up to $\mathcal{O}(\varrho)$ to its first pair of integrals.

By

$$\int\limits_a^b \sqrt{1-q^2}\mathrm{d}q = \frac{1}{2}\left(\arcsin b - \arcsin a + b\sqrt{1-b^2} + a\sqrt{1-a^2}\right), \tag{74}$$

this implies

$$\arcsin\omega_q + \omega_q\sqrt{1-\omega_q^2} = (2q-1)\frac{\pi}{2}, \tag{75}$$

Table 1 Exemplary values of the function θ^{-1}, see text

s	**0.0**	**0.1**	**0.2**	**0.3**	**0.4**	**0.5**
$\theta^{-1}(s)$	0.0000	0.0787	0.1578	0.2379	0.3196	0.4040
s	**0.6**	**0.7**	**0.8**	**0.9**	**1.0**	
$\theta^{-1}(s)$	0.4919	0.5851	0.6871	0.8054	1.0000	

and after back-substitution,

$$z_q = \frac{\varrho\alpha}{\sqrt{1+\alpha^2}}\theta^{-1}(2s-1), \tag{76}$$

with the function θ stated in the theorem. Since $\alpha = \sigma|\nabla u|$, the claim of the theorem is validated. Exemplary values of the function θ^{-1} are given in Table 1. Values for negative arguments are obtained by $\theta^{-1}(-s) = -\theta^{-1}(s)$.

4.4 Analysis of Amoeba M-Smoothers

Let us now consider a corresponding approximation theorem for amoeba M-smoothers.

Theorem 5 *For any $p > 0$, one step of the amoeba M-smoother with exponent p from subsection 3.4, with the amoeba radius ϱ and L^2 amoeba metric, asymptotically approximates for $\varrho \to 0$ a time step of size $\tau = \varrho^2/6$ of an explicit time discretization of the PDE:*

$$u_t = \frac{3}{p+2}\left(\frac{u_{\xi\xi}}{1+\sigma^2|\nabla u|^2} + \frac{\left((p-1)+2\sigma^2|\nabla u|^2\right)u_{\eta\eta}}{\left(1+\sigma^2|\nabla u|^2\right)^2}\right). \tag{77}$$

Proof. To prove Theorem 5, we proceed similarly as in the first proof of Theorem 1, using the Taylor ansatz [Eq. (22)]. Instead of Eq. (24) for the median μ, we have to consider an analogous condition for the minimizer $\mu = \zeta_0/\sigma$ of

$$\int_{z_-}^{z_+} L(\zeta)\frac{\partial x}{\partial z}(\zeta)|\zeta-\zeta_0|^p \mathrm{d}\zeta. \tag{78}$$

Such a condition is obtained by differentiating, as follows:

$$\int_{Z_-}^{\zeta_0} L(\zeta)\frac{\partial x}{\partial z}(\zeta)(\zeta_0-\zeta)^{p-1}\mathrm{d}\zeta = \int_{\zeta_0}^{Z_+} L(\zeta)\frac{\partial x}{\partial z}(\zeta)(\zeta-\zeta_0)^{p-1}\mathrm{d}\zeta. \tag{79}$$

We notice that for $p < 1$, both integrands have poles at the integration boundary ζ_0, but the integrals are convergent.

As in subsection 4.1, the integration bounds $Z_\pm$ can be replaced with $\pm Z_*$, where $Z_* = \varrho/\sqrt{1+1/\alpha^2}$ (note that we are restricted to the L^2 amoeba metric here) with a negligible error.

Starting again from the representation [Eq. (71)] of $L(z)\frac{\partial x}{\partial z}(z)$ as a function of $\omega = z/Z^*$, we obtain, by setting $\omega_0 := \zeta_0/Z_*$,

$$\int_{-Z_*}^{Z^*} L(\zeta)\frac{\partial x}{\partial z}(\zeta)|\zeta-\zeta_0|^{p-1}\operatorname{sgn}(\zeta-\zeta_0)\mathrm{d}\zeta = \frac{2}{\alpha}\left(\frac{\varrho^p\alpha^{p-1}}{(1+\alpha^2)^{\frac{p-1}{2}}}\int_{-1}^{1}\sqrt{1-\omega^2}\right.$$

$$\times\left(1+\frac{(\delta-2\beta)\varrho}{\alpha\sqrt{1+\alpha^2}}\omega\right)|\omega-\omega_0|^{p-1}\operatorname{sgn}(\omega-\omega_0)\mathrm{d}\omega$$

$$\left.+\frac{\varrho^{p+1}\alpha^{p+2}}{(1+\alpha^2)^{\frac{p+2}{2}}}\int_{-1}^{1}\frac{\beta}{\alpha^4}\frac{\omega^3}{\sqrt{1-\omega^2}}|\omega-\omega_0|^{p-1}\operatorname{sgn}(\omega-\omega_0)\mathrm{d}\omega\right). \tag{80}$$

Simplifying the first integral using integration by parts gives

$$\frac{1}{p}\left[\sqrt{1-\omega^2}\left(1+\frac{(\delta-2\beta)\varrho}{\alpha\sqrt{1+\alpha^2}}\omega\right)|\omega-\omega_0|^p\right]_{-1}^{1}$$
$$-\frac{1}{p}\int_{-1}^{1}\frac{\mathrm{d}}{\mathrm{d}\omega}\left(\sqrt{1-\omega^2}\left(1+\frac{(\delta-2\beta)\varrho}{\alpha\sqrt{1+\alpha^2}}\omega\right)\right)\cdot|\omega-\omega_0|^p\mathrm{d}\omega, \tag{81}$$

where the boundary term vanishes due to $\sqrt{1-\omega^2}=0$ at $\omega=\pm1$, such that after carrying out the differentiation by ω in the integrand of the latter expression, our minimality condition reads as

$$0 \stackrel{!}{=} J_1 + \frac{(\delta-2\beta)\varrho}{\alpha\sqrt{1+\alpha^2}}(J_2-J_3) + \frac{p\beta\varrho}{\alpha(1+\alpha^2)^{3/2}}J_4 \tag{82}$$

with the following four integrals:

$$J_1 := \int_{-1}^{1} \frac{\omega}{\sqrt{1-\omega^2}} |\omega - \omega_0|^p \mathrm{d}\omega, \tag{83}$$

$$J_2 := \int_{-1}^{1} \frac{\omega^2}{\sqrt{1-\omega^2}} |\omega - \omega_0|^p \mathrm{d}\omega, \tag{84}$$

$$J_3 := \int_{-1}^{1} \sqrt{1-\omega^2} |\omega - \omega_0|^p \mathrm{d}\omega, \quad \text{and} \tag{85}$$

$$J_4 := \int_{-1}^{1} \frac{\omega^3}{\sqrt{1-\omega^2}} |\omega - \omega_0|^p \operatorname{sgn}(\omega - \omega_0) \mathrm{d}\omega. \tag{86}$$

Provided that $\omega_0 = \mathcal{O}(\varrho)$ (corresponding to $\zeta_0 = \mathcal{O}(\varrho^2)$), all of these integrals can be represented up to higher-order errors in terms of the integrals

$$F(r,s) := \int_0^1 x^s \left(\sqrt{1-x^2}\right)^r \mathrm{d}x, \quad r = \pm 1, \quad s \in \mathbb{R}_0^+. \tag{87}$$

In detail, we have

$$J_1 = -2p\omega_0 F(-1,p) + \mathcal{O}\left(\varrho^{3/2}\right), \tag{88}$$

$$J_2 = 2F(-1,p+2) + \mathcal{O}\left(\varrho^{\frac{p+3}{2}}\right), \tag{89}$$

$$J_3 = 2F(1,p) + \mathcal{O}\left(\varrho^{p/2}\right), \quad \text{and} \tag{90}$$

$$J_4 = 2F(-1,p+2) + \mathcal{O}\left(\varrho^{1/2}\right). \tag{91}$$

(For a detailed proof, see the Appendix.) Inserting these into the minimality condition [Eq. (82)] yields

$$
\begin{aligned}
0 \overset{!}{=} & -p\omega_0 F(-1,p) + \frac{(\delta-2\beta)\varrho}{\alpha\sqrt{1+\alpha^2}}\left(F(-1,p+2) - F(1,p)\right) \\
& + \frac{p\beta\varrho}{\alpha\left(1+\alpha^2\right)^{3/2}} F(-1,p+2) + \mathcal{O}\left(\varrho^{p/2} + \varrho^{1/2}\right),
\end{aligned} \tag{92}
$$

and, thus,

$$
\frac{\zeta_0}{\varrho^2} = \frac{\delta-2\beta}{1+\alpha^2}\,\frac{F(-1,p+2)-F(1,p)}{pF(-1,p)} + \frac{\beta}{(1+\alpha^2)^2}\,\frac{F(-1,p+2)}{F(-1,p)}. \tag{93}
$$

While for most values of p, the integrals $F(\pm 1,p)$ cannot be solved in closed form, one can show that $F(-1,p+2) = \frac{p+1}{p+2}F(-1,p)$ and $F(1,p) = \frac{1}{p+2}F(-1,p)$ for all $p \geq 0$ (for a detailed proof, see the Appendix). Inserting these identities and translating back to the derivatives of u in the ξ, η system yields the claim of Eq. (77).

Remark. In the previous proof, the limits for $\varrho \to 0$ are not uniform for all $p > 0$. As a result, the limit of Eq. (77) for $p \to 0$ does not coincide with a local maximum of $L(z)\partial x/\partial z$, which would be a natural translation of the definition of a mode filter as the most frequent value for the space-continuous setting. We believe, however, that the mode filter as limit $p \to 0$ is the more useful concept since it uses information from the entire structuring element, as opposed to the vicinity of a single level line when maximizing $L(z)\partial x/\partial z$.

4.5 Analysis of Multivariate Median Filtering

In this subsection, we consider multivariate median filtering. Unlike in the preceding sections, where we successfully derived PDEs for amoeba-based filters, we will consider the nonadaptive filter case here (i.e., multivariate median filtering with fixed, disc-shaped structuring elements). Even in this case, it is very tedious to derive a PDE being approximated, and its coefficient functions cannot be stated in closed form but by elliptic integrals, thus marking a limitation of the framework.

This subsection and section A.3 of the Appendix essentially follow subsection 3.5.2 of Welk (2007).

Structure tensor and pseudo-level lines. In order to discuss curvature-based PDEs on multivariate images, it is useful to have some concept that is analogous to the level lines of grayscale images. Note that as soon as the *range* $\mathbb{R}^d$ of the image function $\boldsymbol{u} : \Omega \rightarrow \mathbb{R}^d$ has the same or a higher dimension than its *domain* (i.e., $d \geq 2$ for planar images), one can no longer expect in the generic case to find any point $\boldsymbol{x}$ near a given $\boldsymbol{x}_0 \in \Omega$ with $u(\boldsymbol{x}) = u(\boldsymbol{x}_0)$. Thus, level sets in the strict sense decompose into discrete points. Nevertheless, one can introduce a *pseudo-level set* around a given $\boldsymbol{x}_0 \in \Omega$ as a line of least variation of image values. This notion has been introduced (and given the name *level line*) by Chung and Sapiro (2000).

To define pseudo-level lines, we use the concept of the structure tensor introduced by Förstner and Gülch (1987). For a multivariate image $\boldsymbol{u} \equiv (u_1, u_2, ..., u_d)^{\mathrm{T}} : \Omega \rightarrow \mathbb{R}^d$, the field of structure tensors $J(\nabla u)$ is defined as the sum of tensor products of the channel gradients:

$$J(\nabla u) := \sum_{k=1}^{d} \nabla u_k \, \nabla u_k^{\mathrm{T}} \,. \tag{94}$$

In a planar multivariate image, a pseudo-level line, then, is a curve whose tangent vector is everywhere aligned with the eigenvector for the smallest eigenvalue of $J(\nabla u)$. In an arbitrary image dimension, a pseudo-level set would be a hypersurface orthogonal to the eigenvector for the largest eigenvalue of the structure tensor.

Feddern *et al.* (2006) showed that pseudo-level sets are indeed connected to the multivariate mean curvature motion equation and similar PDEs.

PDE ansatz for multivariate median filtering. Let us consider now a multivariate image $\boldsymbol{u} : \mathbb{R}^2 \rightarrow \mathbb{R}^d$, which is also assumed to be equipped with the Euclidean norm $\| \cdot \|$. We want to analyze the multivariate median [Eq. (19)] with the disc D as the structuring element.

Analogous to the scalar-valued case, we can define a PDE that is associated with the median filter. We say that a PDE

$$\partial_t \boldsymbol{u} = F\big(\partial_x \boldsymbol{u}, \partial_y \boldsymbol{u}, \partial_{xx} \boldsymbol{u}, \partial_{xy} \boldsymbol{u}, \partial_{yy} \boldsymbol{u}, \ldots\big) \tag{95}$$

is associated with the median filter if

$$\lim_{\varrho \rightarrow +0} \frac{\operatorname*{med}_{D_\varrho(x,y)} \boldsymbol{u} - \boldsymbol{u}(x,y)}{\varrho^2/6} = F\big(\partial_x \boldsymbol{u}, \partial_y \boldsymbol{u}, \partial_{xx} \boldsymbol{u}, \partial_{xy} \boldsymbol{u}, \partial_{yy} \boldsymbol{u}, \ldots\big), \tag{96}$$

where $D_\varrho(x,y)$ denotes the disk of radius ϱ around (x,y); i.e., if

$$\operatorname*{med}_{D_\varrho(x,y)} \boldsymbol{u} = \boldsymbol{u}(x,y) + \frac{\varrho^2}{6} F + \mathcal{O}\big(\varrho^{2-\varepsilon}\big) \tag{97}$$

with some positive ε. Note that the right side F of our PDE does not depend on $\boldsymbol{u}$ itself since the shift-invariance of the median rules out such a dependency.

Simplification by symmetries. For simplicity, we investigate the location $x = y = 0$. Exploiting the translational and rotational invariance of the involved filtering procedures, we can assume that $(1, 0)^{\mathrm{T}}$ and $(0,1)^{\mathrm{T}}$ are the eigenvectors for the larger and smaller eigenvalues of the structure tensor [Eq. (94)]. Since it follows, then, that $\langle \partial_x \boldsymbol{u}, \partial_y \boldsymbol{u} \rangle = 0$, we can achieve by rotations and translations in the image range $\mathbb{R}^d$ the fact that $\boldsymbol{u}(0,0) = 0$, $\partial_x \boldsymbol{u}(0,0) = (m, 0, 0, \ldots)^{\mathrm{T}}, \partial_y \boldsymbol{u}(0,0) = (0, n, 0, \ldots)^{\mathrm{T}}$ with $m \geq n \geq 0$.

Minimality condition. We reformulate Eq. (19) by taking partial derivatives with respect to all median entries and conclude that for the multivariate median value $M = (M_1, \ldots, M_d)^{\mathrm{T}}$, the equations

$$\iint\limits_D \frac{u_k(x, y) - M_k}{\sqrt{\sum_{j=1}^d \left(u_j(x, y) - M_j\right)^2}} \mathrm{d}x\mathrm{d}y \tag{98}$$

must hold for $k = 1, \ldots, d$. Given the convexity of the median objective function, fulfillment of Eq. (98) is also sufficient for M to be the sought median. The drawback of using Eq. (98) is its singularity at M, which will make it necessary to split the integrals in the course of our calculations.

Special case. First, let us clarify the case $m \neq 0$, $n = 0$, in which the first derivative of the multivariate image along the local pseudo-level line vanishes. In this case, the argument from the scalar-valued case transfers straightforwardly to the multivariate situation, leading to the following result:

Lemma 6 Let $\boldsymbol{u}$ be analytic and

$$\begin{aligned}
&\boldsymbol{u}(0,0) = 0, \\
&\partial_x \boldsymbol{u}(0,0) = (m, 0, 0, \ldots)^{\mathrm{T}}, \qquad \partial_y \boldsymbol{u}(0,0) = (0, 0, 0, \ldots)^{\mathrm{T}}, \\
&\partial_{yy} \boldsymbol{u}(0,0) = (2\delta, 2\varepsilon, 2\zeta_1, 2\zeta_2, \ldots)^{\mathrm{T}}, \\
&\partial_{xx} \boldsymbol{u}(0,0) \text{ arbitrary}, \qquad \partial_{xy} \boldsymbol{u}(0,0) \text{ arbitrary}
\end{aligned} \tag{99}$$

with $|m| > 0$. Then we have

$$\lim_{\varrho \to +0} \frac{\operatorname*{med}_{D_\varrho(0,0)} \boldsymbol{u} - \boldsymbol{u}(0,0)}{\varrho^2} = \frac{1}{3}(\delta, \varepsilon, \zeta_1, \zeta_2, \ldots)^{\mathrm{T}} = \frac{1}{6} \partial_{yy} \boldsymbol{u}(0,0). \tag{100}$$

This means that in this specific case, iterated median filtering approximates curvature motion even in the multivariate setting.

Generic case. To study the situation if first derivatives in both the pseudo-flow and pseudo-level line directions are different from zero, we introduce three real-valued functions, which will be useful in the following discussion. They arise as quotients of elliptic integrals in the unit disc D_1.

Definition 2 *For* $\lambda \in \mathbb{R} \setminus \{0\}$, let

$$Q_1(\lambda) := \frac{\displaystyle\iint_{D_1} \frac{\xi^2\eta^2}{\left(\xi^2+\lambda^2\xi^2\right)^{3/2}}\,\mathrm{d}\xi\,\mathrm{d}\eta}{\displaystyle\iint_{D_1} \frac{\eta^2}{\left(\xi^2+\lambda^2\eta^2\right)^{3/2}}\,\mathrm{d}\xi\,\mathrm{d}\eta}, \quad Q_2(\lambda) := \frac{\displaystyle\iint_{D_1} \frac{\eta^4}{\left(\xi^2+\lambda^2\eta^2\right)^{3/2}}\,\mathrm{d}\xi\,\mathrm{d}\eta}{\displaystyle\iint_{D_1} \frac{\eta^2}{\left(\xi^2+\lambda^2\eta^2\right)^{3/2}}\,\mathrm{d}\xi\,\mathrm{d}\eta},$$

$$Q_3(\lambda) := \frac{\displaystyle\iint_{D_1} \frac{\eta^2}{\left(\xi^2+\lambda^2\eta^2\right)^{1/2}}\,\mathrm{d}\xi\,\mathrm{d}\eta}{\displaystyle\iint_{D_1} \frac{1}{\left(\xi^2+\lambda^2\eta^2\right)^{1/2}}\,\mathrm{d}\xi\,\mathrm{d}\eta}, \tag{101}$$

as well as

$$Q_1(0) = Q_2(0) = Q_3(0) := \frac{1}{3},$$
$$Q_1(\pm\infty) = Q_2(\pm\infty) = Q_3(\pm\infty) := 0. \tag{102}$$

Remark. Graphs of Q_1, Q_2, and Q_3 are shown in Figure 4. Some of the integrals involved in these definitions can be resolved in closed form, while the remaining ones are reduced to elliptic integrals. We do not pursue this distinction further since for our qualitative discussion, it is more valuable to retain the structural analogy among all the cases. We remark that Q_1, Q_2, and Q_3 are even functions, as follows:

$$Q_1(-\lambda) = Q_1(\lambda), \quad Q_2(-\lambda) = Q_2(\lambda), \quad Q_3(-\lambda) = Q_3(\lambda). \tag{103}$$

We start now by considering important special cases.

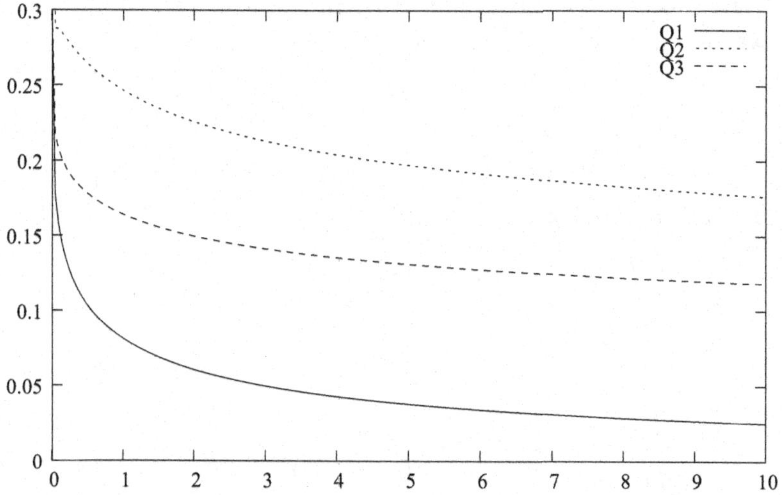

Figure 4 Graphs of Q_1, Q_2, and Q_3 defined in Eq. (101). From Welk (2007).

Lemma 7 *Let* $\boldsymbol{u}$ *be analytic and*

$$\begin{aligned} \boldsymbol{u}(0,0) &= 0, \\ \partial_x \boldsymbol{u}(0,0) &= (m,0,0,\ldots)^{\mathrm{T}}, \quad \partial_y \boldsymbol{u}(0,0) = (0,n,0,\ldots)^{\mathrm{T}}, \\ \partial_{xx} \boldsymbol{u}(0,0) &= (2\alpha,0,0,\ldots)^{\mathrm{T}}, \quad \partial_{xy} \boldsymbol{u}(0,0) = \partial_{yy} \boldsymbol{u}(0,0) = 0 \end{aligned} \tag{104}$$

with $|m| \geq |n| > 0$. *Then we have*

$$\lim_{\varrho \to +0} \frac{\operatorname*{med}_{D_\varrho(0,0)} \boldsymbol{u} - \boldsymbol{u}(0,0)}{\varrho^2} = (a_0, 0, \ldots)^{\mathrm{T}}, \tag{105}$$

where

$$a_0 = \frac{\displaystyle\iint_{D_1} \xi^2 \frac{(n\eta)^2}{\left((m\xi)^2 + (n\eta)^2\right)^{3/2}} \mathrm{d}\xi \mathrm{d}\eta}{\displaystyle\iint_{D_1} \frac{(n\eta)^2}{\left((m\xi)^2 + (n\eta)^2\right)^{3/2}} \mathrm{d}\xi \mathrm{d}\eta} \cdot \alpha = \mathrm{Q}_1\left(\frac{m}{n}\right)\alpha. \tag{106}$$

Proof. In the given case, $\boldsymbol{u}$ is approximated around the origin by

$$\boldsymbol{u}(x,y) = \left(mx + \alpha x^2, ny, 0, \ldots\right)^{\mathrm{T}} + \mathcal{O}\left(x^2 + y^2\right). \tag{107}$$

Clearly, only the first two dimensions of $\mathbb{R}^d$ are relevent in this case, so we will assume that $d = 2$. Assuming that

$$\lim_{\varrho \to +0} \frac{\operatorname*{med}_{D_\varrho(0,0)} \boldsymbol{u} - \boldsymbol{u}(0,0)}{\varrho^2} = \begin{pmatrix} a(\varrho) \\ b(\varrho) \end{pmatrix} = \begin{pmatrix} a \\ b \end{pmatrix}, \tag{108}$$

we obtain from Eq. (98) the conditions

$$\begin{aligned} 0 &= \iint_{D_\varrho} \frac{mx + \alpha x^2 - a\varrho^2}{\sqrt{\left(mx + \left(\alpha x^2 - a\right)\varrho^2\right)^2 + \left(ny - b\varrho^2\right)^2}} \mathrm{d}x\mathrm{d}y \\ 0 &= \iint_{D_\varrho} \frac{ny - b\varrho^2}{\sqrt{\left(mx + \left(\alpha x^2 - a\right)\varrho^2\right)^2 + \left(ny - b\varrho^2\right)^2}} \mathrm{d}x\mathrm{d}y. \end{aligned} \tag{109}$$

Considering the symmetry of the second equation of Eq. (109) with respect to the reflection $y \leftrightarrow -y$ and the convexity of the energy, it is evident that b must be zero. Using this fact and the substitution $\xi := x/\varrho, \eta := y/\varrho$, we find

$$0 = \iint_{D_1} \frac{m\xi + \left(\alpha\xi^2 - a\right)\varrho}{\sqrt{\left(m\xi + \left(\alpha\xi^2 - a\right)\varrho\right)^2 + (n\eta)^2}} \mathrm{d}\xi\mathrm{d}\eta. \tag{110}$$

As proved in section A.3 of the Appendix, this equation transforms into

$$0 = \iint_{D_1} \varrho \frac{(n\eta)^2}{\left((m\xi)^2 + (n\eta)^2\right)^{3/2}} \left(\alpha\xi^2 - a\right)\mathrm{d}\xi\mathrm{d}\eta + \mathcal{O}\left(\varrho^{6/5}\right); \tag{111}$$

thus, after division by ϱ, we get

$$\begin{aligned} & a \cdot \iint_{D_1} \frac{(n\eta)^2}{\left((m\xi)^2 + (n\eta)^2\right)^{3/2}} \mathrm{d}\xi\mathrm{d}\eta \\ &= \alpha \cdot \iint_{D_1} \xi^2 \frac{(n\eta)^2}{\left((m\xi)^2 + (n\eta)^2\right)^{3/2}} \mathrm{d}\xi\mathrm{d}\eta + \mathcal{O}\left(\varrho^{1/5}\right), \end{aligned} \tag{112}$$

from which the assertion follows. This concludes the proof.

Lemma 8 *Let $\boldsymbol{u}$ be analytic and*

$$\begin{aligned} &\boldsymbol{u}(0,0) = 0, \\ &\partial_x \boldsymbol{u}(0,0) = (m,0,0,...)^{\mathrm{T}}, \qquad \partial_y \boldsymbol{u}(0,0) = (0,n,0,...)^{\mathrm{T}}, \\ &\partial_{xx} \boldsymbol{u}(0,0) = (0,2\beta,0,0,...)^{\mathrm{T}}, \quad \partial_{xy} \boldsymbol{u}(0,0) = \partial_{yy} \boldsymbol{u}(0,0) = 0 \end{aligned} \tag{113}$$

with $|m| \geq |n| > 0$. Then we have

$$\lim_{\varrho \to +0} \frac{\operatorname*{med}_{D_\varrho(0,0)} \boldsymbol{u} - \boldsymbol{u}(0,0)}{\varrho^2} = (0, b_0, 0, ...)^{\mathrm{T}}, \tag{114}$$

where

$$b_0 = \frac{\displaystyle\iint_{D_1} \xi^2 \frac{(m\xi)^2}{\left((m\xi)^2 + (n\eta)^2\right)^{3/2}} \mathrm{d}\xi \mathrm{d}\eta}{\displaystyle\iint_{D_1} \frac{(m\xi)^2}{\left((m\xi)^2 + (n\eta)^2\right)^{3/2}} \mathrm{d}\xi \mathrm{d}\eta} \cdot \beta = Q_2\left(\frac{m}{n}\right)\beta. \tag{115}$$

Lemma 9 *Let $\boldsymbol{u}$ be analytic and*

$$\begin{aligned} &\boldsymbol{u}(0,0) = 0, \\ &\partial_x \boldsymbol{u}(0,0) = (m,0,0,...)^{\mathrm{T}}, \qquad \partial_y \boldsymbol{u}(0,0) = (0,n,0,...)^{\mathrm{T}}, \\ &\partial_{xx} \boldsymbol{u}(0,0) = (0,0,2\gamma,0,...)^{\mathrm{T}}, \quad \partial_{xy} \boldsymbol{u}(0,0) = \partial_{yy} \boldsymbol{u}(0,0) = 0, \end{aligned} \tag{116}$$

with $|m| \geq |n| > 0$. Then we have

$$\lim_{\varrho \to +0} \frac{\operatorname*{med}_{D_\varrho(0,0)} \boldsymbol{u} - \boldsymbol{u}(0,0)}{\varrho^2} = (0, 0, c_0, 0, ...)^{\mathrm{T}}, \tag{117}$$

where

$$c_0 = \frac{\displaystyle\iint_{D_1} \frac{\xi^2}{\left((m\xi)^2 + (n\eta)^2\right)^{1/2}} \mathrm{d}\xi \mathrm{d}\eta}{\displaystyle\iint_{D_1} \frac{1}{\left((m\xi)^2 + (n\eta)^2\right)^{1/2}} \mathrm{d}\xi \mathrm{d}\eta} \cdot \gamma = Q_3\left(\frac{m}{n}\right)\gamma. \tag{118}$$

The proofs of Lemmas 8 and 9 are analogous to the one of Lemma 7.

Analogous statements are obtained for cases in which $\partial_{yy}\boldsymbol{u}(0,0) \neq 0$, but $\partial_{xx}\boldsymbol{u}(0,0) = \partial_{xy}\boldsymbol{u}(0,0) = 0$ (note that $|m| \geq |n|$ did not enter the proof of Lemma 7).

If $\partial_{xx}\boldsymbol{u}(0,0) = \partial_{yy}\boldsymbol{u}(0,0) = 0$ and $\partial_{xy}\boldsymbol{u}(0,0) \neq 0$, one easily checks that the evolution is stationary; i.e., $\partial_t \boldsymbol{u} = 0$ is approximated. Since these considerations were based in all cases on linearizations, the contributions of the different second derivative components superpose linearly. Thus, we have the following statement.

Proposition 10 *Let* $\boldsymbol{u}$ *be analytic and*

$$\begin{aligned}
&\boldsymbol{u}(0,0) = 0, \\
&\partial_x \boldsymbol{u}(0,0) = (m,0,0,\ldots)^{\mathrm{T}}, \qquad \partial_y \boldsymbol{u}(0,0) = (0,n,0,\ldots)^{\mathrm{T}}, \\
&\partial_{xx}\boldsymbol{u}(0,0) = (2\alpha, 2\beta, 2\gamma_1, 2\gamma_2, \ldots)^{\mathrm{T}}, \quad \partial_{yy}\boldsymbol{u}(0,0) = (2\delta, 2\varepsilon, 2\zeta_1, 2\zeta_2, \ldots)^{\mathrm{T}}, \\
&\partial_{xy}\boldsymbol{u}(0,0) \quad \textit{arbitrary}
\end{aligned} \tag{119}$$

with $|m| \geq |n| > 0$. Then *we have*

$$\lim_{\varrho \to +0} \frac{\operatorname{med}_{D_\varrho(0,0)} \boldsymbol{u} - \boldsymbol{u}(0,0)}{\varrho^2} = \begin{pmatrix} Q_1(m/n)\alpha + Q_2(n/m)\delta \\ Q_2(m/n)\beta + Q_1(n/m)\varepsilon \\ Q_3(m/n)\gamma_1 + Q_3(n/m)\zeta_1 \\ Q_3(m/n)\gamma_2 + Q_3(n/m)\zeta_2 \\ \vdots \end{pmatrix}, \tag{120}$$

where $Q_1(\lambda)$, $Q_2(\lambda)$, $Q_3(\lambda)$ *are defined as in Eq. (101).*

Transforming back to the general situation leads to the following conclusion.

Corollary 11 *One step of multivariate median filtering with structuring element* D_ϱ *approximates one time step of size* $\tau = \varrho^2$ *of the PDE:*

$$\begin{aligned}
\partial_t \boldsymbol{u} = 3 \Bigg(& Q_1\left(\frac{|\partial_\eta \boldsymbol{u}|}{|\partial_\xi \boldsymbol{u}|}\right) \frac{\langle \partial_{\eta\eta}\boldsymbol{u}, \partial_\eta \boldsymbol{u}\rangle}{\langle \partial_\eta \boldsymbol{u}, \partial_\eta \boldsymbol{u}\rangle} \partial_\eta \boldsymbol{u} + Q_2\left(\frac{|\partial_\xi \boldsymbol{u}|}{|\partial_\eta \boldsymbol{u}|}\right) \frac{\langle \partial_{\xi\xi}\boldsymbol{u}, \partial_\eta \boldsymbol{u}\rangle}{\langle \partial_\eta \boldsymbol{u}, \partial_\eta \boldsymbol{u}\rangle} \partial_\eta \boldsymbol{u} \\
& + Q_2\left(\frac{|\partial_\eta \boldsymbol{u}|}{|\partial_\xi \boldsymbol{u}|}\right) \frac{\langle \partial_{\eta\eta}\boldsymbol{u}, \partial_\xi \boldsymbol{u}\rangle}{\langle \partial_\xi \boldsymbol{u}, \partial_\xi \boldsymbol{u}\rangle} \partial_\xi \boldsymbol{u} + Q_1\left(\frac{|\partial_\xi \boldsymbol{u}|}{|\partial_\eta \boldsymbol{u}|}\right) \frac{\langle \partial_{\xi\xi}\boldsymbol{u}, \partial_\xi \boldsymbol{u}\rangle}{\langle \partial_\xi \boldsymbol{u}, \partial_\xi \boldsymbol{u}\rangle} \partial_\eta \boldsymbol{u} \\
& + Q_3\left(\frac{|\partial_\eta \boldsymbol{u}|}{|\partial_\xi \boldsymbol{u}|}\right) \left(\partial_{\eta\eta}\boldsymbol{u} - \frac{\langle \partial_{\eta\eta}\boldsymbol{u}, \partial_\eta \boldsymbol{u}\rangle}{\langle \partial_\eta \boldsymbol{u}, \partial_\eta \boldsymbol{u}\rangle} \partial_\eta \boldsymbol{u} - \frac{\langle \partial_{\eta\eta}\boldsymbol{u}, \partial_\xi \boldsymbol{u}\rangle}{\langle \partial_\xi \boldsymbol{u}, \partial_\xi \boldsymbol{u}\rangle} \partial_\xi \boldsymbol{u} \right) \\
& + Q_3\left(\frac{|\partial_\xi \boldsymbol{u}|}{|\partial_\eta \boldsymbol{u}|}\right) \left(\partial_{\xi\xi}\boldsymbol{u} - \frac{\langle \partial_{\xi\xi}\boldsymbol{u}, \partial_\eta \boldsymbol{u}\rangle}{\langle \partial_\eta \boldsymbol{u}, \partial_\eta \boldsymbol{u}\rangle} \partial_\eta \boldsymbol{u} - \frac{\langle \partial_{\xi\xi}\boldsymbol{u}, \partial_\xi \boldsymbol{u}\rangle}{\langle \partial_\xi \boldsymbol{u}, \partial_\xi \boldsymbol{u}\rangle} \partial_\xi \boldsymbol{u} \right) \Bigg)
\end{aligned} \tag{121}$$

where $\boldsymbol{\eta}$ *and* $\boldsymbol{\xi}$ *are unit eigenvectors for the major and minor eigenvalues of* $\sum_k \nabla u_k \nabla u_k^{\mathrm{T}}$, *respectively; i.e.,* $\boldsymbol{\xi}$ *is tangential and* $\boldsymbol{\eta}$ *perpendicular to the local pseudo-level line.*

To summarize, we have seen that only if the first derivative of the multichannel image in the pseudo-level line direction vanishes, the multivariate median filter provides an approximation to the curvature motion equation (see Lemma 6). The corollary reveals, however, that upon transition to the generic case $m \neq 0$, $n \neq 0$, one obtains not only a mixture of the second derivatives in the pseudo-level line direction $\boldsymbol{\xi}$ and pseudo-flow line direction $\boldsymbol{\eta}$, but the influence of $\partial_{\xi\xi} u$ is subject to different decay characteristics for its components parallel to $\partial_{\xi} \boldsymbol{u}$, $\partial_{\eta} \boldsymbol{u}$ and perpendicular to both, as described by $Q_1(n/m)$, $Q_2(n/m)$, and $Q_3(n/m)$. An analogous decomposition applies to $\partial_{\eta\eta} \boldsymbol{u}$.

The different decay characteristics preclude a comparably simple PDE approximation property for the multivariate median filter, as in the scalar-valued case, even in the case of nonadaptive filtering. Therefore, we do not pursue a generalization of this analysis to the amoeba filtering case.

5. PRESMOOTHING AND AMOEBA FILTERS

Theorem 1 establishes the relationship between the self-snakes PDE [Eq. (2)] and AMF. As discussed in the Introduction, this PDE is ill posed since it incorporates a shock-filter component that leads to staircasing in numerical solutions. In order to eliminate this ill-posedness, it is often recommended to employ a presmoothing of the gradient data that is used for steering the edge-stopping function; i.e.,

$$u_t = |\nabla u| \operatorname{div}\left(g(|\nabla u_\sigma|) \frac{\nabla u}{|\nabla u|} \right), \tag{122}$$

with $u_\sigma = K_\sigma * u$.

In this section, we discuss how presmoothing can take place in an AMF context and how this compares to presmoothed self-snakes. This section essentially follows section 5 of Welk (2013a).

Note that throughout this section, σ denotes the standard deviation of a Gaussian kernel used for presmoothing. The contrast-scale parameter, which is called σ in the remainder of the chapter, is fixed to 1 here.

The translation of the presmoothing procedure to AMF is straightforward. One simply has to compute the amoeba-structuring elements from the presmoothed image u_σ instead of u. This means that there is a decoupling between the image to which the filter is applied and the blurred image employed for amoeba construction. This is analogous to the AAC model, with $f \equiv u_\sigma/\lambda$. Therefore, the asymptotic approximation result from Theorem 3 applies, which implies that for $\varrho \to 0$, the procedure does not approximate exactly Eq. (122). Nevertheless, the resulting PDE can still be interpreted as a self-snakes process but with a way of presmoothing different from the simple Gaussian convolution of Eq. (122).

Considering a practical computation of discrete data, however, one always employs a positive amoeba radius ϱ. For instance, using the median filter on an amoeba means that equivalently, the amoeba itself acts like a regularizer since it determines the data used for the filtering. A filtering with amoebas derived from $f = u_\sigma/\lambda$, therefore, would involve two spatial scale parameters, σ and ϱ, each of which has the meaning of a spatial averaging step.

Employing this logic, it seems evident that the amoeba radius itself acts similarly as a presmoothing step. We investigate this conjecture by experimentally comparing presmoothed self-snakes with $g(s^2) = 1/(1+s^2)$ to AMF, where $f = u$ with a positive amoeba radius for a very simple example. In the rest of this section, we describe our test setup and then discuss the results of self-snakes and AMF filtering.

5.1 The Test Case

The test situation that we employ allows us to study staircasing. The idea is to model a smooth transition that would be reproduced by filtering with both methods of interest, perturbed by fluctuations with a small amplitude that we can control. This is realized via the function $u : \mathbb{R}^2 \to \mathbb{R}$ with

$$u(x, y) = x + \varepsilon \cos(kx), \quad \varepsilon \ll 1. \tag{123}$$

We will analyze the response of the self-snakes and AMF filters to that perturbation in dependence on the frequency parameter k.

Since both of the filters of interest are nonlinear, there is no superposition property for solutions corresponding to the two components of Eq. (123). Let us note that sufficiently small perturbations will interact with each other, but this will only be in the sense of higher-order terms $\mathcal{O}(\varepsilon^2)$. Thus, the experiment will give an intuition of the behavior of the filters.

5.2 The Self-Snakes Test

In this test case, all level lines are parallel. Therefore, we can simplify the 2-D self-snakes PDE $u_t = g \cdot |\nabla u| \operatorname{div}(\nabla u/|\nabla u|) + \langle \nabla g, \nabla u \rangle$ since the first summand vanishes, while the second one simplifies to $g_x u_x$. From Eq. (123), we can compute $u_x = 1 - \varepsilon k \sin(kx)$. Using a Perona-Malik-type edge-stopping function g, as in Eq. (3), we obtain $g_x = \frac{1}{2}\varepsilon k^2 \cos(kx) + \mathcal{O}(\varepsilon^2)$. Thus, we have

$$u_t = g_x u_x = \frac{\varepsilon k^2}{2} \cos(kx) + \mathcal{O}(\varepsilon^2), \tag{124}$$

from which it can be interpreted that at higher frequencies, amplification factors are unbounded. To worsen the problem, the higher-order terms resulting from nonlinearity instantaneously propagate any perturbation from a given frequency k to higher frequencies. As a consequence, even a single-frequency perturbation causes arbitrarily high frequencies with arbitrarily high amplification ratios to appear within a short evolution period. Therefore, the initial function is evolved into states of lesser regularity.

Presmoothing. Let us now replace $g \equiv g(|\nabla u|)$ with its presmoothed version $g_\sigma \equiv g(|\nabla u_\sigma|)$. In the test case, we have $u_\sigma = x + \varepsilon e^{-k^2\sigma^2/2} \cos(kx)$, and thus $\partial_x g_\sigma = \frac{\varepsilon k^2}{2} e^{-k^2\sigma^2/2} \cos(kx) + \mathcal{O}(\varepsilon^2)$, such that we obtain

$$u_t = \partial_x g_\sigma \cdot \partial_x u = \frac{\varepsilon k^2}{2} e^{-k^2\sigma^2/2} \cos(kx) + \mathcal{O}(\varepsilon^2). \tag{125}$$

Since the exponential functions decays more rapidly in the parameter k than the other contributions, the amplification ratio $k^2 \exp(-k^2\sigma^2/2)$ is bounded and reaches a maximum for $k = \sqrt{2}/\sigma$. This implies that in contrast to the use of unsmoothed data in g, the regularity of the evolving function is kept.

5.3 The Amoeba Test

Let us recall that in the second proof of Theorem 1, the effect of AMF with amoeba radius ϱ is determined via the area difference $\Delta := |\mathcal{A}_1| - |\mathcal{A}_2|$. Now we proceed analogously.

As in our test, the level lines are straight, only the asymmetry contributions Δ_1 need be considered. In this setup, the amoeba around $\boldsymbol{x}_0 = (x_0, y_0)$ is symmetric with respect to the line $y = y_0$, which is parallel to the x-axis. We parameterize this line via $(x(s), y_0)$, where s has the meaning of an arc-length parameter in the amoeba metric; i.e.,

$$\int_0^{x(s)} \sqrt{1 + u_x^2(z, y_0)} \mathrm{d}z = s. \tag{126}$$

Let us now consider the level line through $(x(s), y_0)$ which is parallel to the y-axis. Intersecting this level line with the amoeba cuts out a piece with length $2\sqrt{\varrho^2 - s^2}$. The area difference under consideration is, therefore,

$$\Delta(\boldsymbol{x}_0) = \int_0^{\varrho} \frac{2\sqrt{\varrho^2 - s^2}}{\sqrt{1 + u_x^2(x(s), y_0)}} \mathrm{d}s - \int_{-\varrho}^{0} \frac{2\sqrt{\varrho^2 - s^2}}{\sqrt{1 + u_x^2(x(s), y_0)}} \mathrm{d}s$$

$$= 2\int_0^{\varrho} \sqrt{\varrho^2 - s^2} \left(\frac{1}{\sqrt{1 + u_x^2(x(s), y_0)}} - \frac{1}{\sqrt{1 + u_x^2(x(-s), y_0)}} \right) \mathrm{d}s \tag{127}$$

with $u_x(x, y) = 1 - k\varepsilon \sin(kx)$.

Analogous to the median calculated in Eq. (62), one can compute the resulting median as $u(\boldsymbol{x}_0) + \Delta(\boldsymbol{x}_0)/(4\varrho)$. A numerical integration of Eq. (127) confirms that $\Delta(x, y_0)$ itself is approximately a multiple of the perturbation function $\varepsilon \cos(kx)$.

For easy comparison of these analytical results with Eq. (125), we divide the amplification factor $\Delta(x, y_0)/(4\varrho \cos(kx))$ by the evolution time $\varrho^2/6$ corresponding to amoeba radius ϱ in the asymptotic approximation.

Figure 5 is a plot of the numerically computed factor $\Delta(x, y_0)/(4\varrho \cos(kx))\cdot 6/\varrho^2$ compared to the factor $k^2 \exp(-k^2\sigma^2/2)/2$ from

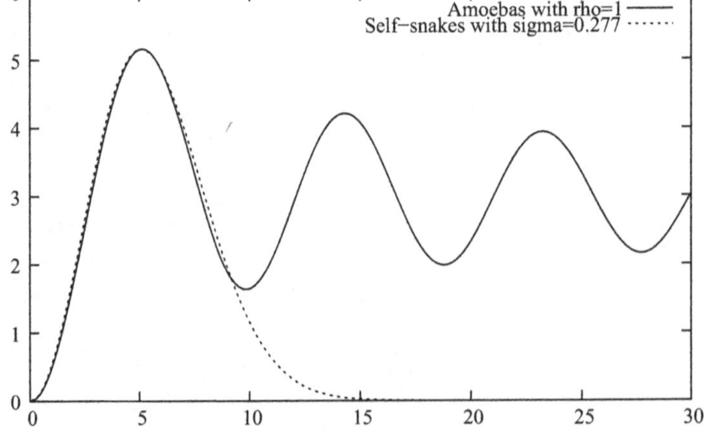

Figure 5 Comparison of amplification factors depending on the frequency parameter k for presmoothed self-snakes and AMF with fixed amoeba size. The horizontal axis shows k, and the vertical axis shows amplification factors. From Welk (2013a).

Eq. (125) as functions of the frequency parameter k. Here, ϱ and σ were chosen to give an optimal fit of the first maximum (i.e., for low frequencies). Obviously, the first lobe of the amplification functions is very similar. For higher frequency parameter values k, presmoothed self-snakes with their exponential dampening factor are superior to the amoeba filter whose amplification factor oscillates around a nonzero value. However, when working with discretized images, spatial discretization automatically cuts off or strongly dampens higher-frequency contributions. So long as the amoeba radius does not exceed a value of approximately $10/\pi \approx 3$, the higher lobes of the amplification function in Figure 5 will disappear completely.

6. EXPERIMENTS

In this section, we present experiments that illuminate the relationship between amoeba filters and corresponding differential models. Moreover, we consider the influence of amoeba parameters and show results for the new amoeba filters.

6.1 Experiments on Amoeba Median Filtering

In this subsection, we consider iterated AMF and its PDE counterpart, self-snakes. As these two filters are denoising methods, we use a test image perturbed by Gaussian noise in this subsection; see Figure 6(a).

We begin our experiments with a study of adaptivity and different amoeba metrics (see Figure 6). To counteract the decrease in the number of pixels in a structuring element of fixed radius with increasing σ, we choose an amoeba radius of $\varrho = 3$ for the nonadaptive case, $\varrho = 4$ for $\sigma = 0.03$, and $\varrho = 5$ for $\sigma = 0.1$. We observe the typical behavior of classical median filtering, where corners are rounded and edges of small tonal contrast are not kept. In comparison, we observe the edge-enhancing properties when denoising with AMF with L^1 and L^2 amoeba metric, respectively. We also observe in this example that at the same contrast scale, the use of the L^2 amoeba metric tends to result in more homogeneous regions and fewer peaks from the noisy input being retained. On the other hand, the results of AMF with the L^1 amoeba metric preserve some medium-scale contrasts in a more pronounced way, which is evident e.g. in the floor region of the office test image. This is related to a slightly stronger trend toward staircasing effects than with the L^2 amoeba metric. However, the same contrast parameter values ($\sigma = 0.03$ and $\sigma = 0.1$) have been employed with both

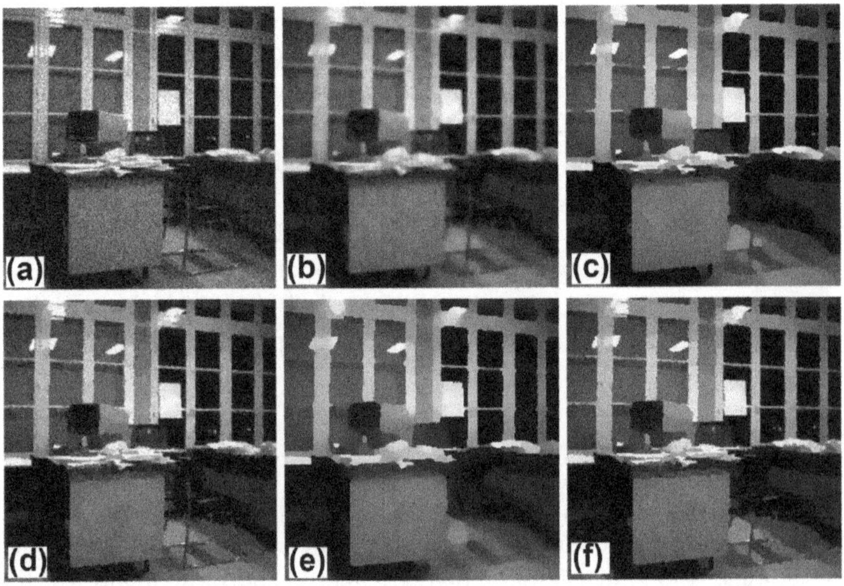

Figure 6 Iterated nonadaptive median filtering versus AMF of a noisy image. (a) Test image (256 x 256 pixels, gray-value range [0, 255]) with Gaussian noise (standard deviation 20). (b) Filtered by nonadaptive median filtering using a 3 x 3 stencil, for 10 iterations. (c) Iterated AMF with the L^1 amoeba metric, contrast scale $\sigma = 0.03$, and amoeba radius $\varrho = 4$, for 10 iterations. (d) Iterated AMF with the L^1 amoeba metric, contrast scale $\sigma = 0.1$, and amoeba radius $\varrho = 5$, for 10 iterations. (e) Same as (c), but with the L^2 amoeba metric. (f) Same as (d), but with the L^2 amoeba metric.

amoeba metrics in this experiment. The results show that often smaller amoebas arise for the L^1 amoeba metric.

In the next experiment, we discuss the influence of two crucial parameters of iterated AMF–namely, the amoeba radius and number of iterations (see Figure 7). As suggested by analytical computations, halving an original amoeba radius ϱ leads to similar results when performing four times the original number of iterations. However, when using very large amoebas, a higher variation of tonal difference is automatically allowed, which introduces a blurring effect. This factor, along with the higher computational load, makes very large amoebas less attractive for computations than a higher number of iterations with smaller amoebas.

To illustrate the approximation statement from Theorem 1, Figure 8 shows the results from curvature motion and self-snakes filtering of the noisy test image from Figure 6(a). Both PDE filters were implemented via explicit (Euler forward) finite-difference schemes. Herein, the right side of the

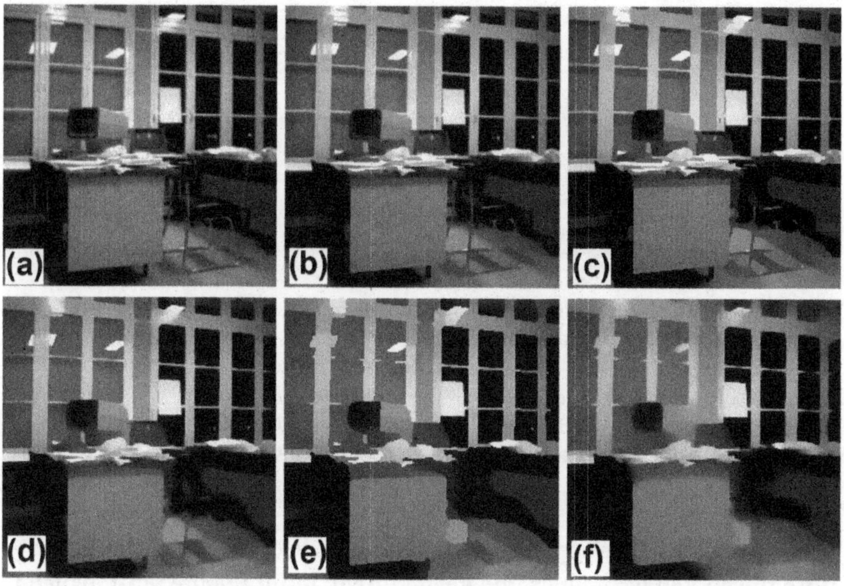

Figure 7 Iterated AMF with different parameters. The L^2 amoeba metric with contrast scale $\sigma = 0.1$ is used throughout. (a) Test image (256 x 256 pixels, gray-value range [0, 255]). (b) Amoeba radius $\varrho = 5$, for 1 iteration. (c) $\varrho = 5$, for 4 iterations. (d) $\varrho = 10$, for 1 iteration. (e) $\varrho = 10$, for 4 iterations. (f) $\varrho = 20$, for 1 iteration.

MCM equation [Eq. (1)], as well as the first summand in the right side of the GAC equation [Eq. (6)], were discretized by central differences. For the second summand of Eq. (6), an upwind discretization (Osher & Sethian, 1988) was employed.

The parameters of Figures 8(a-c) are chosen such that they correspond via Theorem 1 to the parameters of Figures 6(b), (e), and (f), respectively. The evolution time of curvature motion of 20/6 corresponds to 10 iterations of median filtering with structuring element radius $\varrho = \sqrt{2}$, while the self-snakes evolution times of 160/6 and 250/6 belong to $\varrho = 4$ and $\varrho = 5$, respectively. The choice of the Perona-Malik edge-stopping function [Eq. (3)] in the self-snakes evolution corresponds to the L^2 amoeba metric. The Perona-Malik parameters (λ) are the inverses of the σ values from Figure 6.

The results of the PDE schemes indeed look largely comparable to their iterated median filtering/AMF counterparts from Figures 6(b), (e), and (f), in particular when comparing the remaining noise level in the first image pair and the degree of structure simplification in all pairs. The PDE scheme tends to remove small details more aggressively, which can be attributed to the numerical dissipation that is inherent to finite differences.

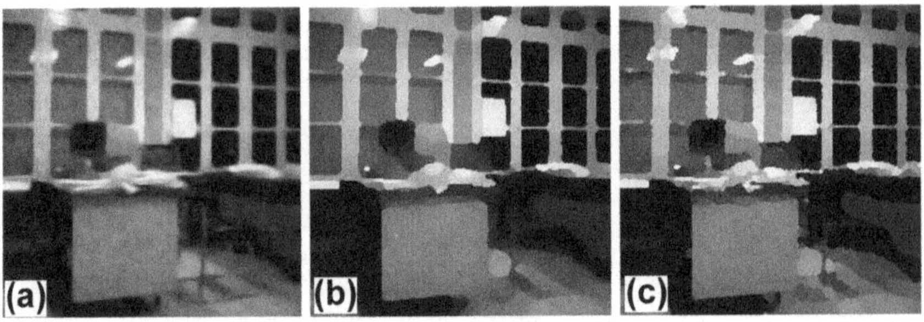

Figure 8 Curvature motion and self-snakes filtering of the noisy test image from Figure 6(a) with different parameters. (a) Finite-difference scheme for curvature motion; time step size $\tau = 1/6$, for 20 iterations. Compare to the median filtering result in Figure 6(b). (b) Finite-difference scheme for self-snakes with Perona-Malik edge-stopping function [Eq. (3)]; $\lambda = 33.3$ and $\tau = 1/6$, for 160 iterations. Compare to Figure 6(e). (c) Same as (b), but $\lambda = 10$ and $\tau = 1/6$, for 250 iterations. Compare to Figure 6(f).

A limitation of the approximation statement becomes visible in Figure 9. With the chosen parameters, the iterated AMF result [Figure 9(b)], which is obtained by 100 iterations with $\varrho = 3$, and the outcome of self-snakes filtering up to evolution time 900/6 [Figure 9(c)] could be expected to correspond. However, Figure 9(b) is drastically less filtered than Figure 9(c); indeed, it hardly differs from Figure 9(a), where only 20 AMF iterations have been carried out. This is caused by a property already known from nonadaptive iterated median filtering: Median filters possess nontrivial steady states, so-called root signals, as studied by Eckhardt (2003). Particularly for small structuring elements, the iterated median filter often locks in at root

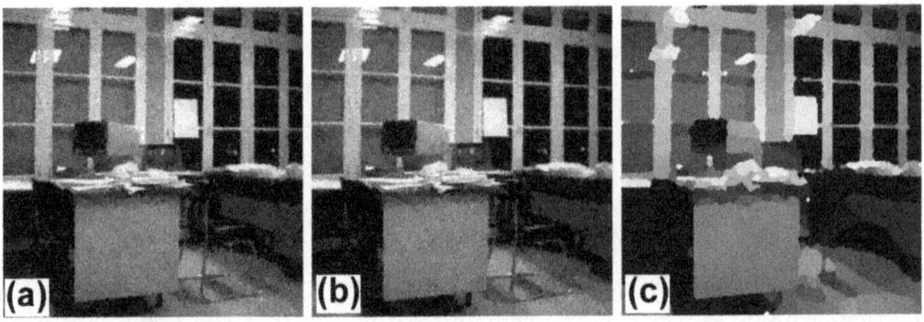

Figure 9 Iterated AMF and self-snakes filtering of the noisy test image from Figure 6(a). (a) Iterated AMF with L^2 amoeba metric, $\sigma = 0.1$, $\varrho = 3$, 20 iterations. (b) Iterated AMF, same parameters as in (a) but 100 iterations. (c) Finite-difference scheme for self-snakes with Perona-Malik edge-stopping function [Eq. (3)], $\lambda = 10$, $\tau = 1/6$, 900 iterations.

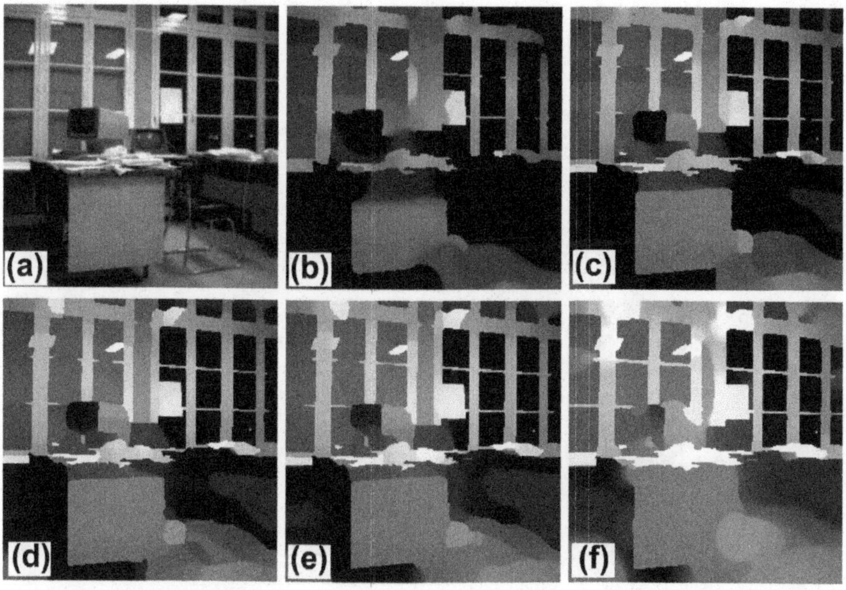

Figure 10 (a) Test image, same as Figure 7(a). (b) Iterated amoeba quantile filtering of the test image from (a), with $\sigma = 0.1$ and $\varrho = 10$, for 5 iterations and $q = 0.2$. (c) Same as (b), but $q = 0.4$. (d) Same as (b), but $q = 0.5$ (identical to median). (e) Same as (b), but $q = 0.6$. (f) Same as (b), but $q = 0.8$.

signals after a few iterations. To avoid this undesired effect, sufficiently large structuring elements need to be used in median filtering. The same is true for AMF. In practice, ϱ should not be set to be smaller than about 5.

6.2 Experiments on Amoeba Quantile Filtering

Our experiments for the remaining amoeba filters were carried out on a test image without noise [Figure 10(a)], because not all of these filters are equally suitable as denoising filters.

We start by amoeba quantile filtering in the framework of the L^2 amoeba metric. It is of interest to demonstrate here the effects of using several quantiles q. As per the discussion in section 2.3, it is evident that a small q should induce an erosion-like effect, a medium-range q should result in a median-like filtering, and a large q should yield dilation-like effects. This is confirmed via the experiments documented in Figure 10. Interestingly, many strong edges still are not displaced, as would have been the case with classical dilation/erosion.

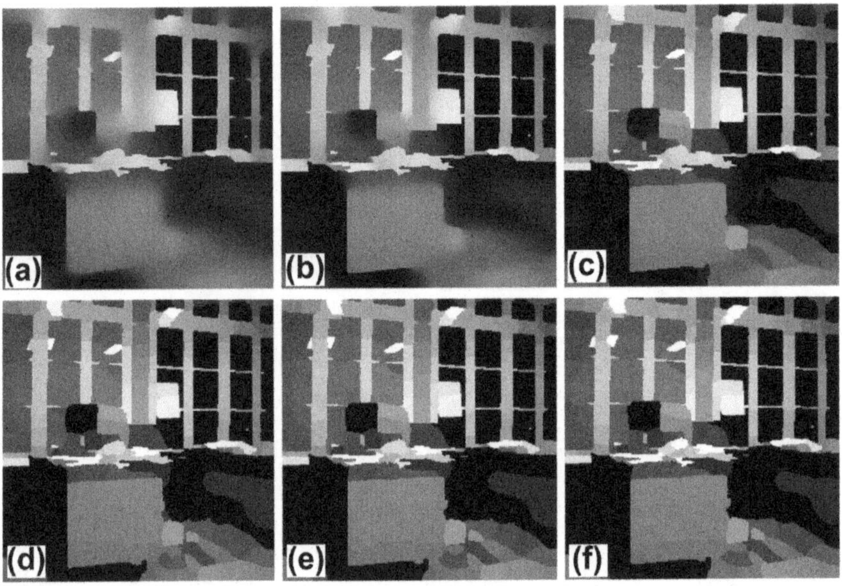

Figure 11 Iterated amoeba M-smoothing of the test image from Figure 7(a), using the L^2 amoeba metric; $\sigma = 0.1$ and $\varrho = 10$, for 5 iterations. (a) Filtered with $p = 2$, equivalent to averaging within structuring elements. (b) Same as (a), but $p = 1.5$. (c) Same as (a), but $p = 1$ (identical to median). (d) Same as (a), but $p = 0.5$. (e) Same as (a), but $p = 0.3$. (f) Same as (a), but $p = 0.1$.

6.3 Experiments on Amoeba Mode Filtering

In Figure 11, we demonstrate the iterated M-smoothers from subsection 2.4 with different values of p. Again, the L^2 amoeba metric is used.

For $p = 2$ [Figure 11(a)], the M-smoother comes down to averaging gray-values within each structuring element, a diffusion-like process. In this case, a strong blurring effect of the filter is observed. This remains true so long as $p > 1$ [see Figure 11(b) for $p = 1.5$]. The choice $p = 1$ [see Figure 11(c)] is identical to AMF. For $p < 1$ (i.e., nonconvex objective functions), the image is processed into sharply contoured patches. The more p is decreased, the more pronounced staircasing and oversegmentation are observed; see Figure 11(d–f).

6.4 Experiments on Amoeba Active Contours

We now demonstrate the AAC algorithm and compare it to its GAC counterpart, discretized by a finite-difference scheme. We use a biomedical test image showing a human head (Figure 12) and aim at segmenting the cerebellum.

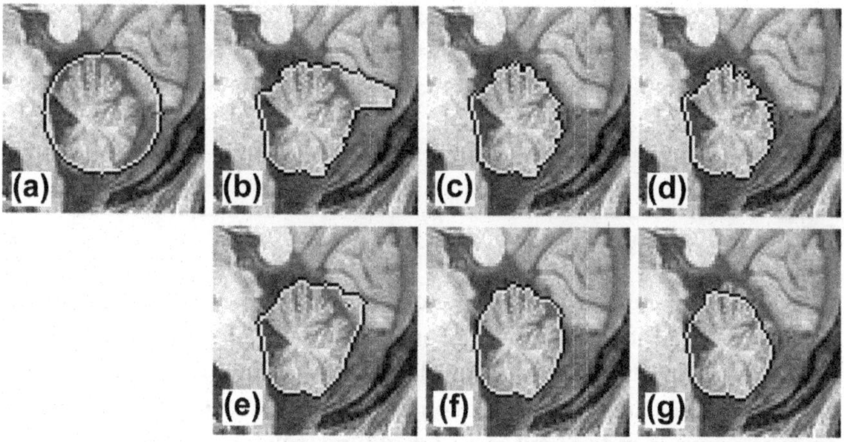

Figure 12 Segmentation by AAC. (a) MR image with initial contour (detail). (b) AAC (unbiased) with the L^2 amoeba metric; contrast scale $\sigma = 0.1$, amoeba radius $\varrho = 10$, and $\sigma = 0.1$, for 20 iterations. (c) Same as (b), but $\sigma = 0.1$ and $\varrho = 12$, for 10 iterations. (d) Same as (b), but $\sigma = 0.1$ and $\varrho = 12$, for 60 iterations. (e) GAC (4), $\sigma = 0.1$ and $\tau = 0.25$, for 960 iterations. (f) Same as (e), but for 3,000 iterations. (g) Same as (e), but for 57,600 iterations. From Welk (2012). (See the color plate.)

As our initial contour is not very precise, the AAC with amoeba radius $\varrho = 10$ gets locked at a high contrast away from the desired region. With a slightly enlarged amoeba radius ($\varrho = 12$), a fairly good segmentation is obtained. Running the algorithm for some more iterations shows that the latter AAC becomes stationary at a contour that neglects some small details.

Let us now turn to PDE-based segmentation. Running a finite difference method for GAC up to evolution time $T = 240$, which matches the final stationary state reached by the amoebas in this experiment, still does not segment the cerebellum well. A reasonable segmentation is achieved here only after a considerably longer evolution time. The stationary solution of the GAC scheme cuts off some details and is also depicted here.

Our experiments confirm the similarity of the AAC and GAC methods. A stronger rounding of contours with the latter method can be explained by the inherent numerical dissipation of the finite difference approach. In contrast, the amoeba-based algorithm has a higher tendency to get locked at small-scale details; this also can be confirmed by other experiments and is related to the existence of root signals for the amoeba median filter; compare this to the discussion in subsection 6.1. On the one hand, this property contributes to stabilizing a fine segmentation result. On the other hand, it means that a minimal amoeba size is needed to obtain reasonable segmentation results. Experiments suggest that the ϱ value for AAC should not be smaller than 10.

7. CONCLUSION

In this chapter, we have collected and extended our previous studies on discrete amoeba filters and their relation to PDE-based processes. This theory extends the work by Guichard and Morel (1997), who proved that iterated nonadaptive median filtering in a continuous domain approximates (mean) curvature motion in the limit case of vanishing structuring element size.

The cornerstone of the work presented here is the result from Welk, Breuß, and Vogel (2011), which warrants an analogous approximation property for iterated AMF and the self-snakes equation. In this paper, we have provided two different proofs for this result. The first goes back to Welk, Breuß, and Vogel (2011) and relies on an analysis of the lengths and density of level lines within an amoeba structuring element, while the other proof, based on an idea introduced in Welk (2013a), captures structuring elements via a polar coordinate representation.

These proofs open the way to different generalizations. The principle of the first proof can be adapted to analyze PDE limits of dilation, erosion, quantile filters, and M-smoothers with amoeba structuring elements. The analysis of these filters has been presented here for the first time. The concept of the second proof, in contrast, is more helpful when investigating an active contour segmentation approach based on amoeba median filters that was introduced in Welk (2012). Results cited from Welk (2012, 2013a) establish the correspondence of such an AAC method to a PDE similar to GAC, although the exact GAC evolution occurs as limit only in special configurations like rotationally symmetric images.

Furthermore, we have presented a first step toward an analysis of the relation between presmoothing in the context of the self-snakes method and iterated AMF with a positive structuring element. Our frequency-response analysis of a test case, following Welk (2013a), indicates that the execution of iterated AMF with a nonzero structuring element radius, as is inevitable in practical computation, already possesses a regularizing effect closely related to that of Gaussian presmoothing in the self-snakes PDE.

To complement the theoretical exposition, we have demonstrated the basic properties of the amoeba filters in question by experimenting on a test image. In the case of AMF, a comparison with the corresponding self-snakes PDE has been provided that confirms the validity of the PDE approximation result. One of the experiments showed the limitations of the approximation induced by the appearance of root signals in AMF.

The immediate outcome of this investigation is a deeper understanding of similarities and differences between discrete morphological filters and PDE methods as two originally disparate image filter classes. While this is a theoretical achievement in its own right, it also has possible practical benefits: On such a basis, discrete morphological filters can be interpreted as numerical realizations of PDE methods; compare this with the work by Jalba and Roerdink (2009). This is of particular interest for those PDEs that have interesting theoretical properties, but due to their ill-posedness, they cause substantial difficulties to conventional numerical approaches such as finite difference schemes.

Future work. At this point, some open questions remain for further research. On the theoretical side, the relationship between the regularizations on the PDE and the discrete filter side (i.e., presmoothing versus structuring element radius) calls for a more general analysis than has been done for the single test case considered so far. Further consideration of this question should also envision the other discrete filters with their corresponding PDEs.

Specifically for the amoeba median filter, it would be of interest to analyze root signals in a similar way as Eckhardt (2003) for the classical median filter.

From the perspective of applications, a more detailed consideration of algorithmic complexity of the different filters and more efficient implementations will be of interest. In the long run, this work should lead to the integration of discrete morphological amoeba filters as numerical realizations of PDE-filtering steps into practical application scenarios.

Another starting point for further investigation comes from the process of computing amoeba structuring elements. This is typically done with a variant of Dijkstra's algorithm, which establishes a tree of the shortest paths to pixels within the amoeba. Whereas amoeba image filters use just the resulting set of pixels, it appears worthwhile to exploit the tree structure itself in order to extract additional information about the image. For example, such an approach could be beneficial in texture analysis.

APPENDIX

A.1 Derivation of I_3 from Subsection 4.1

In the first proof of Theorem 1, the expression $sg'(s) - h(s)$, with g and h given by Eqs. (46) and (47), needs to be calculated, paying appropriate attention to the singularity of the integrands of Eqs. (40) and (41) at $\xi = 1$.

One can show that the derivative of the integrand of I_1 with regard to α is bounded over the closed rectangular domain $[a, b] \times [0, 1]$, and therefore is uniformly continuous, and its integral converges uniformly with regard to α. In calculating $I_1'(\alpha)$, therefore, it is possible to differentiate under the integral:

$$I_1'(\alpha) = \int_0^1 \xi^2 \frac{\partial}{\partial \alpha} R_\psi(\xi, \alpha) \mathrm{d}\xi. \tag{128}$$

By virtue of

$$\begin{aligned}
\frac{\partial}{\partial \alpha} R_\psi(\xi, \alpha) &= \frac{1}{2R_\psi(\xi, \alpha)} \frac{\partial}{\partial \alpha} \left(\psi^{-2}\left(\frac{1}{\xi}\psi\left(\frac{1}{\alpha}\right)\right) - \frac{1}{\alpha^2} \right) \\
&= \frac{1}{2R_\psi(\xi, \alpha)} \left(2\psi^{-1}\left(\frac{1}{\xi}\psi\left(\frac{1}{\alpha}\right)\right) \cdot \frac{\partial}{\partial \alpha}\left(\psi^{-1}\left(\frac{1}{\xi}\psi\left(\frac{1}{\alpha}\right)\right)\right) + \frac{2}{\alpha^3} \right) \\
&= \frac{1}{R_\psi(\xi, \alpha)} \left(\psi^{-1}\left(\frac{1}{\xi}\psi\left(\frac{1}{\alpha}\right)\right) \cdot \frac{1}{\psi'\left(\psi^{-1}\left(\frac{1}{\xi}\psi\left(\frac{1}{\alpha}\right)\right)\right)} \right. \\
&\qquad \left. \times \frac{1}{\xi}\psi'\left(\frac{1}{\alpha}\right) \cdot \left(-\frac{1}{\alpha^2}\right) + \frac{1}{\alpha^3} \right) \\
&= -\frac{1}{R_\psi(\xi, \alpha)} \left(\frac{\xi^{-1}\psi^{-1}\left(\frac{1}{\xi}\psi\left(\frac{1}{\alpha}\right)\right)\psi'\left(\frac{1}{\alpha}\right)}{\alpha^2 \psi'\left(\psi^{-1}\left(\frac{1}{\xi}\psi\left(\frac{1}{\alpha}\right)\right)\right)} - \frac{1}{\alpha^3} \right),
\end{aligned} \tag{129}$$

we obtain

$$I_1'(\alpha) = \int_0^1 \frac{\xi}{R_\psi(\xi,\alpha)} \left(\frac{\psi^{-1}\left(\frac{1}{\xi}\psi\left(\frac{1}{\alpha}\right)\right)\psi'\left(\frac{1}{\alpha}\right)}{\alpha^2\psi'\left(\psi^{-1}\left(\frac{1}{\xi}\psi\left(\frac{1}{\alpha}\right)\right)\right)} - \frac{\xi}{\alpha^3} \right) \mathrm{d}\xi. \tag{130}$$

Using Eq. (46), the quotient rule, Eqs. (47) and (130), and the relation $\alpha = \sigma s$, we calculate

$$\begin{aligned} sg'(s) - h(s) &= 3\left(\frac{1}{\alpha\psi^3\left(\frac{1}{\alpha}\right)} I_1'(\alpha) - \frac{2}{\alpha^2\psi^3\left(\frac{1}{\alpha}\right)} I_1(\alpha) + \frac{3\psi'\left(\frac{1}{\alpha}\right)}{\alpha^3\psi^4\left(\frac{1}{\alpha}\right)} I_1(\alpha) \right. \\ &\qquad \left. - \frac{1}{\alpha^4\psi^3\left(\frac{1}{\alpha}\right)} I_2(\alpha) + \frac{2}{\alpha^2\psi^3\left(\frac{1}{\alpha}\right)} I_1(\alpha) \right) \\ &= \frac{3}{\alpha^4\psi^4\left(\frac{1}{\alpha}\right)} \left(\alpha^3\psi\left(\frac{1}{\alpha}\right) I_1'(\alpha) + 3\alpha\psi'\left(\frac{1}{\alpha}\right) I_1(\alpha) - \psi\left(\frac{1}{\alpha}\right) I_2(\alpha) \right) \\ &= \frac{3}{\alpha^4\psi^4\left(\frac{1}{\alpha}\right)} \int_0^1 \frac{\xi^2}{R_\psi(\xi,\alpha)} \left(\left(- \frac{\alpha\xi^{-1}\psi\left(\frac{1}{\alpha}\right)\psi^{-1}\left(\frac{1}{\xi}\psi\left(\frac{1}{\alpha}\right)\right)\psi'\left(\frac{1}{\alpha}\right)}{\psi'\left(\psi^{-1}\left(\frac{1}{\xi}\psi\left(\frac{1}{\alpha}\right)\right)\right)} \right. \right. \\ &\qquad \left. \left. + \psi\left(\frac{1}{\alpha}\right) - \psi\left(\frac{1}{\alpha}\right) \right) + 3\xi^2\alpha\psi'\left(\frac{1}{\alpha}\right) R_\psi(\xi,\alpha) \right) \mathrm{d}\xi \\ &= \frac{3\psi'\left(\frac{1}{\alpha}\right)}{\alpha^3\psi^4\left(\frac{1}{\alpha}\right)} I_3(\alpha), \end{aligned} \tag{131}$$

where I_3 is given by Eq. (49). This completes the proof.

A.2 Integrals from Subsection 4.4

The integrals J_1, J_2, J_3, and J_4 from Eqs. (83)–(86) can be represented in the following form:

$$J_{r,s,p,t} = (-1)^t \int_{-1}^{\omega_0} \omega^s \left(1-\omega^2\right)^{r/2} (\omega_0-\omega)^p \mathrm{d}\omega + \int_{\omega_0}^{1} \omega^s \left(1-\omega^2\right)^{r/2} (\omega-\omega_0)^p \mathrm{d}\omega, \tag{132}$$

with $s \in \{0,1,2,3\}, r \in \{1,-1\}, t \in \{0,1\}$. The exponent p is nonnegative in J_1, J_2, J_3 and greater than -1 in J_4.

Assume that $\omega_0 = \mathcal{O}(\varrho)$. The following transformations of $J_{r,s,p,t}$ hold, therefore, if ϱ is small enough. Without loss of generality, let ω_0 be positive. Then

$$\begin{aligned} J_{r,s,p,t} &= (-1)^t \int_{-1}^{-\omega_0} \omega^s \left(1-\omega^2\right)^{r/2} (\omega_0-\omega)^p \mathrm{d}\omega \\ &\quad + (-1)^t \int_{-\omega_0}^{\omega_0} \omega^s \left(1-\omega^2\right)^{r/2} (\omega_0-\omega)^p \mathrm{d}\omega \\ &\quad + \int_{\omega_0}^{1} \omega^s \left(1-\omega^2\right)^{r/2} (\omega-\omega_0)^p \mathrm{d}\omega \\ &= (-1)^t \int_{-\omega_0}^{\omega_0} \omega^s \left(1-\omega^2\right)^{r/2} (\omega_0-\omega)^p \mathrm{d}\omega \\ &\quad + \int_{\omega_0}^{1} \omega^s \left(1-\omega^2\right)^{r/2} \left((\omega-\omega_0)^p + (-1)^{t+s}(\omega+\omega_0)^p\right) \mathrm{d}\omega \end{aligned}$$

$$= \underbrace{(-1)^t \int_{-\omega_0}^{\omega_0} \omega^s \left(1-\omega^2\right)^{r/2} (\omega_0-\omega)^p \mathrm{d}\omega}_{K_1}$$

$$+ \underbrace{\int_{\omega_0}^{\sqrt{\omega_0}} \omega^s \left(1-\omega^2\right)^{r/2} \left((\omega-\omega_0)^p + (-1)^{t+s}(\omega+\omega_0)^p\right) \mathrm{d}\omega}_{K_2}$$

$$+ \underbrace{\int_{\sqrt{\omega_0}}^{1} \omega^s \left(1-\omega^2\right)^{r/2} \left((\omega-\omega_0)^p + (-1)^{t+s}(\omega+\omega_0)^p\right) \mathrm{d}\omega}_{K_3}. \tag{133}$$

For $p \geq 0$ (as is always the case in J_1, J_2, and J_3), one has $K_1 = \mathcal{O}(\varrho^{1+s+p})$; for $-1 < p < 0$ (as can occur in J_4), K_1 is of order $\mathcal{O}(\varrho^{1+s})$ because the integral $\int_{-\omega_0}^{\omega_0} (\omega_0 - \omega)^p \mathrm{d}\omega$ is finite. Similarly, K_2 is of order $\mathcal{O}\left(\varrho^{\frac{1+s+p}{2}}\right)$ if $p \geq 0$, or $\mathcal{O}\left(\varrho^{\frac{1+s}{2}}\right)$ for $-1 < p < 0$. In all cases, therefore, K_3 dominates $J_{r,s,p,t}$.

In analyzing K_3 further, notice that throughout its integration domain, $\omega_0/\omega = \mathcal{O}(\sqrt{\varrho})$, and thus $(1-\omega_0/\omega)^p = 1 + \mathcal{O}(\sqrt{\varrho})$ for any fixed p. For even $t+s$ (which is the case in J_2, J_3, and J_4), we have

$$K_3 = \int_{\sqrt{\omega_0}}^{1} \omega^s \left(1-\omega^2\right)^{r/2} \omega^p \left(\left(1-\frac{\omega_0}{\omega}\right)^p + \left(1+\frac{\omega_0}{\omega}\right)^p\right) \mathrm{d}\omega$$

$$= \left(1+\mathcal{O}(\sqrt{\varrho})\right) \int_{\sqrt{\omega_0}}^{1} 2\omega^s \left(1-\omega^2\right)^{r/2} \omega^p \mathrm{d}\omega$$

$$= 2\int_0^1 \omega^{s+p} \left(1-\omega^2\right)^r \mathrm{d}\omega - 2\int_0^{\sqrt{\omega_0}} \omega^{s+p} \left(1-\omega^2\right)^r \mathrm{d}\omega + \mathcal{O}(\sqrt{\varrho}), \tag{134}$$

and since the subtracted integral is $\mathcal{O}(\sqrt{\varrho})$, it follows that in this case,

$$J_{r,s,p,t} = 2F(r, s+p) + \mathcal{O}\left(\varrho^{1/2} + \varrho^{1+s}\right), \tag{135}$$

where F is defined as in Eq. (87).

For odd $t+s$ (applies to J_1), we use the fact that $\frac{(\omega+\omega_0)^p - (\omega-\omega_0)^p}{2\omega_0}$ is a first-order approximation of $\mathrm{d}x^p/\mathrm{d}x$ at $x = \omega - \omega_0$, which implies the following:

$$\begin{aligned} K_3 &= \int_{\sqrt{\omega_0}}^{1} \omega^s \left(1-\omega^2\right)^{r/2} (-2p\omega_0)(\omega-\omega_0)^{p-1} \left(1 + \mathcal{O}(\sqrt{\varrho})\right) \mathrm{d}\omega \\ &= -2p\omega_0 \int_{\sqrt{\omega_0}}^{1} \omega^s \left(1-\omega^2\right)^{r/2} \omega^{p-1} \left(1 - \frac{\omega_0}{\omega}\right)^{p-1} \mathrm{d}\omega + \mathcal{O}(\sqrt{\varrho}) \\ &= -2p\omega_0 \int_{\sqrt{\omega_0}}^{1} \omega^{s+p-1} \left(1-\omega^2\right)^{r/2} \mathrm{d}\omega + \mathcal{O}(\sqrt{\varrho}). \end{aligned} \tag{136}$$

For $s+p-1>0$, the integral over ω^{s+p-1} is finite even at 0. Therefore, we can proceed as follows:

$$\begin{aligned} K_3 = {} & -2p\omega_0 \int_0^1 \omega^{s+p-1} \left(1-\omega^2\right)^{r/2} \mathrm{d}\omega + 2p\omega_0 \int_0^{\sqrt{\omega_0}} \omega^{s+p-1} \left(1-\omega^2\right)^{r/2} \mathrm{d}\omega \\ & + \mathcal{O}(\sqrt{\varrho}), \end{aligned} \tag{137}$$

where again the subtracted integral itself is $\mathcal{O}(\sqrt{\varrho})$, yielding

$$J_{r,s,p,t} = -2p\omega_0 F(r, s+p-1) + \mathcal{O}(\sqrt{\varrho}). \tag{138}$$

From Eqs. (135) and (138), Eqs. (88)–(91) can be read immediately by inserting the appropriate parameters (r, s, t, and p).

A.3 Integral Approximation from Subsection 4.5

In this section, we will prove that Eq. (110) is approximated by Eq. (111) for sufficiently small values of ϱ.

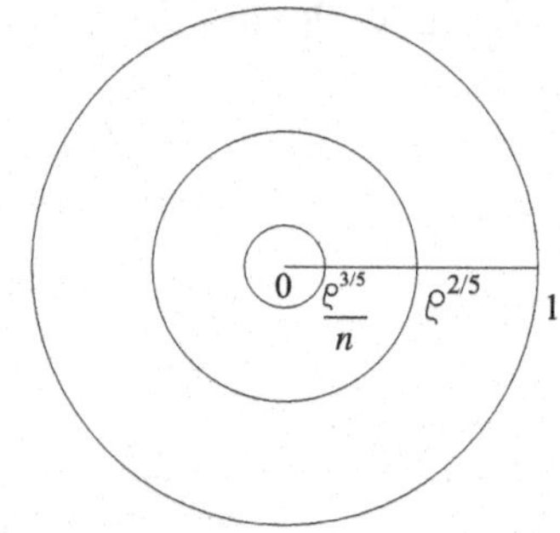

Figure 13 Partition of the integration domain D_1 into three concentric regions. From Welk (2007).

Assuming that $\varrho \ll 1$, we split the integration domain D_1 into an inner disk of radius $\varrho^{3/5}/n$, an annulus with inner radius $\varrho^{3/5}/n$ and outer radius $\varrho^{2/5}$, and the remaining annulus (see Figure 13).

1. For the inner disk, we estimate

$$\iint\limits_{D_{\varrho^{3/5}/n}} \frac{m\xi + (\alpha\xi^2 - a)\varrho}{\sqrt{\left(m\xi + (\alpha\xi^2 - a)\varrho\right)^2 + (n\eta)^2}} \mathrm{d}\xi\mathrm{d}\eta = \mathcal{O}(\varrho^{6/5}) \tag{139}$$

since the absolute value of the integrand is bounded by 1 and is integrated over an integration domain of area $\mathcal{O}(\varrho^{6/5})$.

2. For the middle annulus, we find

$$\begin{aligned} &\iint\limits_{D_{\varrho^{2/5}}\setminus D_{\varrho^{3/5}/n}} \frac{m\xi + (\alpha\xi^2 - a)\varrho}{\sqrt{\left(m\xi + (\alpha\xi^2 - a)\varrho\right)^2 + (n\eta)^2}} \mathrm{d}\xi\mathrm{d}\eta \\ &= \iint\limits_{\left(D_{\varrho^{2/5}}\setminus D_{\varrho^{3/5}/n}\right)\cap(\mathbb{R}_0^+\times\mathbb{R})} \left(\frac{m\xi + (\alpha\xi^2 - a)\varrho}{\sqrt{\left(m\xi + (\alpha\xi^2 - a)\varrho\right)^2 + (n\eta)^2}} \right. \\ &\qquad \left. + \frac{-m\xi + (\alpha\xi^2 - a)\varrho}{\sqrt{\left(-m\xi + (\alpha\xi^2 - a)\varrho\right)^2 + (n\eta)^2}} \right) \mathrm{d}\xi\mathrm{d}\eta \\ &= \mathcal{O}\left(\varrho^{6/5}\right) \end{aligned} \tag{140}$$

since the integrand is bounded by $\mathcal{O}(\varrho^{2/5})$ according to

$$
\begin{aligned}
&\frac{m\xi+(\alpha\xi^2-a)\varrho}{\sqrt{\left(m\xi+(\alpha\xi^2-a)\varrho\right)^2+(n\eta)^2}}+\frac{-m\xi+(\alpha\xi^2-a)\varrho}{\sqrt{\left(-m\xi+(\alpha\xi^2-a)\varrho\right)^2+(n\eta)^2}}\\
&\quad=m\xi\left(\frac{1}{\sqrt{\left(m\xi+(\alpha\xi^2-a)\varrho\right)^2+(n\eta)^2}}-\frac{1}{\sqrt{\left(-m\xi+(\alpha\xi^2-a)\varrho\right)^2+(n\eta)^2}}\right)\\
&\quad\quad+(d\xi^2-a)\varrho\left(\frac{1}{\sqrt{\left(m\xi+(\alpha\xi^2-a)\varrho\right)^2+(n\eta)^2}}+\frac{1}{\sqrt{\left(-m\xi+(\alpha\xi^2-a)\varrho\right)^2+(n\eta)^2}}\right)\\
&\quad=\underbrace{\left(\frac{1}{\sqrt{\left(1+(\alpha\xi^2-a)\frac{\varrho}{m\xi}\right)^2+\left(\frac{n\eta}{m\xi}\right)^2}}-\frac{1}{\sqrt{\left(1-(\alpha\xi^2-a)\frac{\varrho}{m\xi}\right)^2+\left(\frac{n\eta}{m\xi}\right)^2}}\right)}_{\mathcal{O}(\varrho^{2/5})}\\
&\quad+(d\xi^2-a)\underbrace{\frac{\varrho}{m\xi}}_{\mathcal{O}(\varrho^{2/5})}\underbrace{\left(\frac{1}{\sqrt{\left(1+(\alpha\xi^2-a)\frac{\varrho}{m\xi}\right)^2+\left(\frac{n\eta}{m\xi}\right)^2}}+\frac{1}{\sqrt{\left(1-(\alpha\xi^2-a)\frac{\varrho}{m\xi}\right)^2+\left(\frac{n\eta}{m\xi}\right)^2}}\right)}_{\mathcal{O}(1)}\\
&\quad=\mathcal{O}(\varrho^{2/5}),
\end{aligned}
\tag{141}
$$

whereas the area of the integration domain is $\mathcal{O}(\varrho^{4/5})$.

3. For the outer annulus, we use the expansion

$$
(p+q)^{-1/2}=p^{-1/2}\left(1-\frac{q}{2p}+\mathcal{O}\left(\frac{q^2}{p^2}\right)\right) \tag{142}
$$

to obtain

$$\iint_{D_1\backslash D_{\varrho^{2/5}}} \frac{m\xi + (\alpha\xi^2 - a)\varrho}{\sqrt{(m\xi + (\alpha\xi^2 - a)\varrho)^2 + (n\eta)^2}} \mathrm{d}\xi \mathrm{d}\eta$$

$$= \underbrace{\iint_{D_1\backslash D_{\varrho^{2/5}}} \frac{m\xi}{\sqrt{(m\xi)^2 + (n\eta)^2}} \mathrm{d}\xi \mathrm{d}\eta}_{=0} + \iint_{D_1\backslash D_{\varrho^{2/5}}} \frac{(\alpha\xi^2 - a)\varrho}{((m\xi)^2 + (n\eta)^2)^{3/2}}$$

$$\times ((m\xi)^2 + (n\eta)^2 - (m\xi)^2) \mathrm{d}\xi \mathrm{d}\eta + \mathcal{O}\left(\varrho^{6/5}\right)$$

$$= \iint_{D_1\backslash D_{\varrho^{2/5}}} \varrho \frac{(n\eta)^2}{((m\xi)^2 + (n\eta)^2)^{3/2}} (\alpha\xi^2 - a) \mathrm{d}\xi \mathrm{d}\eta + \mathcal{O}\left(\varrho^{6/5}\right). \quad (143)$$

4. Finally, we have that

$$\iint_{D_{\varrho^{2/5}}} \varrho \frac{(n\eta)^2}{((m\xi)^2 + (n\eta)^2)^{3/2}} (\alpha\xi^2 - a) \mathrm{d}\xi \mathrm{d}\eta$$

$$= \iint_{D_{\varrho^{2/5}}} \varrho \underbrace{\frac{(n\eta)^2}{(m\xi)^2 + (n\eta)^2}}_{\mathcal{O}(1)} \cdot \frac{1}{\sqrt{(m\xi)^2 + (n\eta)^2}} (\alpha\xi^2 - a) \mathrm{d}\xi \mathrm{d}\eta$$

$$= \int_0^{\varrho^{2/5}} \int_0^{2\pi} \mathcal{O}(\varrho)\mathcal{O}(r^{-1}) \cdot r \, \mathrm{d}\varphi \mathrm{d}r \quad (144)$$

$$= \mathcal{O}(\varrho) \cdot \int_0^{\varrho^{2/5}} \mathrm{d}r = \mathcal{O}\left(\varrho^{7/5}\right).$$

5. Combining Eqs. (110), (139), (140), (143), and (144), the approximation [Eq. (111)] follows.

REFERENCES

Adalsteinsson, D., & Sethian, J. A. (1995). A fast level set method for propagating interfaces. *Journal of Computational Physics, 118*(2), 269–277.

Alvarez, L., Lions, P.-L., & Morel, J.-M. (1992). Image selective smoothing and edge detection by nonlinear diffusion II. *SIAM Journal on Numerical Analysis, 29*, 845–866.

Angulo, J. (2011). Morphological bilateral filtering and spatially-variant adaptive structuring functions. In P. Soille, M. Pesaresi, & G. Ouzounis (Eds.), *Mathematical Morphology and Its Applications to Image and Signal Processing* (pp. 212–223). Vol. 6671 of Lecture Notes in Computer Science. Berlin: Springer.

Astola, J., Haavisto, P., & Neuvo, Y. (1990). Vector median filters. *Proceedings of the IEEE, 78*(4), 678–689.

Austin, T. L. (1959). An approximation to the point of minimum aggregate distance. *Metron, 19*, 10–21.

Barash, D. (2001). Bilateral filtering and anisotropic diffusion: towards a unified viewpoint. In M. Kerckhove (Ed.), *Scale-Space and Morphology in Computer Vision* (pp. 273–280). Berlin: Springer. Vol. 2106 of Lecture Notes in Computer Science.

Barnett, V. (1976). The ordering of multivariate data. *Journal of the Royal Statistical Society A, 139*(3), 318–355.

Barni, M., Buit, F., Bartolini, F., & Cappellini, V. (2000). A quasi-Euclidean norm to speed up vector median filtering. *IEEE Transactions on Image Processing, 9*(10), 1704–1709.

Barral Souto, J. (1938). *El modo y otras medias, casos particulares de una misma expresión matemática. Technical Report 3, Cuadernos de Trabajo, Instituto de Biometria.* Argentina: Universidad Nacional de Buenos Aires.

Blum, M., Floyd, R. W., Pratt, V., Rivest, R., & Tarjan, R. (1973). Time bounds for selection. *Journal of Computers and System Sciences, 7*, 448–461.

Borgefors, G. (1986). Distance transformations in digital images. *Computer Vision, Graphics, and Image Processing, 34*, 344–371.

Borgefors, G. (1996). On digital distance transforms in three dimensions. *Computer Vision and Image Understanding, 64*(3), 368–376.

Braga-Neto, U. M. (1996). Alternating sequential filters by adaptive neighborhood structuring functions. In P. Maragos, R. W. Schafer, & M. A. Butt (Eds.), *Mathematical Morphology and Its Applications to Image and Signal Processing* (pp. 139–146). Vol. 5 of Computational Imaging and Vision. Kluwer, Dordrecht.

Breuß, M., & Welk, M. (2007). Analysis of staircasing in semidiscrete stabilised inverse linear diffusion algorithms. *Journal of Computational and Applied Mathematics, 206*(1), 520–533.

Breuß, M., Burgeth, B., & Weickert, J. (2007). Anisotropic continuous-scale morphology. In J. Martí, J. M. Benedí, A. Mendonça, & J. Serrat (Eds.), *Pattern Recognition and Image Analysis* (pp. 515–522). Berlin: Springer. Vol. 4478 of Lecture Notes in Computer Science.

Burgeth, B., Pizarro, L., Breuß, M., & Weickert, J. (2011). Adaptive continuous scale morphology for matrix fields. *International Journal of Computer Vision, 92*(2), 146–161.

Carmona, R., & Zhong, S. (1998). Adaptive smoothing respecting feature directions. *IEEE Transactions on Image Processing, 7*(3), 353–358.

Caselles, V., Catté, F., Coll, T., & Dibos, F. (1993). A geometric model for active contours in image processing. *Numerische Mathematik, 66*, 1–31.

Caselles, V., Kimmel, R., & Sapiro, G. (1995). Geodesic active contours. In *Proceedings of the Fifth International Conference on Computer Vision* (pp. 694–699). Cambridge, MA: IEEE Computer Society Press.

Caselles, V., Morel, J.-M., & Sbert, C. (1998). An axiomatic approach to image interpolation. *IEEE Transactions on Image Processing, 7*(3), 376–386.

Caselles, V., Sapiro, G., & Chung, D. H. (2000). Vector median filters, inf-sup operations, and coupled PDEs: Theoretical connections. *Journal of Mathematical Imaging and Vision, 8*, 109–119.

Chui, C. K., & Wang, J. (2009). PDE models associated with the bilateral filter. *Advances in Computational Mathematics, 31*(1-3), 131–156.

Chung, D. H., & Sapiro, G. (2000). On the level lines and geometry of vector-valued images. *IEEE Signal Processing Letters, 7*(9), 241–243.

Cohen, L. D. (1991). On active contour models and balloons. *CVGIP: Image Understanding, 53*(2), 211–218.

Ćurić, V., Hendriks, C. L., & Borgefors, G. (2012). Salience adaptive structuring elements. *IEEE Journal of Selected Topics in Signal Processing, 6*(7), 809–819.

Debayle, J., & Pinoli, J.-C. (2006). General adaptive neighborhood image processing. Part I: Introduction and theoretical aspects. *Journal of Mathematical Imaging and Vision, 25*(2), 245–266.

Didas, S., & Weickert, J. (2007). Combining curvature motion and edge-preserving denoising. In F. Sgallari, F. Murli, & N. Paragios (Eds.), *Scale Space and Variational Methods in Computer Vision* (pp. 568–579). Berlin: Springer. Vol. 4485 of Lecture Notes in Computer Science.

Dougherty, E. R., & Astola, J. (Eds.), (1999). *Nonlinear Filters for Image Processing*. Bellingham, WA: SPIE Press.

Eckhardt, U. (2003). Root images of median filters. *Journal of Mathematical Imaging and Vision, 19*, 63–70.

Fabbri, R., Da F. Costa, L., Torelli, J. C., & Bruno, O. M. (2008). 2D Euclidean distance transform algorithms: A comparative survey. *ACM Computing Surveys, 40*(1), art. 2.

Feddern, C., Weickert, J., Burgeth, B., & Welk, M. (2006). Curvature-driven PDE methods for matrix-valued images. *International Journal of Computer Vision, 69*(1), 91–103.

Förstner, W., & Gülch, E. (1987). A fast operator for detection and precise location of distinct points, corners, and centres of circular features. In *Proceedings of the ISPRS Intercommission Conference on Fast Processing of Photogrammetric Data* (pp. 281–305). Interlaken, Switzerland.

Grazzini, J., & Soille, P. (2009). Edge-preserving smoothing using a similarity measure in adaptive geodesic neighbourhoods. *Pattern Recognition, 42*, 2306–2316.

Guichard, F., & Morel, J.-M. (1997). Partial differential equations and image iterative filtering. In I. S. Duff, & G. A. Watson (Eds.), *The State of the Art in Numerical Analysis* (pp. 525–562). Oxford, UK: Clarendon Press. No. 63 in IMA Conference Series (New Series).

Heijmans, H. J. A. M. (1994). *Morphological Image Operators*. Boston: Academic Press.

Huber, P. J. (1981). *Robust Statistics*. New York: Wiley.

Hyman, J. M., & Shashkov, M. (1997). Natural discretizations of the divergence, gradient, and curl on logically rectangular grids. *International Journal of Computers and Mathematics with Applications, 33*(4), 81–104.

Hyman, J., Morel, J., Shashkov, M., & Steinberg, S. (2002). Mimetic finite difference methods for diffusion equations. *Computational Geosciences, 6*, 333–352.

Ikonen, L. (2007). Priority pixel queue algorithm for geodesic distance transforms. *Image and Vision Computing, 25*(10), 1520–1529.

Ikonen, L., & Toivanen, P. (2005). Shortest routes on varying height surfaces using gray-level distance transforms. *Image and Vision Computing, 23*(2), 133–141.

Jalba, A. C., & Roerdink, J. B. T. M. (2009). An efficient morphological active surface model for volumetric image segmentation. In M. H. F. Wilkinson, & J. B. T. M. Roerdink (Eds.), *Mathematical Morphology and Its Application to Signal and Image Processing* (pp. 193–204). Berlin: Springer. Vol. 5720 of Lecture Notes in Computer Science.

Kass, M., Witkin, A., & Terzopoulos, D. (1988). Snakes: Active contour models. *International Journal of Computer Vision, 1*(4), 321–331.

Kichenassamy, S., Kumar, A., Olver, P., Tannenbaum, A., & Yezzi, A. (1995). Gradient flows and geometric active contour models. In *Proceedings of the Fifth International Conference on Computer Vision. Cambridge, MA* (pp. 810–815). Washington, DC: IEEE Computer Society Press.

Kimmel, R. (2003). Fast edge integration. In S. Osher, & N. Paragios (Eds.), *Geometric Level Set Methods in Imaging, Vision and Graphics* (pp. 59–77). New York: Springer.

Kimmel, R., Sochen, N., & Malladi, R. (1997). Images as embedding maps and minimal surfaces: movies, color, and volumetric medical images. In *Proceedings of the 1997 IEEE Computer Society Conference on Computer Vision and Pattern Recognition, San Juan, Puerto Rico* (pp. 350–355). Washington, DC: IEEE Computer Society Press.

Klette, R., & Zamperoni, P. (1996). *Handbook of Image Processing Operators.* New York: Wiley.

Koschan, A., & Abidi, M. (2001). A comparison of median filter techniques for noise removal in color images. In *Proceedings of the 7th German Workshop on Color Image Processing* (pp. 69–79). Germany: Erlangen.

Kuhn, H. W. (1973). A note on Fermat's problem. *Mathematical Programming, 4*, 98–107.

Kuwahara, M., Hachimura, K., Eiho, S., & Kinoshita, M. (1976). Processing of RI-angiocardiographic images. In K. Preston, Jr., & M. Onoe (Eds.), *Digital Processing of Biomedical Images* (pp. 187–202). New York: Plenum.

Lerallut, R., Decencière, E., & Meyer, F. (2005). Image processing using morphological amoebas. In C. Ronse, L. Najman, & E. Decencière (Eds.), *Mathematical Morphology: 40 Years On* (pp. 13–22). Dordrecht: Springer. Vol. 30 of Computational Imaging and Vision.

Lerallut, R., Decencière, E., & Meyer, F. (2007). Image filtering using morphological amoebas. *Image and Vision Computing, 25*(4), 395–404.

Malladi, R., Sethian, J. A., & Vemuri, B. C. (1993). A topology-independent shape modeling scheme. In B. Vemuri (Ed.), *Geometric Methods in Computer Vision* (pp. 246–258). Bellingham, WA: SPIE Press. Vol. 2031 of Proceedings of SPIE.

Maragos, P., & Vachier, C. (2009). Overview of adaptive morphology: Trends and perspectives. In *Proceedings of the 2009 IEEE International Conference on Image Processing* (pp. 2241–2244).

Matheron, G. (1967). *Eléments pour une théorie des milieux poreux.* Paris: Masson.

Mickens, R. E. (1994). *Nonstandard Finite Difference Models of Differential Equations.* Singapore: World Scientific.

Nagao, M., & Matsuyama, T. (1979). Edge preserving smoothing. *Computer Graphics and Image Processing, 9*(4), 394–407.

Osher, S., & Rudin, L. I. (1990). Feature-oriented image enhancement using shock filters. *SIAM Journal on Numerical Analysis, 27*, 919–940.

Osher, S., & Sethian, J. A. (1988). Fronts propagating with curvature-dependent speed: Algorithms based on Hamilton-Jacobi formulations. *Journal of Computational Physics, 79*, 12–49.

Perona, P., & Malik, J. (1990). Scale space and edge detection using anisotropic diffusion. *IEEE Transactions on Pattern Analysis and Machine Intelligence, 12*, 629–639.

Pierpaoli, C., Jezzard, P., Basser, P. J., Barnett, A., & Di Chiro, G. (1996). Diffusion tensor MR imaging of the human brain. *Radiology, 201*(3), 637–648.

Roerdink, J. (2009). Adaptive and group invariance in mathematical morphology. In *Proceedings of the 2009 IEEE International Conference on Image Processing* (pp. 2253–2256).

Rosin, P., & West, G. (1995). Salience distance transforms. *CVGIP: Graphical Models and Image Processing, 57*(6), 483–521.

Sapiro, G. (1996). *Vector (self) snakes: A geometric framework for color, texture, and multiscale image segmentation*. In *Proceedings of the 1996 IEEE International Conference on Image Processing* (vol. 1) (pp. 817–820). Switzerland: Lausanne.

Serra, J. (1982). *Image Analysis and Mathematical Morphology* (vol. 1). London: Academic Press.

Serra, J. (1988). *Image Analysis and Mathematical Morphology* (vol. 2). London: Academic Press.

Seymour, D. R. (1970). Note on Austin's "An approximation to the point of minimum aggregate distance." *Metron, 28*, 412–421.

Shih, F. Y., & Cheng, S. (2004). Adaptive mathematical morphology for edge linking. *Information Sciences, 167*(1-4), 9–21.

Spence, C., & Fancourt, C. (2007). *An iterative method for vector median filtering*. In *Proceedings of the 2007 IEEE International Conference on Image Processing* (vol. 5) (pp. 265–268).

Spira, A., Kimmel, R., & Sochen, N. (2007). A short-time Beltrami kernel for smoothing images and manifolds. *IEEE Transactions on Image Processing, 16*(6), 1628–1636.

Tomasi, C., & Manduchi, R. (1998). Bilateral filtering for gray and color images. In *Proceedings of the 6th International Conference on Computer Vision, Bombay, India* (pp. 839–846). New Delhi, India: Narosa Publishing House.

Torroba, P. L., Cap, N. L., Rabal, H. J., & Furlan, W. D. (1994). Fractional order mean in image processing. *Optical Engineering, 33*(2), 528–534.

Tukey, J. W. (1971). *Exploratory Data Analysis*. Menlo Park: Addison-Wesley.

van den Boomgaard, R. (2002). Decomposition of the Kuwahara-Nagao operator in terms of linear smoothing and morphological sharpening. In H. Talbot, & R. Beare (Eds.), *Mathematical Morphology: Proceedings of the 6th International Symposium, Sydney, Australia* (pp. 283–292). Collingwood, Victoria, Australia: CSIRO Publishing.

Verdú-Monedero, R., Angulo, J., & Serra, J. (2011). Anisotropic morphological filters with spatially-variant structuring elements based on image-dependent gradient fields. *IEEE Transactions on Image Processing, 20*(1), 200–212.

Verly, J. G., & Delanoy, R. L. (1993). Adaptive mathematical morphology for range imagery. *IEEE Transactions on Image Processing, 2*(2), 272–275.

Weickert, J., Welk, M., & Wickert, M. (2013). L^2-stable nonstandard finite differences for anisotropic diffusion. In A. Kuijper, T. Pock, K. Bredies, & H. Bischof (Eds.), *Scale Space and Variational Methods in Computer Vision* (pp. 380–391). Berlin: Springer. vol. 7893 of Lecture Notes in Computer Science.

Weiss, B. (2006). Fast median and bilateral filtering. In *Proceedings of SIGGRAPH 2006* (pp. 519–526).

Weiszfeld, E. (1937). Sur le point pour lequel la somme des distances de *n* points donnés est minimum. *Tôhoku Mathematics Journal, 43*, 355–386.

Welk, M. (2007). *Dynamic and geometric contributions to digital image processing*. Habilitation thesis. Saarbrücken, Germany: Saarland University.

Welk, M. (2012). Amoeba active contours. In A. M. Bruckstein, B. ter Haar Romeny, A. M. Bronstein, & M. M. Bronstein (Eds.), *Scale Space and Variational Methods in Computer Vision* (pp. 374–385). Berlin: Springer. vol. 6667 of Lecture Notes in Computer Science.

Welk, M. (2013a). Relations beween amoeba median algorithms and curvature-based PDEs. In A. Kuijper, T. Pock, K. Bredies, & H. Bischof (Eds.), *Scale Space and Variational Methods in Computer Vision* (pp. 392–403). Berlin: Springer. Vol. 7893 of Lecture Notes in Computer Science.

Welk, M. (2013b). *Analysis of amoeba active contours*. Technical Report cs:1310.0097, http://arxiv.org/abs/1310.0097.

Welk, M., Feddern, C., Burgeth, B., & Weickert, J. (2003). Median filtering of tensor-valued images. In B. Michaelis, & G. Krell (Eds.), *Pattern Recognition* (pp. 17–24). Berlin: Springer. Vol. 2781 of Lecture Notes in Computer Science.

Welk, M., Weickert, J., Becker, F., Schnörr, C., Feddern, C., & Burgeth, B. (2007). Median and related local filters for tensor-valued images. *Signal Processing, 87*, 291–308.

Welk, M., Breuß, M., & Vogel, O. (2009). Differential equations for morphological amoebas. In M. H. F. Wilkinson, & J. B. T. M. Roerdink (Eds.), *Mathematical Morphology and Its Application to Signal and Image Processing* (pp. 104–114). Berlin: Springer. Vol. 5720 of Lecture Notes in Computer Science.

Welk, M., Breuß, M., & Vogel, O. (2011). Morphological amoebas are self-snakes. *Journal of Mathematical Imaging and Vision, 39*, 87–99.

Whitaker, R. T., & Xue, X. (2001). Variable-conductance, level-set curvature for image denoising. In *Proceedings of the 2001 IEEE International Conference on Image Processing, Thessaloniki, Greece* (pp. 142–145).

Winkler, G., Aurich, V., Hahn, K., Martin, A., & Rodenacker, K. (1999). Noise reduction in images: Some recent edge-preserving methods. *Pattern Recognition and Image Analysis, 9*(4), 749–766.

Yang, G., Burger, P., Firmin, D., & Underwood, S. (1995). Structure adaptive anisotropic image filtering. *Image and Vision Computing, 14*, 135–145.

Yezzi, A., Jr. (1998). Modified curvature motion for image smoothing and enhancement. *IEEE Transactions on Image Processing,* 7(3), 345–352.

You, Y.-L., Kaveh, M., Xu, W., & Tannenbaum, A. (1994). *Analysis and design of anisotropic diffusion for image processing.* In *Proceedings of the 1994 IEEE International Conference on Image Processing* (Vol. 2) (pp. 497–501). Texas: Austin.

Contents of Volumes 151-184

Volume 151[1]

Volume 152

Volume 153

[1] Lists of the contents of volumes 100–149 are to be found in volume 150; the entire series can be searched on ScienceDirect.com

Volume 154

Volume 155

Volume 156

Volume 157

Volume 158

Volume 159

Volume 160

Volume 167

Volume 168

Volume 169

Volume 170

Volume 171

Volume 172

Volume 173

Volume 174

Volume 175

Volume 176

Volume 177

Volume 178

Volume 179

Volume 180

Volume 181

Volume 182

Volume 183

Volume 184

INDEX

Note: Page numbers followed by f indicate figures.

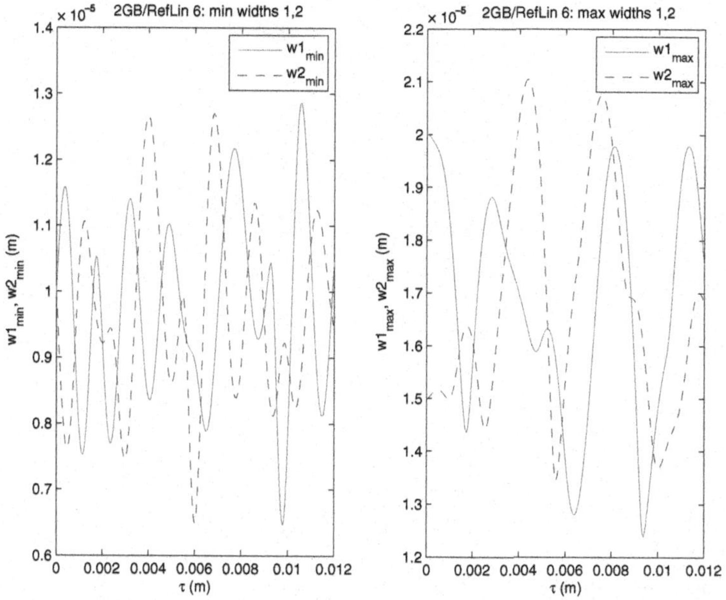

PLATE 1 (Figure 31 on page 51 of this Volume)

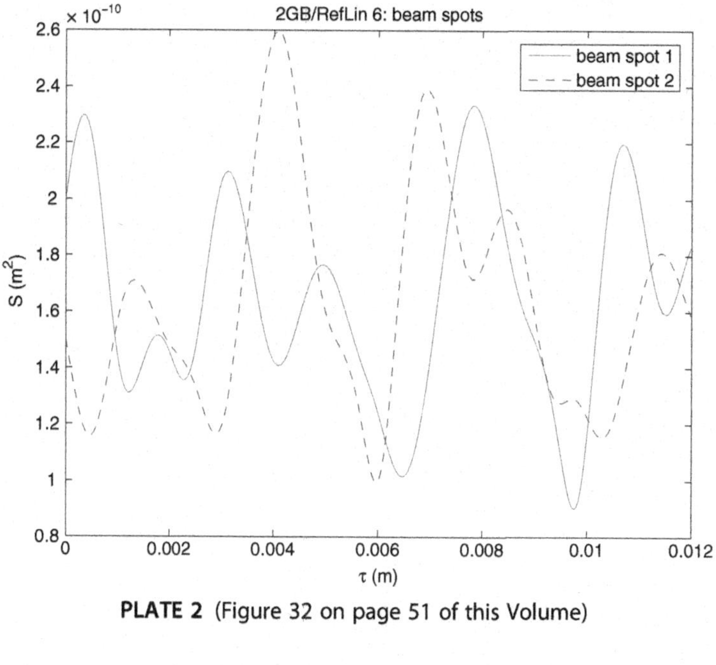

PLATE 2 (Figure 32 on page 51 of this Volume)

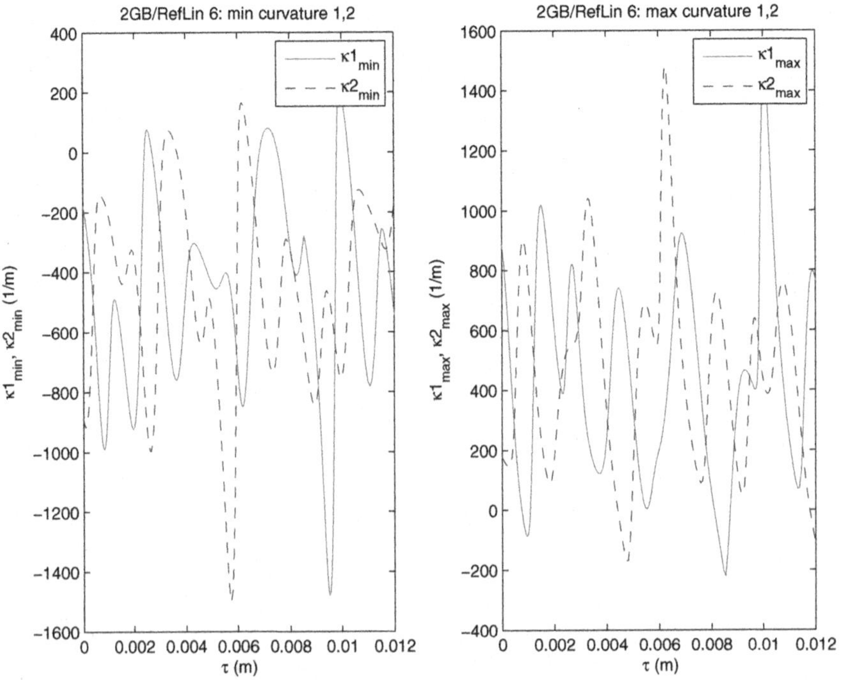

PLATE 3 (Figure 33 on page 52 of this Volume)

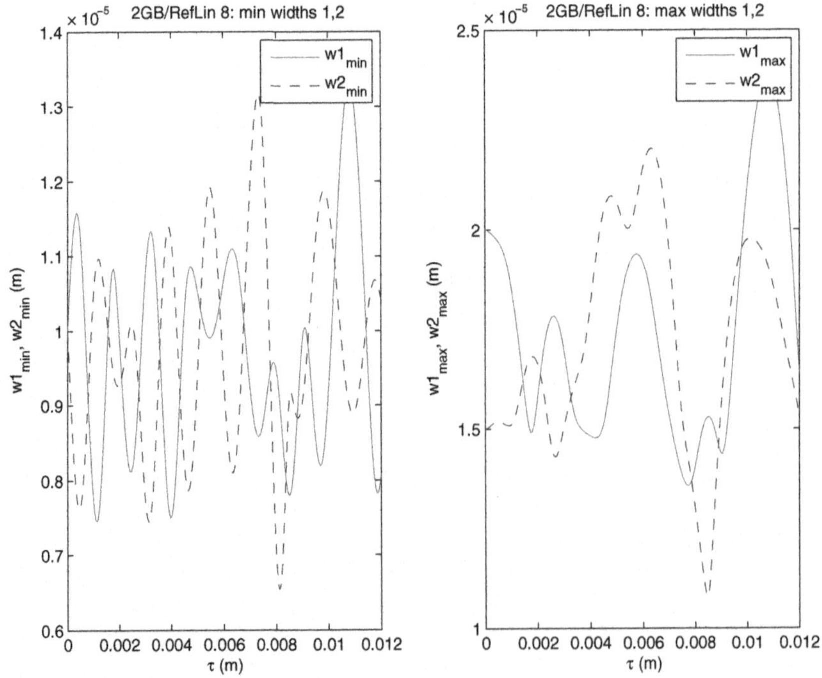

PLATE 4 (Figure 36 on page 54 of this Volume)

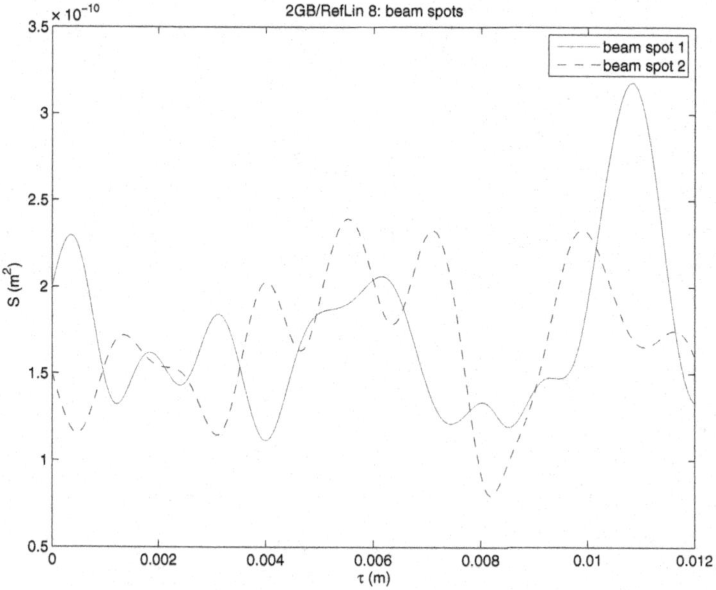

PLATE 5 (Figure 37 on page 55 of this Volume)

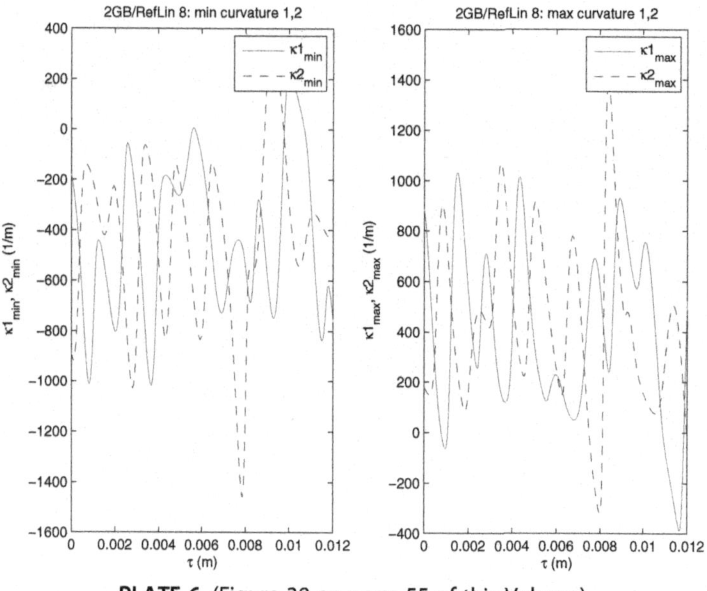

PLATE 6 (Figure 38 on page 55 of this Volume)

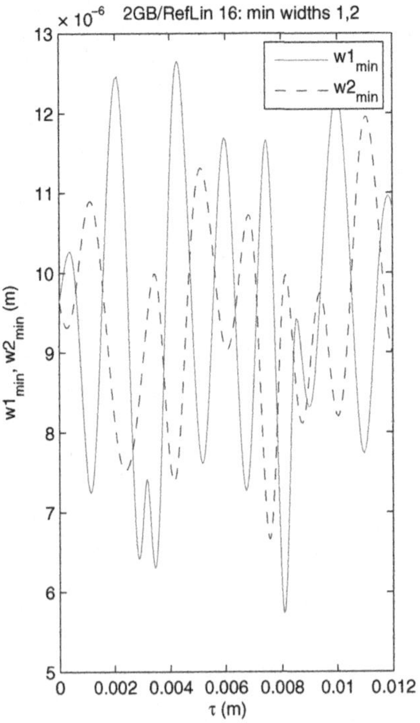

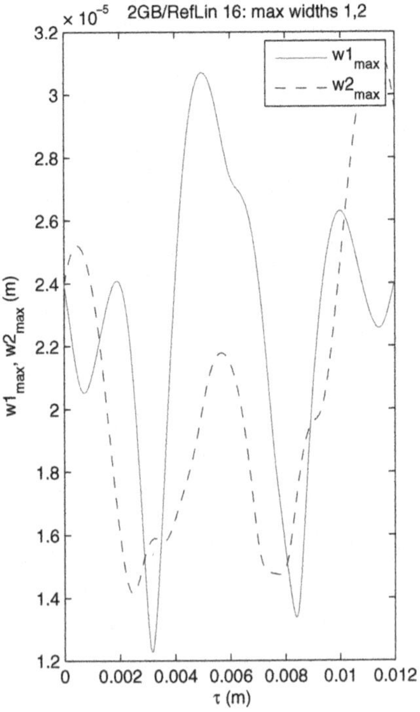

PLATE 7 (Figure 41 on page 57 of this Volume)

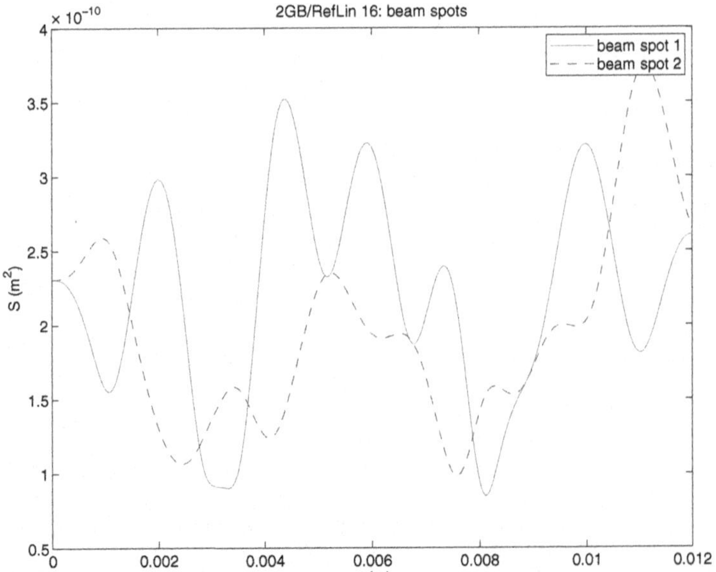

PLATE 8 (Figure 42 on page 58 of this Volume)

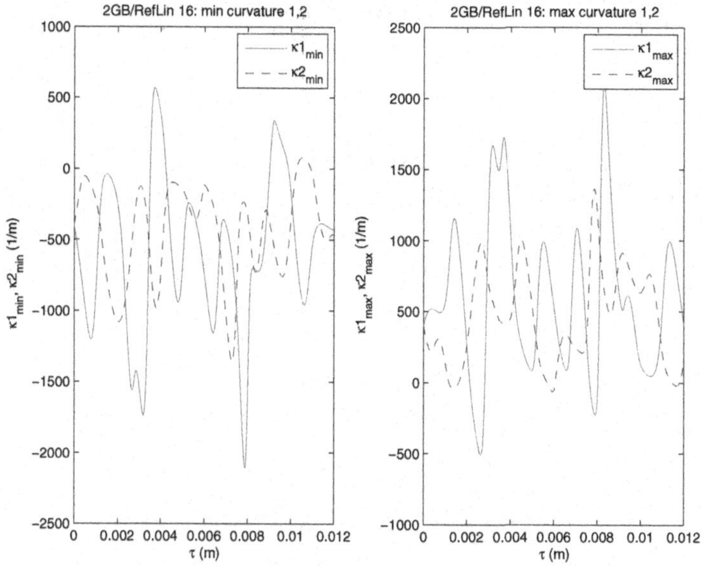

PLATE 9 (Figure 43 on page 58 of this Volume)

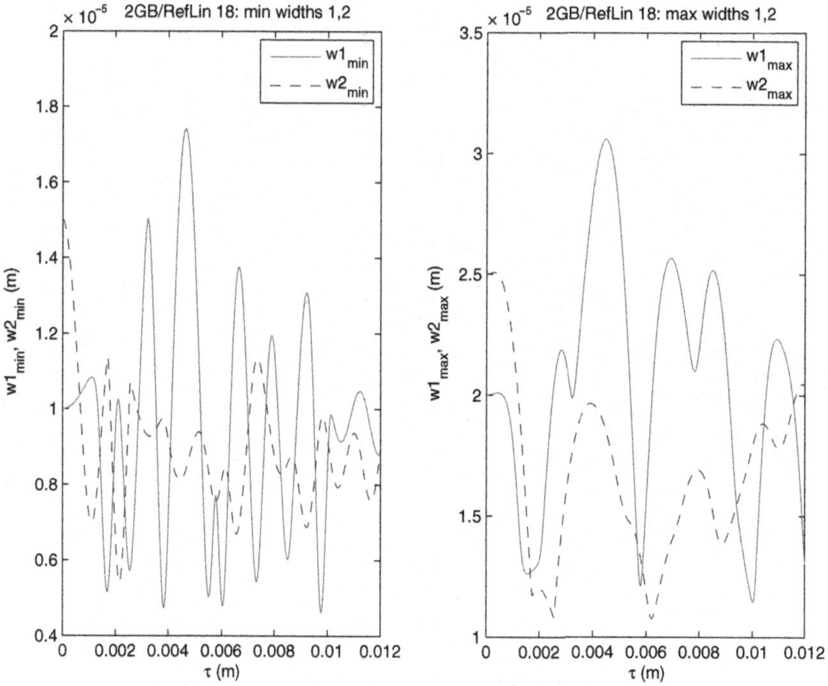

PLATE 10 (Figure 46 on page 60 of this Volume)

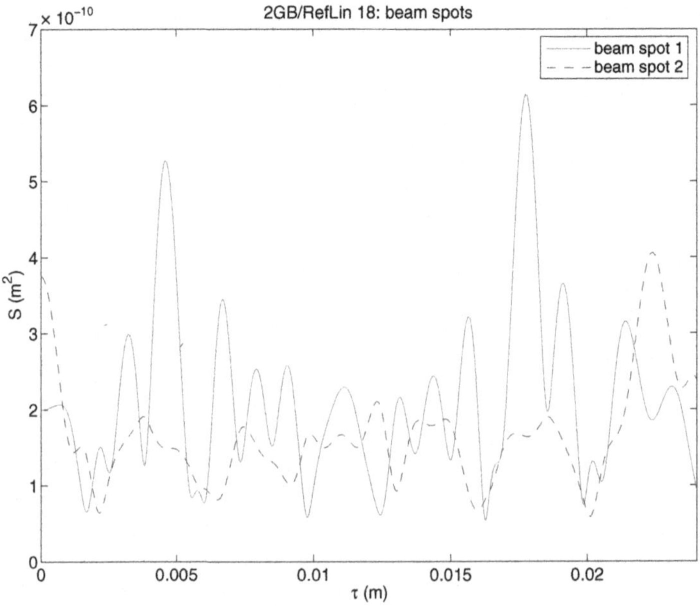

PLATE 11 (Figure 47 on page 61 of this Volume)

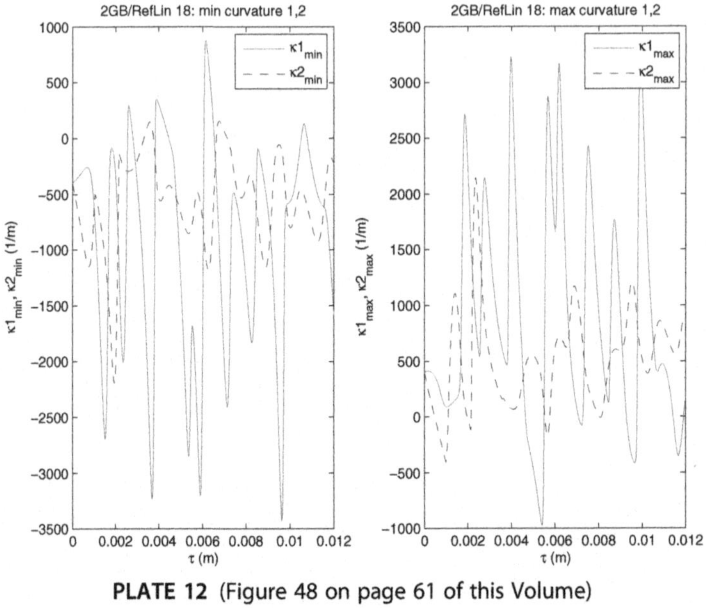

PLATE 12 (Figure 48 on page 61 of this Volume)

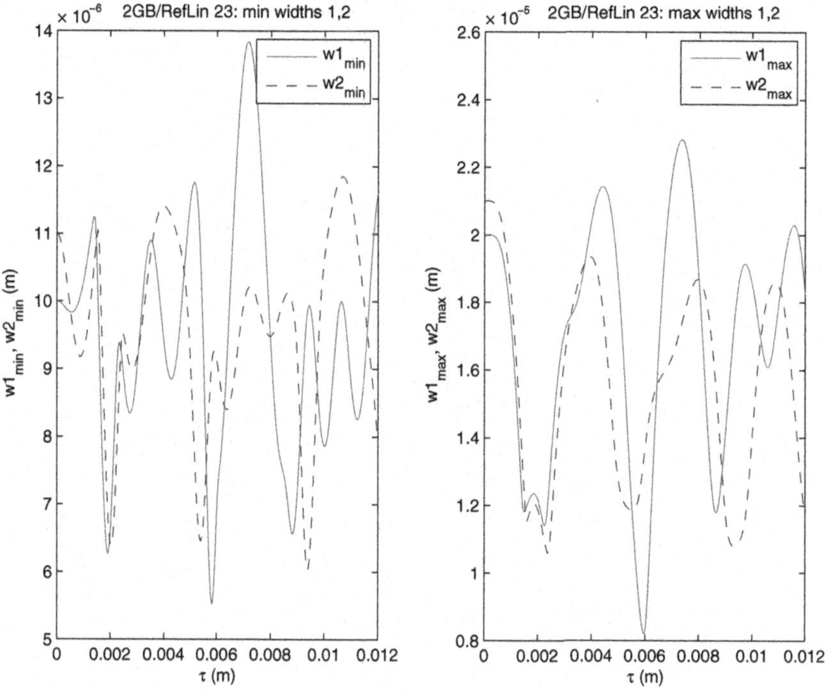

PLATE 13 (Figure 51 on page 63 of this Volume)

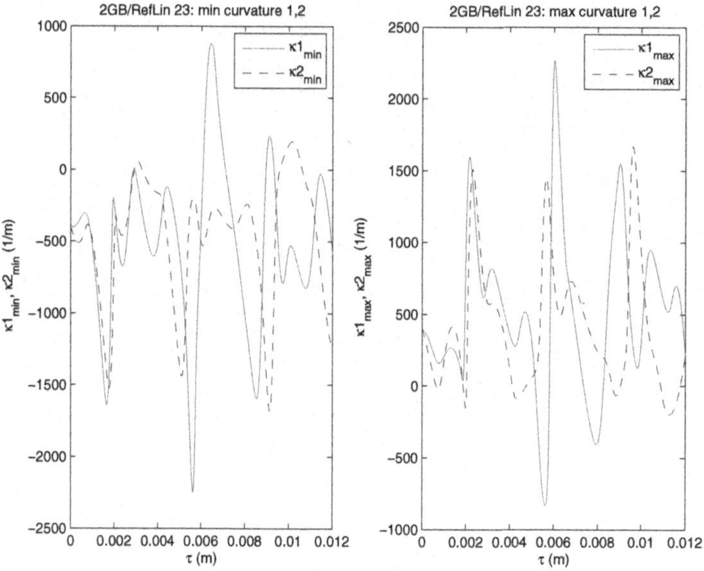

PLATE 14 (Figure 52 on page 64 of this Volume)

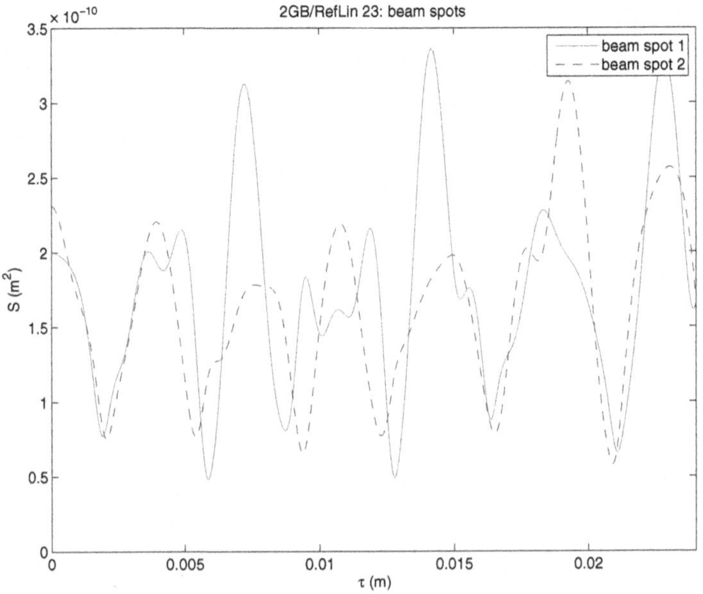

PLATE 15 (Figure 53 on page 64 of this Volume)

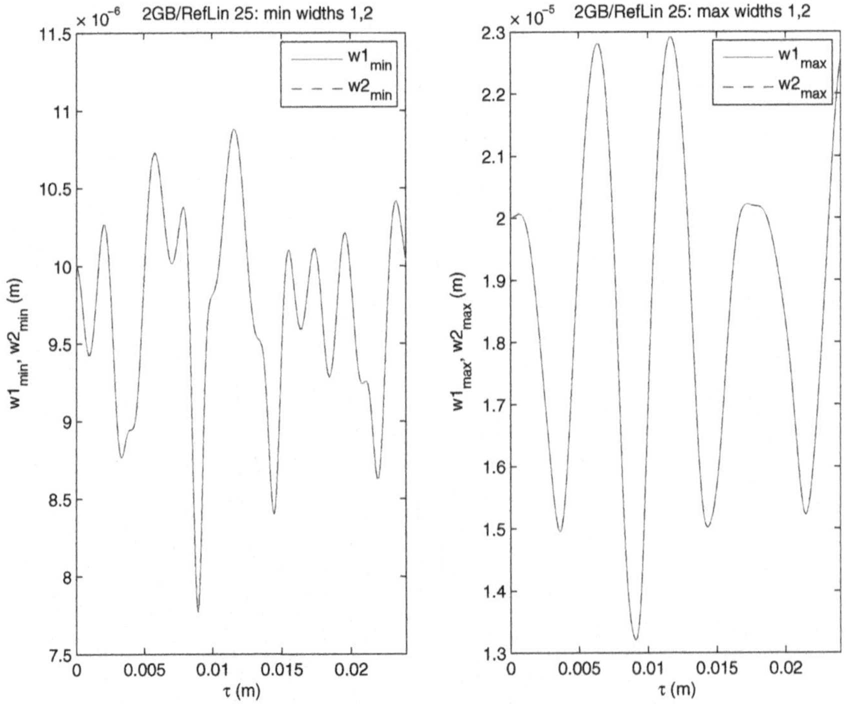

PLATE 16 (Figure 56 on page 66 of this Volume)

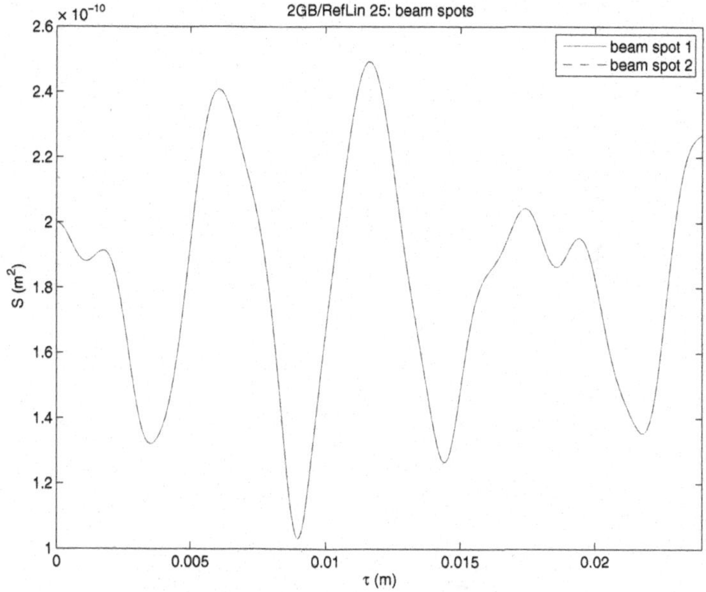

PLATE 17 (Figure 57 on page 67 of this Volume)

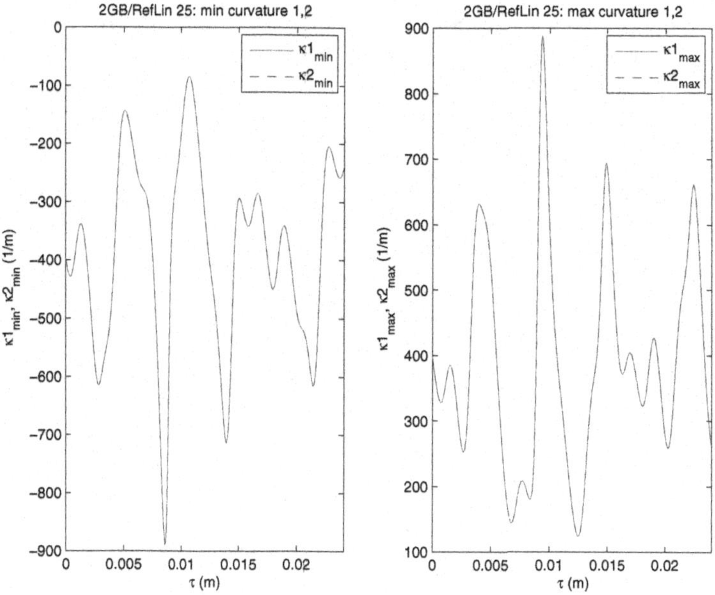

PLATE 18 (Figure 58 on page 67 of this Volume)

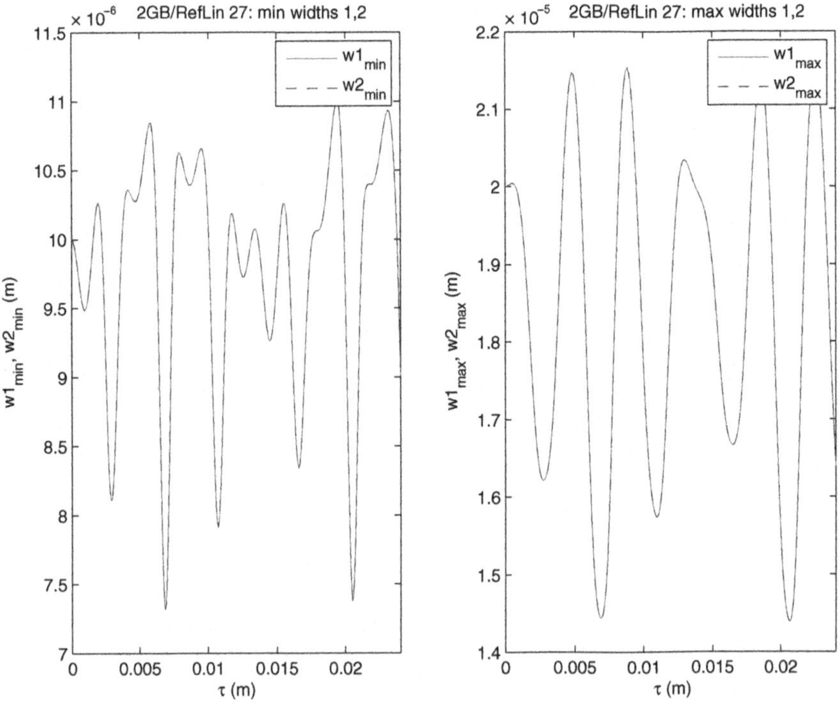

PLATE 19 (Figure 61 on page 69 of this Volume)

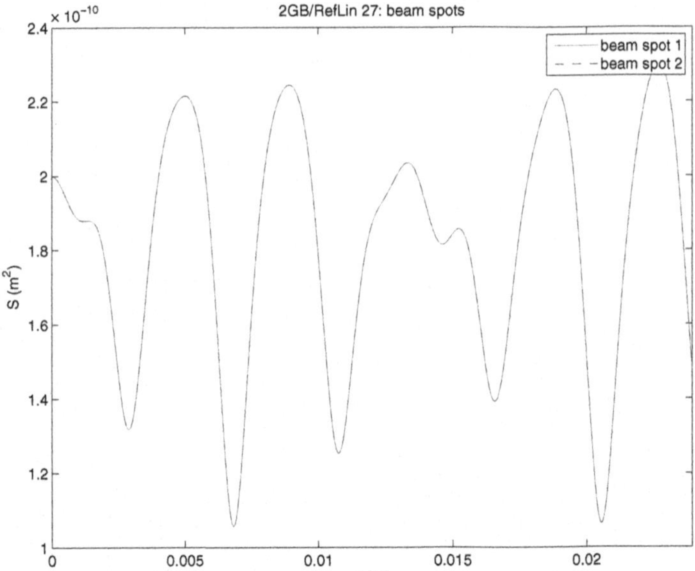

PLATE 20 (Figure 62 on page 70 of this Volume)

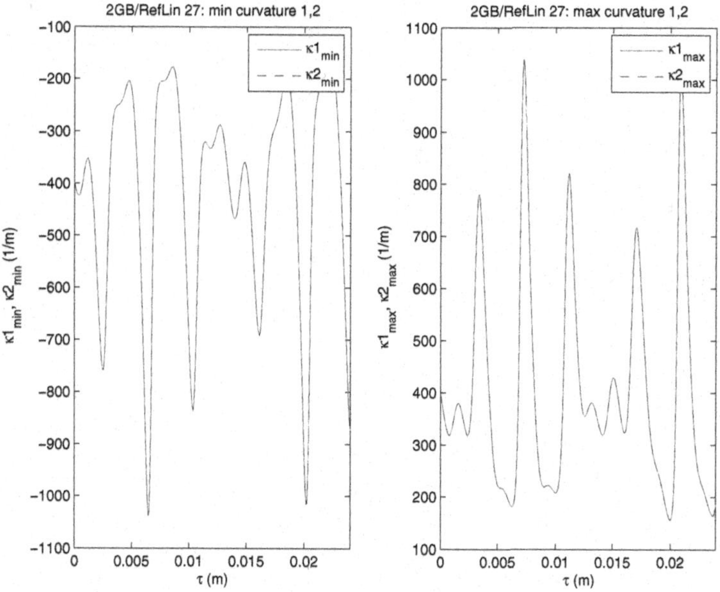

PLATE 21 (Figure 63 on page 70 of this Volume)

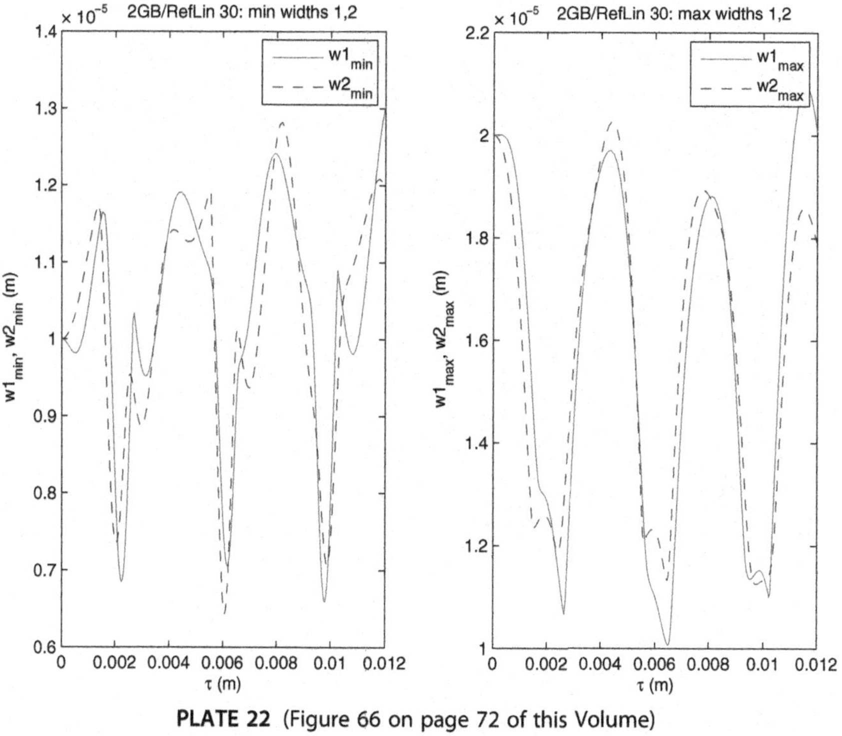

PLATE 22 (Figure 66 on page 72 of this Volume)

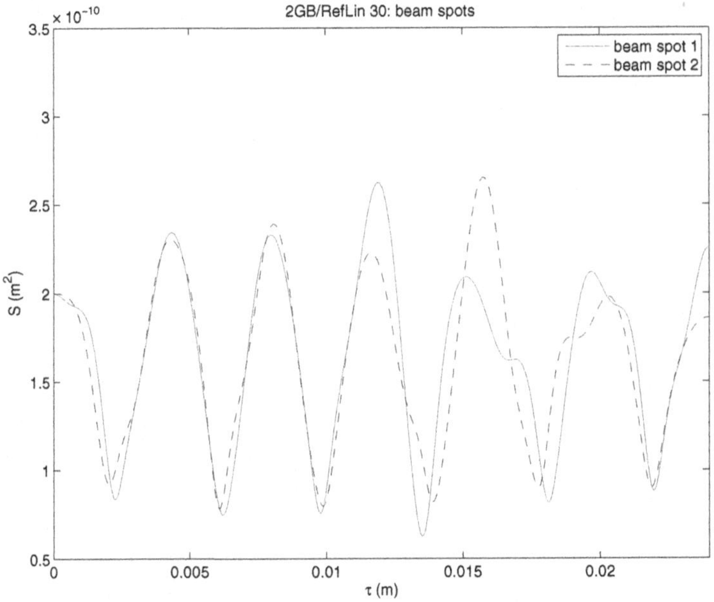

PLATE 23 (Figure 67 on page 73 of this Volume)

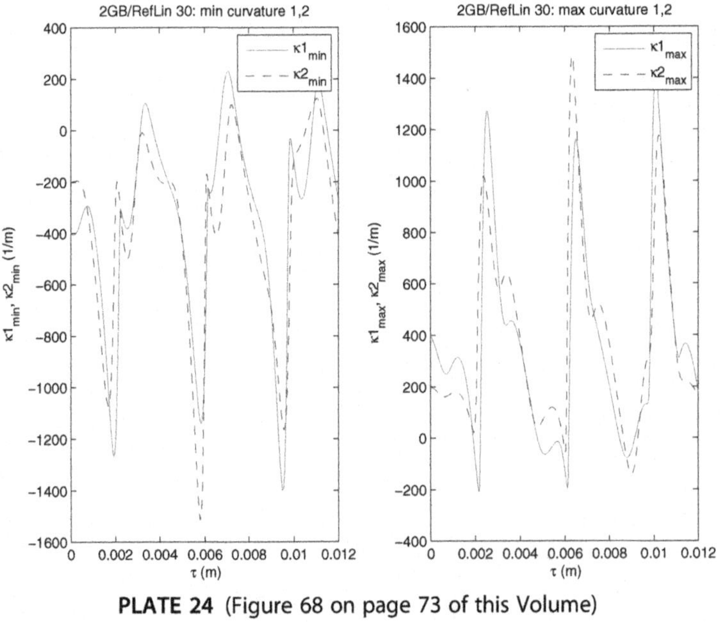

PLATE 24 (Figure 68 on page 73 of this Volume)

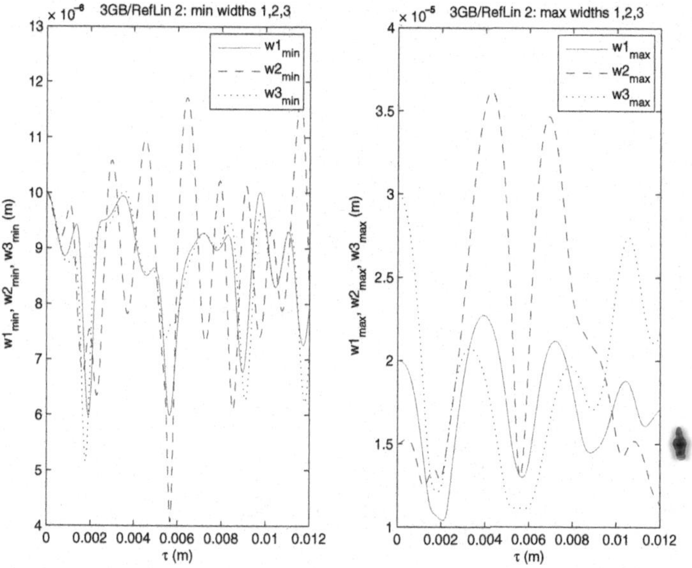

PLATE 25 (Figure 71 on page 76 of this Volume)

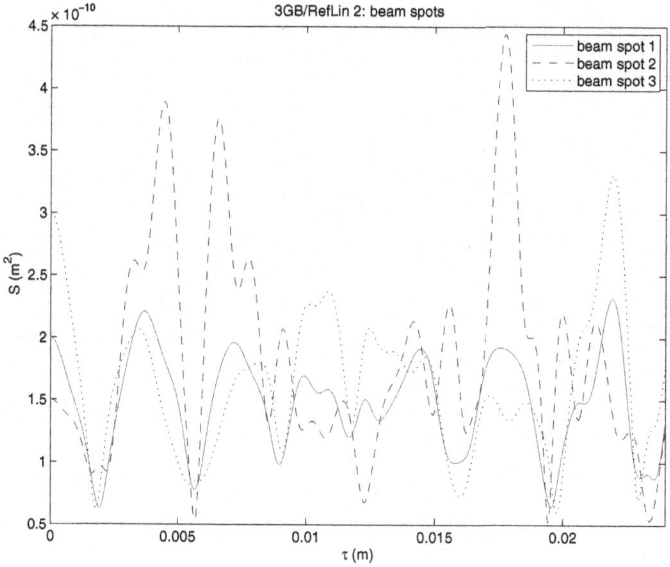

PLATE 26 (Figure 72 on page 76 of this Volume)

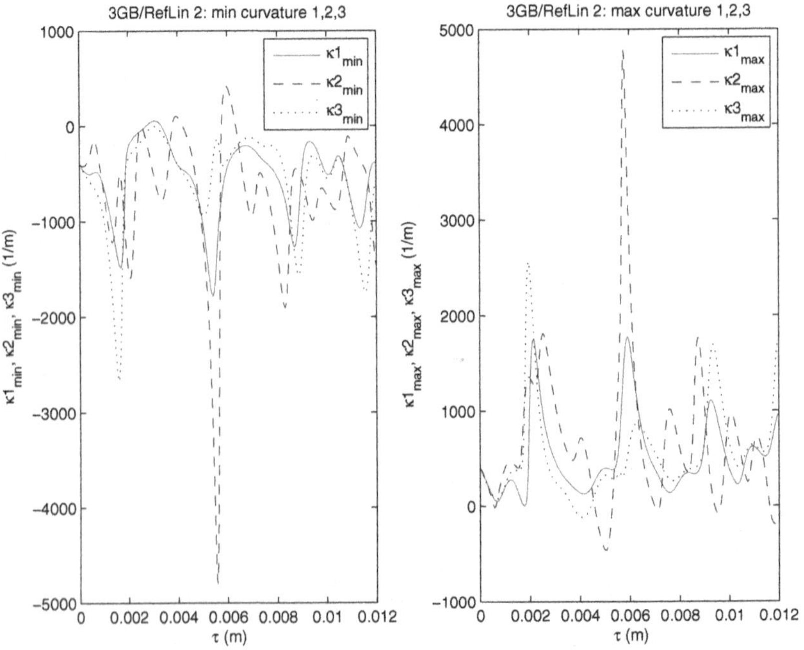

PLATE 27 (Figure 73 on page 77 of this Volume)

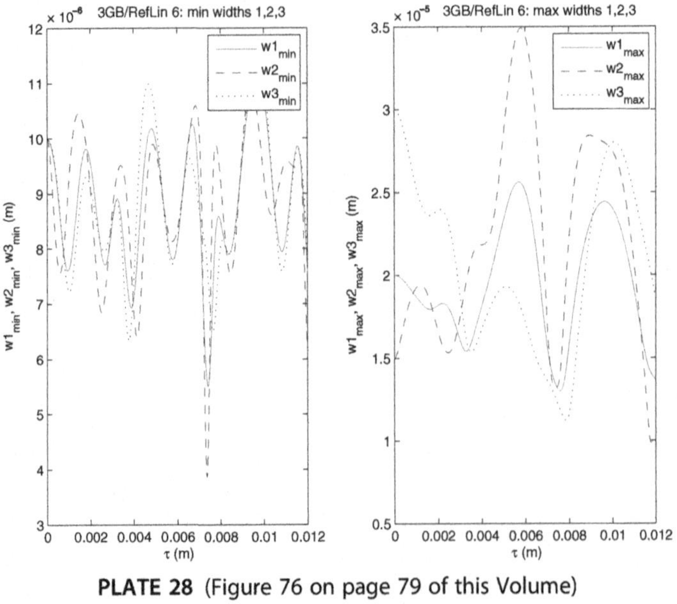

PLATE 28 (Figure 76 on page 79 of this Volume)

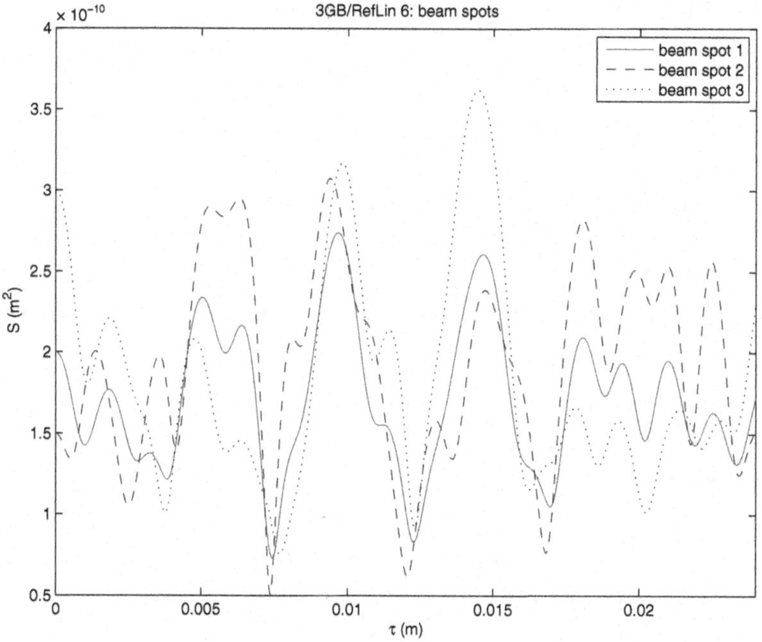

PLATE 29 (Figure 77 on page 80 of this Volume)

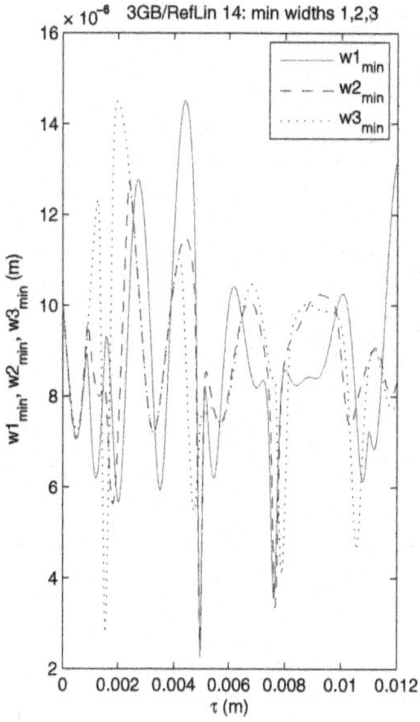

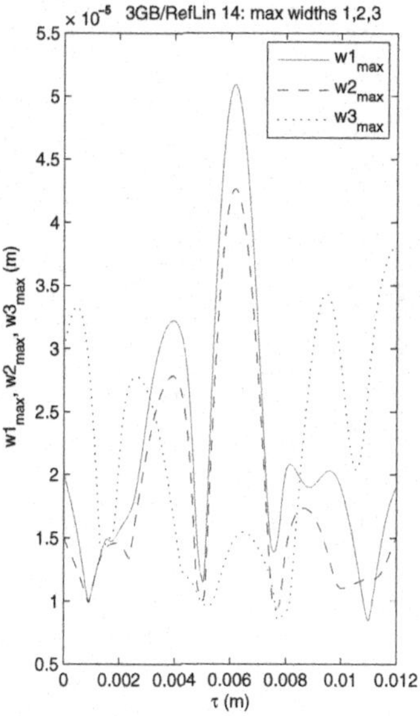

PLATE 30 (Figure 80 on page 82 of this Volume)

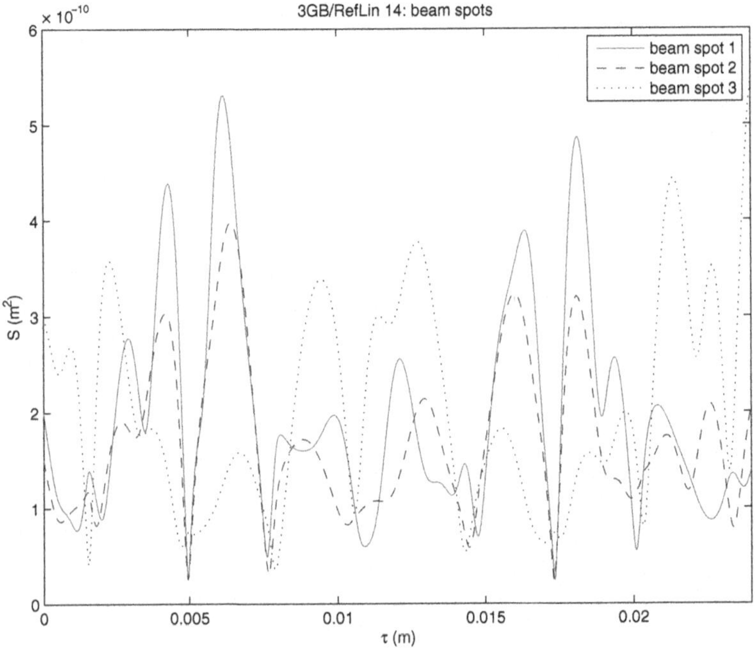

PLATE 31 (Figure 81 on page 83 of this Volume)

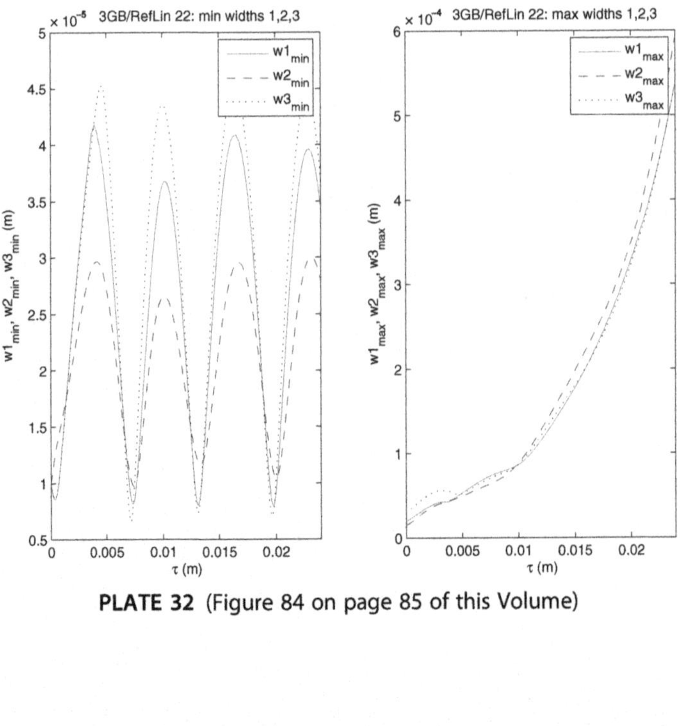

PLATE 32 (Figure 84 on page 85 of this Volume)

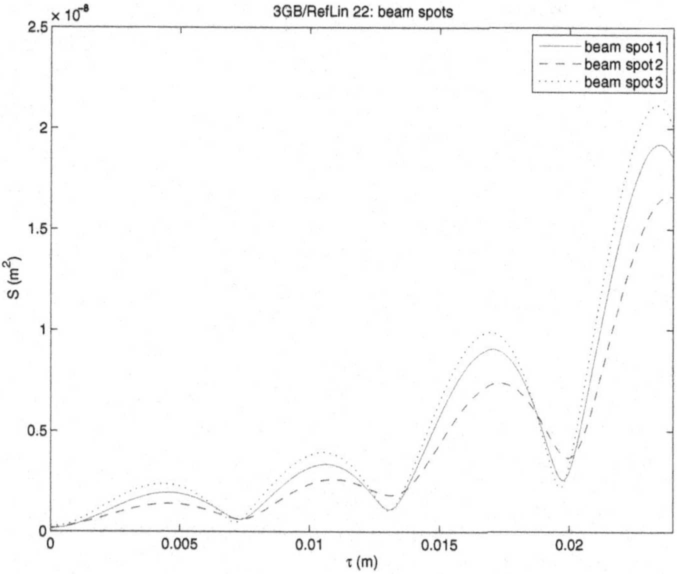

PLATE 33 (Figure 85 on page 86 of this Volume)

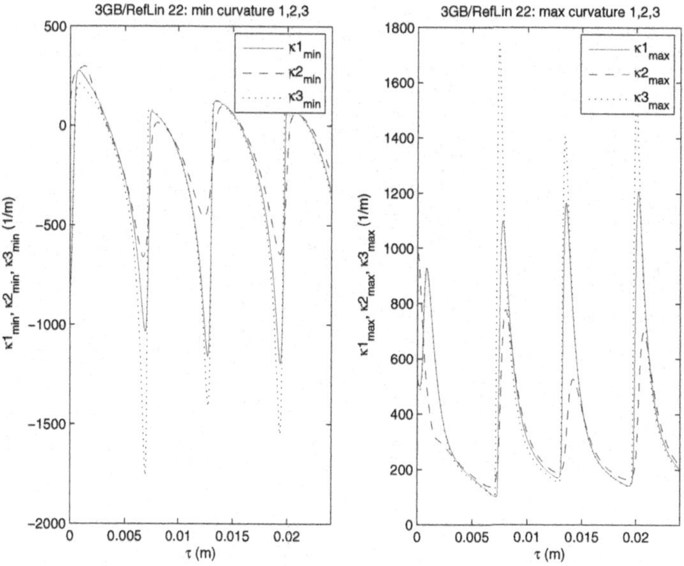

PLATE 34 (Figure 86 on page 86 of this Volume)

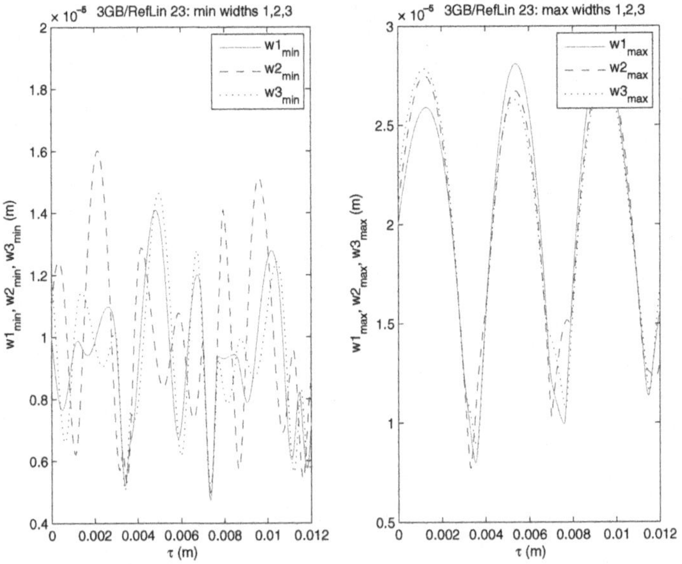

PLATE 35 (Figure 89 on page 89 of this Volume)

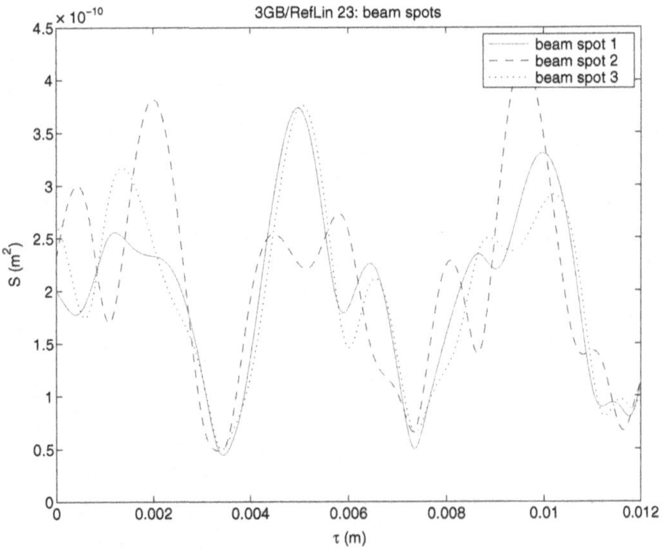

PLATE 36 (Figure 90 on page 89 of this Volume)

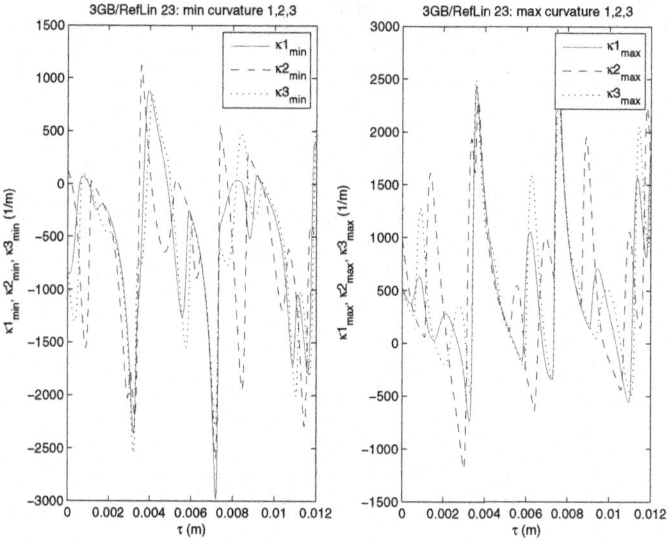

PLATE 37 (Figure 91 on page 90 of this Volume)

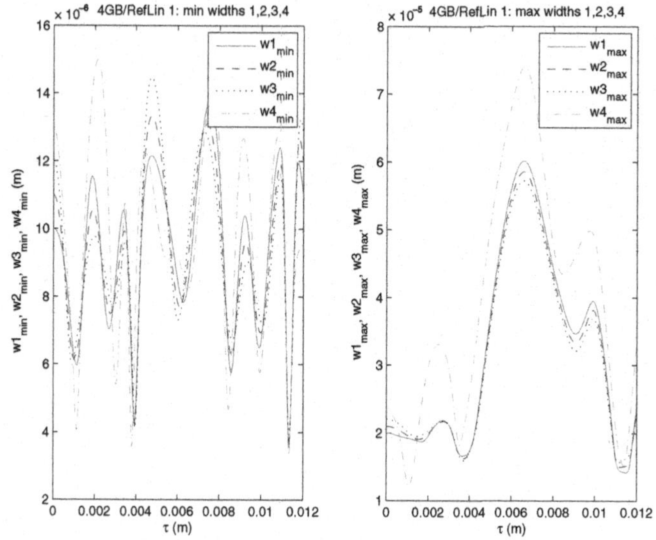

PLATE 38 (Figure 94 on page 92 of this Volume)

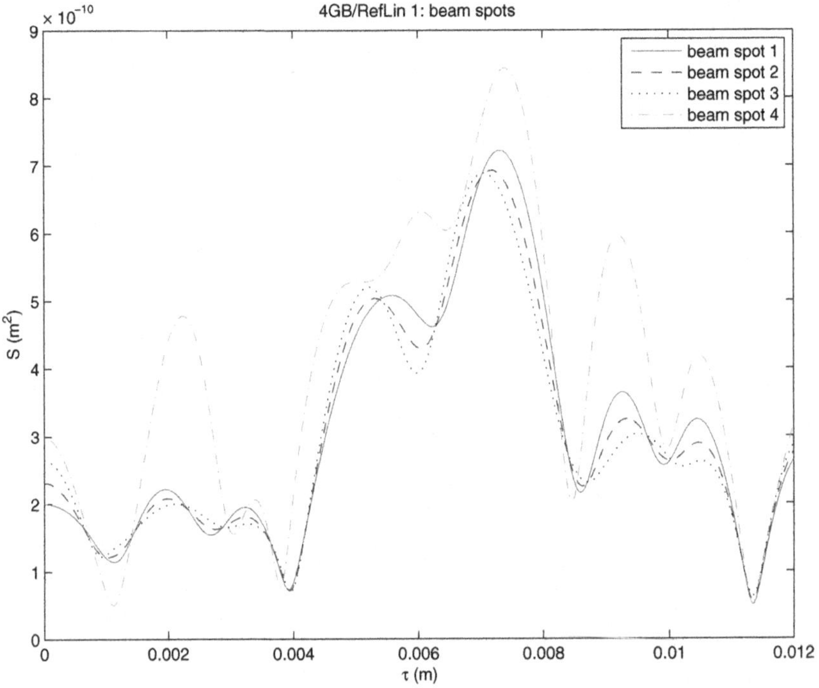

PLATE 39 (Figure 95 on page 93 of this Volume)

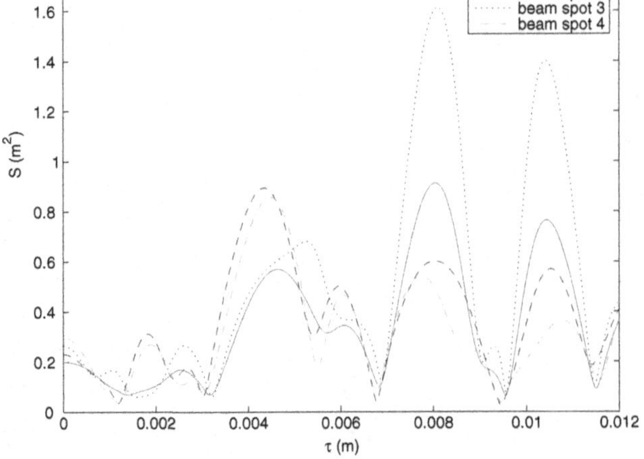

PLATE 40 (Figure 100 on page 99 of this Volume)

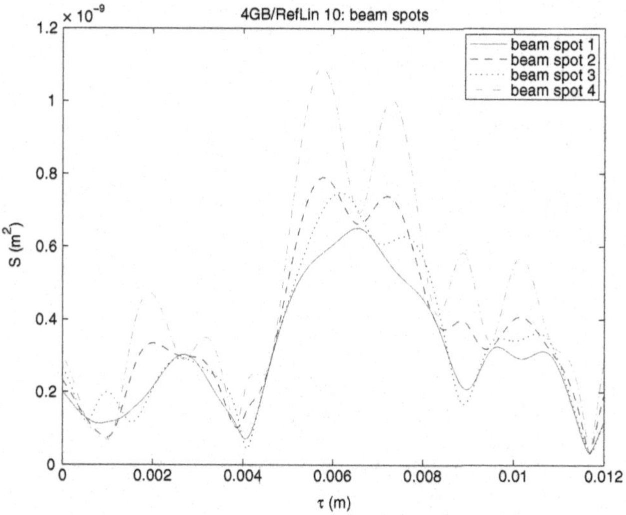

PLATE 41 (Figure 103 on page 102 of this Volume)

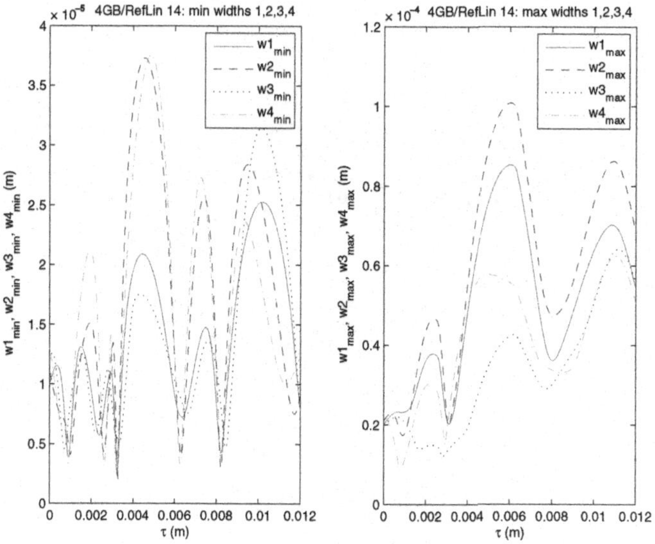

PLATE 42 (Figure 106 on page 105 of this Volume)

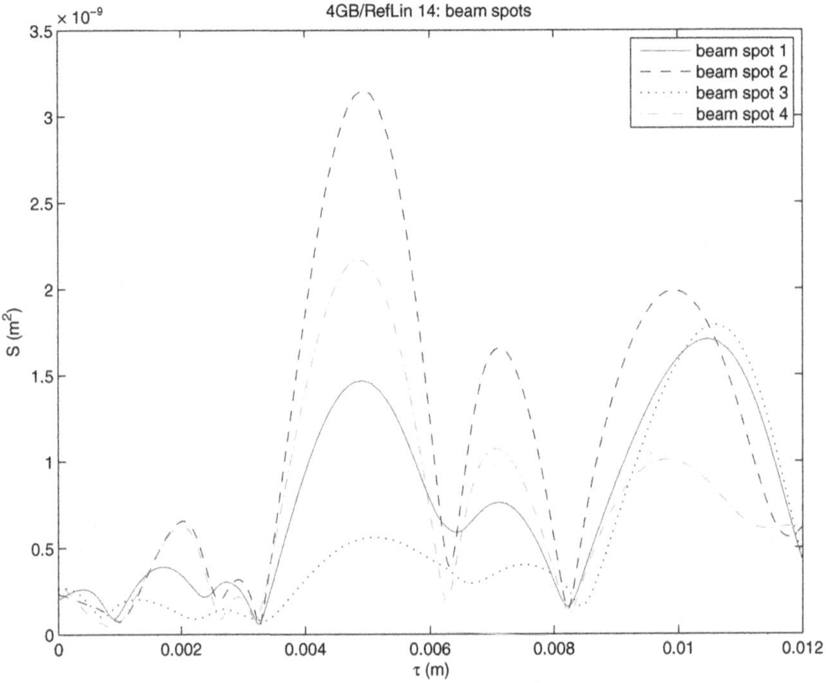

PLATE 43 (Figure 107 on page 106 of this Volume)

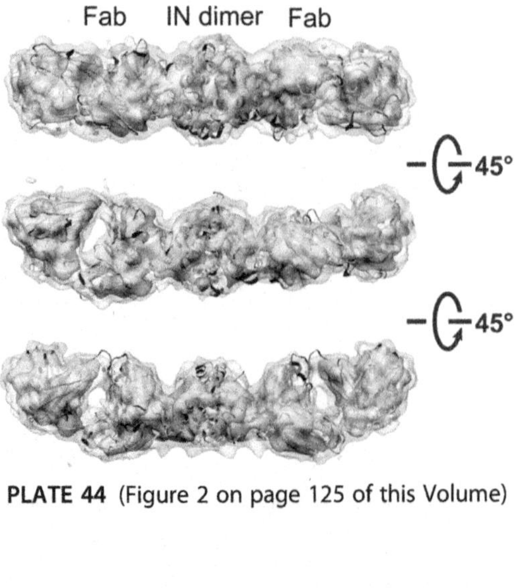

PLATE 44 (Figure 2 on page 125 of this Volume)

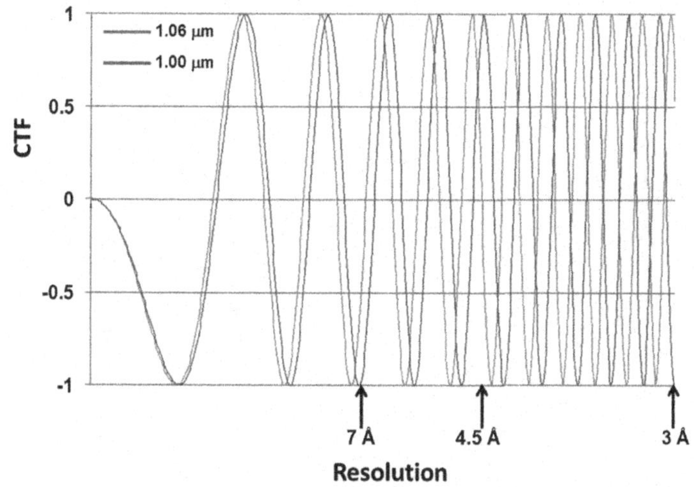

PLATE 45 (Figure 4 on page 130 of this Volume)

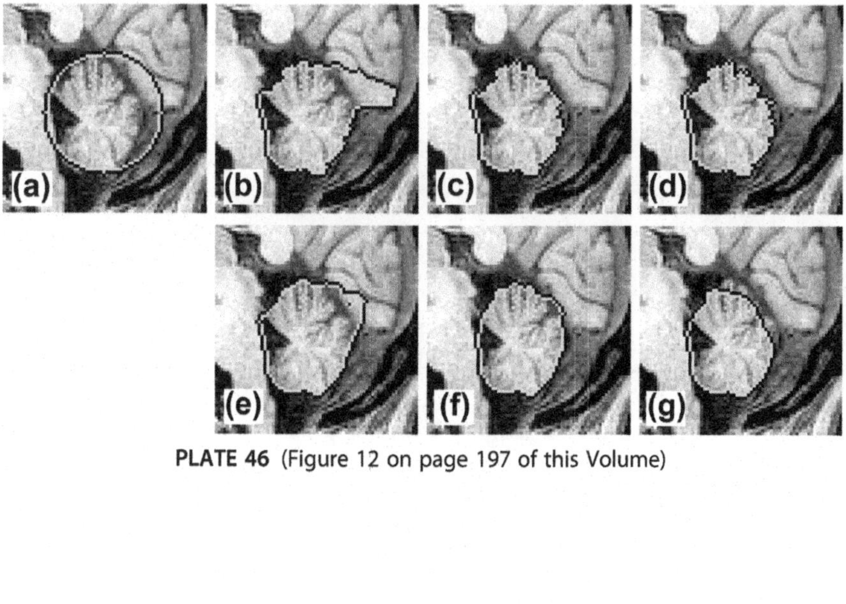

PLATE 46 (Figure 12 on page 197 of this Volume)